BASIC STATISTICS

[For students of Engineering Studies, Agricultural Sciences, BBA, MBA, Biotechnology, Applied Microbiology, Pharmaceutical Sciences, Forestry and Environmental Sciences]

(2ND EDITION)

Operations Research

P.K. Mohanty & S. K. Patel

ISBN: 9789386237996

The book covers clear and crisp pedagogy in the field of decision making process, which pervades the activities of every business manager. Modest attempt has been made to discuss some of the commonly used quantitative techniques in a wide spectrum of decision-making situations. It presents the application of various techniques through a large number of examples and review illustrations. A number of problems from various examinations have also been incorporated. Simplicity in explaining complex phenomena and lucidity in style are the twin objectives of the authors' in organizing the chapters of the book so that students of Civil, Production, Mechanical, Electrical and Electronics Engineering, Commerce, Management, CA and ICWA can derive maximum benefit.

Contents:

1. Introduction
2. Linear Programming- Graphical Approach
3. Linear Programming - Simplex Method
4. Duality and Sensitivity Analysis
5. Transportation Problem
6. Assignment Problem
7. Sequencing Problem
8. Decision Theory
9. Game Theory
10. Simulation
11. Queuing Theory
12. CPM / Pert
13. Inventory Control
14. Integer Programming
Selected References

Related Books

Title	Author	ISBN
Communication Skills and Personality Development	*J.R. Kadam*	9789387741652
Entrepreneurship Development and Communication Skills	*R.R. Chole*	9788172338008
Financial Management for Agribusiness	*W.J. Obst*	9788172336516
Fundamentals of Agricultural Statistics	*K.P. Dhamu*	9788172335632
Fundamentals of Human Ethics and Agriculture Extension	*D.K. Meena*	9789388172653
Indian Agriculture & Agribusiness Management	*Smita Diwase*	9788190938426
Objective Agribusiness Management	*S.R. Panigrahy*	9789386347206

For complete details visit www.scientificpub.com

BASIC STATISTICS

[For students of Engineering Studies, Agricultural Sciences, BBA, MBA, Biotechnology, Applied Microbiology, Pharmaceutical Sciences, Forestry and Environmental Sciences]

2ND EDITION

DR. P. K. MOHANTY
Ph.D., F.I.E.
Chief Engineer,
Deptt of Water Resources,
Govt. of Odisha,
BHUBANESWAR.

DR. S. K. PATEL
Ph.D., F.I.E.
Professor,
Deptt of Mechanical Engineering
National Institute of Technology,
ROURKELA.

Published by

SCIENTIFIC PUBLISHERS (INDIA)
5 A, New Pali Road, P.O. Box 91
Jodhpur 342 001, INDIA
E-mail: info@scientificpub.com
Website: http://www.scientificpub.com

ISBN: 978-93-88812-16-0
eISBN: 978-93-88812-17-7

Printed in India

FOREWORD

In the present world, Statistics is finding applications in almost all branches of study–social as well as physical–such as biology, psychology, education, economics, business management and the list is endless. It is hardly possible to enumerate even a single department of human activity where statistics is not applied. It has rather become indispensable in all phases of human life. Keeping these in view, the endeavor of the authors in bringing out this textbook deserves commendation.

I have found four basic goals which have been achieved by the book- a strong real world focus, a structured approach to the subject matter, lucidity of style and simplicity of expression of many complex mathematical formulae and intuitiveness.

I am sure the book will prove to be involuable students, researchers and academicians alike.

Prof. B.K. Nanda

Prof. Bijoy Kumar Nanda

Formet- Professor, Department of Mechanical Engineering,

National Institute of Technology, Rourkela

Former Vice Chancellor,

Veer Surendra Sai University of Technology, Burla

PREFACE TO THE SECOND EDITION

It is our profound pleasure and pride, as the time is now ripe, to bring out a revised and updated edition of the book, which has been greatly metamorphosed with addition of new chapter on *Statistical Methods in Environmental Studies* and many worked out examples, by which the student can get a solid foundation for understanding. The broad plan essentially remains the same as the first edition. In order to cater to the long-felt need of the students, this has been made possible on our part due to numerous feedbacks and responses received from many readers via e-mails, phone calls and postal correspondences. The enthusiasms exhibited by such readers has prompted us to bring out this newer edition, which, we hope, will no doubt fill the gap that have possibly been created in the earlier one. We will also not spare all our readers without expressing our thanks to them, who have given their precious time by going through the book and giving valuable suggestions and we wish them to continue to do so for the betterment of the book. The cover design of the book has been purposefully changed to give it a new look distinguishable from the first edition.

Dr. P. K. Mohanty
Dr. S. K. Patel

Bhubaneswar
19-Oct-2018
Vijaya Dashami

PREFACE TO THE FIRST EDITION

Writing a textbook is always an exciting experience. In the last one year that it took to complete the manuscript, we got to interact with a number of stakeholders, we benefitted from many insightful comments of professors, who reviewed the manuscript and above all our students, who always have a lot of ideas. Every effort has been made to make it most teacher- and student-friendly textbook in business statistics. The chapters in this book have been so organized to cater to the needs of the research scholars and students of Engineering, Commerce, Economics, BCA, MCA, BBA and MBA, Biotechnology, Applied Microbiology, Environmental Science etc., who are required to learn the subject and to apply its techniques and methods in their sphere of work. The questions and problems have all been solved in a very simple, easy and comprehensive manner to suit the readers' requirements. Wherever possible real world data, particularly from biology, industrial, agricultural and census data, etc. have been taken for explaining various statistical techniques.

The book content has been organized into seventeen chapters. To begin with the Chapter 1 contains the introduction. Chapter 2 contains collection and classification of data, where various methods of sampling, qualitative and quantitative data classification, population, sample, types of variables etc. have been described. Chapter 3 is all about *representation of data*, where frequency distribution, histogram, frequency polygon, frequency curve, relative frequency curve, cumulative frequency curve or ogive have been described with suitable diagrams. A clear understanding of class, class interval, class boundary, class limits, class magnitude, mid value, class frequency, etc. can be had from this chapter. The *Measures of Central Tendency* have been dealt with in Chapter 4. Detailed discussion of mode, median, arithmetic and geometric mean have been made here. Chapter 5 describes *Measures of Dispersion* like Range, Quartile Deviation, Mean Deviation and Standard Deviation and Chapter 6 explains about skewness and kurtosis. *Probability, Laws of Probabilities* and *Set Theory, Permutation and Combination* have been described in Chapter 7. Chapter 8 is devoted to the *Theoretical Distributions* like Binomial, Poisson and Normal Distribution. Chapter 9 deals with *Design of Experiments*. Sampling distributions is dealt with in Chapter 10. The all-important chapter in Statistics, *Tests of Hypothesis* and *Tests of Significance* have been dealt with in an extensive and exhaustive way in Chapter 11. It includes a meaning of Standard Error, Central Limit Theorem, difference between Large and Small Samples, and One Tailed & Two Tailed Hypothesis Testing at various significance levels. Comparison of means of two samples (both Large & Small), Significance tests for dependent samples or paired observations have also found place in this chapter. One paragraph describing *Degrees of Freedom* in a layman's language is the specialty of this chapter. The lucidity of style and simplicity of expression have been our twin objectives to get rid of the difficulty, which is usually associated with statistical hypothesis testing procedures. Chapter 12 deals with *Analysis of Variance or F-Test*. Variance Analyses in both One Way & Two Way Classifications have been described in this chapter. Seeing the complexity of the calculations in preparing the ANOVA Table, the shortcut method has been explained for ease. Chapter 13 and Chapter 14 is devoted in explaining the Regression and Correlation with briefings on Method of Least Square and Standard Error of Estimate. Non Parametric Tests have been discussed in the Chapter 15, where very sophisticated statistical tests like Sign Test, Mann-Whitney U Test, Kruskal Wallis Test and Kolmogorov-Smirnov Test mostly used in statistical experiments are explained in a very simple language with adequate number of examples. Various statistical tables showing t-values, z-values,^{2} values and F values etc. have also been incorporated in the book. Chapters 16 and 17 contain subjects on *Time Series Analysis* and *Index Numbers* respectively.

We record our deep sense of approbation to Sri Krishna Mohan Mohanty, former-Reader in Economics, who himself is an institution of self-discipline and a man of letters, for his pedagogic guidance during this uphill task. God could not be everywhere and so He created mother. Any woman can be a mother but it takes someone special to be called "Maa" and she is none other than Smt. Sarojini Mohanty, whom we express our knee bent veneration for her oceanic love and affection during the ordeal. *Maa Tujhe Pranam*.

Sri Sanmay Patel and Ms. Indrani Mohanty, students of NIT, Rourkela deserve special mention for making the cover design of the book.

Lastly, we record our sincere thanks to the publishers for the energetic efforts they have put forth in bringing out this book.

Dr. P. K. Mohanty

Dr. S. K. Patel

SYMBOLS USED

D_i	Decile
σ	Population Standard Deviation
	Estimated Population Standard Deviation
σ^2	Population Variance
	Estimated Population Variance
μ	Population Mean
μ_1	First Moment
μ_2	Second Moment
μ_3	Third Moment
μ_4	Fourth Moment
s	Sample Standard Deviation
s^2	Sample Variance
s_k	Coefficient of Skewness
x	Sample Mean
d	Deviations from the mean
δ	Mean Deviation (M.D.)
d	Modulus of d
Σ	Summation
Π	Pi = 3.141
e	Logarithmic Base = 2.718
∞	Infinity
$\pm$	Plus or Minus
$\sqrt{}$	Square Root
$\neq$	Not Equal to
$\leq$	Less than Equal to
$\geq$	More than Equal to
$\cap$	Intersection of Sets
U	Union of Sets
P_i	Percentile
P(X)	Probability of happening of X
p	Probability of Success
q	Probability of Failure = 1 – p
f	Frequency
M	Median
N	Number of observations = Σ f
A.M.	Arithmetic Mean
G.M.	Geometric Mean
H.M.	Harmonic Mean
m	Mid Value of Class
l_1	Lower Class Boundary
l_2	Upper Class Boundary
i	Class Width
Q	Quartile Deviation
Q_1	Lower Quartile
Q_3	Upper Quartile
Z	Mode

CONTENTS

Foreword (v)

Preface to 1st Edition (vii)

Preface to 2nd Edition (ix)

Symbols Used (x)

1. **Introduction** 1
2. **Data Collection** 5

   Types of Data- Population- Sample- Population Survey- Sampling Methods- Classification of Data- Variable Types
3. **Representation of Data** 19

   Frequency Distribution-Histogram-Frequency Polygon- Frequency Curve- Normal Frequency Curve-Relative Frequency Curve-Cumulative Frequency Curve or Ogive
4. **Measures of Central Tendency** 33

   Objectives-Arithmetic Mean-Geometric Mean-Harmonic Mean-Mode-Median
5. **Measures of Dispersion** 53

   Range- Quartile Deviation-Mean Deviation- Standard Deviation-Coefficient of Variation
6. **Skewness- Moments- Kurtosis** 71

   Definition- Purpose- Measures- Karl Pearson- Bowley coefficients
7. **Probability** 85

   Basic Concepts- Measure of Probability- Probability Approaches-Classical, Relative Frequency, Subjective, Axiometric-Terminology- Permutations and Combinations- Laws of Probability and Set Theory- Bayes' Theorem
8. **Probability Distribution** 125

   Introduction-Expected Value-Binomial Distribution-Poisson Distribution-Normal Distribution- Standard Normal Distribution
9. **Design of Experiments** 161

   Basic Principle-Randomization- Control- Informal Experimental Designs- Formal Experimental Designs

10. Sampling Distributions- Statistical Inference 181

Central Limit Theorem-Statistic-Parameter-Standard Error-Statistical inference-Point Estimation-Interval Estimation-Confidence Interval

11. Test of Hypothesis 217

Tests of Significance-Null Hypothesis-Alternative Hypothesis-Types of Error-One Tailed test-Two tailed Test-Chi Square Tests-F Test (Test of Hypothesis on equality of variances)

12. Analysis of Variance (F- Test) 249

Analysis of Variance (ANOVA)-One Way Classification- Two Way Classification- Least Significant Difference (LSD)

13. Regression 267

Introduction- Definition-Scatter diagram- Method of least square-Standard error of estimate- Comparison of correlation and regression- Properties

14. Correlation 291

Introduction- Definition-Utility-Types of Correlation- Coefficient of Determination- Coefficient of Correlation

15. Non Parametric Test 309

Introduction-Characteristics-Sign Test-Sign Test (Paired Data)-Mann-Whitney U Test-Kruskal Walis Test-Kolmogorov Smirnov Test(K-S test)- Rank Correlation

16. Time Series 333

Reason- Fitting Linear Trend-Second Degree Trend-Secular Trend- Seasonal Variation-Cyclical Variation- Irregular Variation

17. Index Numbers 385

Types of Index Numbers- Wholesale Price Index- Consumer Price Index-Good Index Number criteria- Cost of Living Index Number-Quantity Index Number- Chain Index Number

18. Statistical/ Mathematical Models in Environmental Studies 413

Air Pollution Concentration Models-Fixed Box Model-Gaussian Plume Model-Population Model-Leslie Matrix-Lokta Voltera Predator Prey Model

Glossary of Terms 437

INTRODUCTION

India, being the largest democracy in the world, goes to the poll every five years for the members to be elected to the parliament. There is a system of exit-poll by many TV broadcasters, who inform the viewers, soon after the poll is over and prior to declaration of results, about the expected number of seats that will be won by various political parties with a certain degree of confidence. But for the general public, what does this statement mean? What is this degree of confidence? What is the basis of arriving at such conclusions by the broadcasters? How many people were interviewed by the broadcasters? Is it worth believing such assertions? Take another example- A drug is manufactured by a pharmaceutical company. Before it is marketed, it is first subjected to rigorous laboratory tests. The drug is made public only after the company becomes 99 % sure that it actually addresses the problem and has no side effects. These are few instances, where Statistics plays vital role in arriving at any meaningful conclusions. The word *statistics* means different to different people. To the TV broadcaster, it may mean the proportion of voters in favour of certain political party. To the pharmaceutical company, it is the number of cases, where the drug has proved its effectiveness. But each of these people is using the word correctly, yet each person uses it in a different way. After all Statistics is used by all to help them in making a decision.

Though Statistics has been defined differently by different authors from time to time, the word *statistik* originated from the Latin word *statista* (meaning "statesman"). It was first used by Gottfried Achenwall (1719 – 1772), professor at Marlborough and Gottingen Unweixtes. Dr. E.A.W. Zimmerman introduced the word 'statistics' into England and then it was popularized by Sir John Sinclair in his work *Statistical Account of Scotland* (1791 – 1799). The history of the development of statistical theory and practice is a lengthy one. In the early years, statistics was meant for mere collection of facts about the state or people in the state for political purposes. It was then used in collecting data about economic and social conditions of people living in different parts of the country, for proper functioning of the State. In this way statistics developed as "*Political Arithmetic*". In course of time, the scope of statistics widened, which was once viewed as a branch of Economics, is now regarded as a full-fledged subject. Now it includes all types of quantitative analyses too. Modern concept of Statistics evolved in 1749. Sir R.A.Fisher treated the science of Statistics as "*Mathematics to observational data*".

In modern times, Statistics is viewed not as a mere tool for collecting numerical data but as a means of developing sound techniques for their handling and analysis, and drawing valid inferences from them. It is now finding applications in almost all sciences–social as well as physical–such as biology, psychology, education, economics, business management, etc. It is hardly possible to enumerate even a single department of human activity where statistics is not in use. It has rather become indispensable in all spheres of human activities.

Some writers define it as "statistical data", i.e. numerical statements of facts, while others define it as "statistical methods", i.e. complete body of the principles and techniques used in collecting and analyzing the data.

According to the Oxford Dictionary, "Statistics has two meanings, as in plural sense and singular sense". In plural sense, it means a systematic collection of numerical facts and in singular sense, it is the science of collecting, classifying and using statistics". Webster Dictionary defines statistics as "classified facts representing the conditions of the people in a state… especially those facts which can be stated in numbers or in any other tabular or classified arrangement". This definition, since it confines statistics only to the data pertaining to state, is inadequate as the domain of statistics is much wider.

A more exhaustive definition is given by Prof. Horace Secrist as follows:

"By Statistics we mean aggregates of facts affected by a marked extent by multiplicity of causes numerically expressed, enumerated or estimated according to reasonable standard of accuracy, collected in a systematic manner for a pre-determined purpose and placed in relation to each other."

Perhaps the best definition seems to be one given by Croxton and Cowden, according to whom statistics may be defined as "the science, which deals with the collection, analysis and interpretation of numerical data".

Management professionals and scientific investigators apply statistical technique to virtually every branch of public as well as private enterprises. But the techniques being very diverse, statisticians separate them into two broad categories: *Descriptive Statistics* and *Inferential Statistics*. Suppose an investigator wants to know the voting preference in a particular village, as to which political party the villagers want to come to power. He then makes a door-to-door survey, puts questions to the individual villager and tabulates the data. He prepares graphs, tables and charts, etc., to display the data so that it becomes easy to understand. This falls in the category of descriptive statistics. On the contrary, if the investigator wants to know the voters' preference in favour of certain political party in the entire district, he puts questions to some voters picked up at random and find its average, which is then used as an inference for estimating the voters' preference for the entire district. This process of estimating the average would be a problem in *inferential statistics*. Statisticians also refer to this category as *statistical inference*. Admittedly, any conclusion that the investigator makes about the voting preference for the district is based on the data collected from few voters picked up at random, which is then generalized, which may not be actually valid. Hence, the investigator must state how likely it is to be true. So also, statistical inference involves generalizations and statements about the probability of their validity.

The methods and techniques of statistical inference are also used in a branch of statistics called *Decision Theory*. As managerial decisions are often taken under conditions of uncertainty, knowledge of decision theory becomes helpful. For example, a manufacturer of a new product can not precisely specify its demand in the market beforehand. Similarly, it is nearly impossible to say how much exactly the share price would fall in the event of some unprecedented happenings.

In business processes like financial analysis, production, costing, market research, distribution system, product research and quality control etc., in biological research works like DNA or protein sequence alignments, statistics play a vital role in its use and objective solution of most problems. For prediction of future course of business activities, businessmen, industrialists and traders make use of past records and data only. It may be noted that probability theory of statistics is the backbone of all insurance business.

Keeping this background, the chapters of the book have been organized to cater to the needs of students from all disciplines and professionals working in practical field.

data. He prepares graphs, tables and charts, etc., to display the data so that it becomes easy to understand. This falls in the category of descriptive statistics. On the contrary, if the investigator wants to know the voters' preferences in favour of certain political party in the entire district, he puts questions to some voters picked up at random and finds its average, which is then used as an inference for estimating the voters' preference for the entire district. This process of estimating the average would be a problem in *inferential statistics*. Statisticians also refer to this category as *statistical inference*. Admittedly any conclusion that the investigator makes about the voting preference for the district is based on the data collected from few voters picked up at random, which is then generalized, which may not be actually valid. Hence, the investigator must state how likely it is to be true. So statistical inference involves generalizations and statements about the probability of their validity.

The methods and techniques of statistical inference are also used in a branch of statistics called *Decision Theory*. As managerial decisions are often taken under conditions of uncertainty, knowledge of decision theory becomes helpful. For example, a manufacturer of a new product can not precisely specify its demand in the market beforehand. Similarly, it is nearly impossible to say how much exactly the share price would fall in the event of some unprecedented happenings.

In business processes like financial analysis, production, costing, market research, distribution system, product research and quality control etc., in biological research works like DNA or protein sequence analysis, statistics play a vital role in its use and objective solution of most problems. For prediction of future course of business activities businessmen, industrialists and traders make use of past records and data only. It may be noted that probability theory of statistics is the backbone of all these aspects business.

Keeping this background, the chapters of the book have been organized to cater to the needs of students from all disciplines and professionals working in practical field.

COLLECTION OF DATA

Data are collection of any number of related observations and it is the first step in any statistical investigation. Extreme care must be taken in the collection of data, faulty data leads to unreliable conclusions. Data refers to a set of measurements on one or more observational units that happen to interest us. For example, if we are interested to know the acceptability of a new product in the market by the consumers, then the sales records of this product, say daily, weekly or monthly, at various outlet shops are collected and this constitutes data, which is then subjected to various statistical analyses to arrive at meaningful conclusions.

Depending on the source, statistical data is classified into two types:

1. Primary data and 2. Secondary data.

PRIMARY DATA

Primary data are those, which are collected for the first time and are thus original in nature. These are in the shape of raw materials to which statistical methods are applied for the purpose of analysis and interpretation; e.g. a study is undertaken to find the labour productivity of Odisha and Jharkhand. For this purpose, selected labourers in both the states are contacted and their productivity figures are noted. This constitutes primary data.

Advantages

1. Where secondary data is not available, primary data is to be collected.
2. It depicts the data in greater detail.
3. It does not require extra caution.
4. Degree of accuracy is high.

Disadvantages

1. It is a labour intensive job.

2. It requires much skill.
3. It is time consuming and expensive.
4. Sometimes, primary data collection is not possible.

SECONDARY DATA

Secondary data are those data which have already been collected by some other persons and which have passed through the statistical machine at least once. These are usually in the shape of finished products since they have been treated statistically in some form or the other. Data obtained from Labour and Employment Department is being used by the Water Resources, Forest, Agriculture, Industry Departments etc. Data collected by the National Sample Survey are being used by economists, population scientists, sociologists, etc., for their further study.

Advantages

1. It saves time and money.
2. Use of secondary data is very convenient.
3. In some situations, primary data collection becomes difficult.
4. Reliable secondary data can be collected for many investigations.

Disadvantages

1. These are not available for all types of enquiry.
2. Extra care is required to use secondary data.
3. Many times, need of present investigation is not exactly fulfilled by secondary data.
4. Sufficient accurate secondary data may not always be available.

Before any further discussion on data collection and representation, it is worth knowing two terms very frequently used in statistics i.e. population and sample.

POPULATION

- The term 'Population' may be defined as an aggregate of items or individuals under study for a definite purpose.
- In statistics, population means whole.
- Population refers to the whole of the observations, which comes under the purview of a statistical investigation.
- It is the totality of the observations, which are studied in order to find some of their characteristics.

- If the population is countable then it is known as "finite population" otherwise it is known as "infinite population". Suppose a researcher wants to give certain comments on average income of daily labourers in a district, then the entire labour-force of the district constitutes to form a population, more specifically finite population, as the total number of labourers is countable.
- On the contrary, if the investigator is interested to know the environmental effect of setting up of new industries on a particular variety of bacteria, then the entire bacteria of that variety present in the universe constitute together to form a population, more specifically infinite population.

SAMPLE

- When a portion of the population is examined in order to find the characteristics of the whole population or universe, it is called a Sample.
- Population is the whole and sample is its part. A sample is a selected portion of the population. A sample provides valuable information about its population.
- In most cases it gives almost the same results as that of the whole mass. It is the representative of the whole mass.
- It gives fairly accurate results and a reliable picture of the total observations under investigation.
- It is used to measure and estimate the corresponding characteristics of the population.

For collection of data, either the entire population is surveyed like census survey in every 10 years in India, or a part of the population, called sample, is studied.

POPULATION SURVEY vs SAMPLE SURVEY

- All items in any field of enquiry constitute 'Universe' or 'Population'.
- Complete enumeration of all items in the population is known as census survey.
- In census survey, all items are covered, no element of chance is left and highest accuracy is obtained.
- In practice, it may not be possible and any bias in census survey cannot be checked.
- Census survey involves huge amounts money, time and energy.

- Hence, a part of the population is studied called sample survey.
- Sample should be as representative of the total population as far as possible.

ADVANTAGES OF SAMPLING

- Complete enumeration or complete census is not always feasible.
- Census survey is expensive and time consuming.
- Census survey involves trained personnel and other facilities.
- Sampling helps to avoid destructive tests e.g. quality of products.
- But sampling involves sampling error, which is non-existent in complete enumeration.
- On the contrary, non-sampling error (e.g. observational error, error due to non-observation, non-response, tabulation error etc) is lesser in sampling than that in census survey.
- Sampling is more flexible.

STEPS IN SAMPLING DESIGN

The following attention must be paid in sampling design:

Type of Universe:

- Clearly define the set of objects.
- Finite or Infinite Universe.

Sampling Unit:

- Geographical unit like state, district, village etc.
- Construction unit like house, flat etc.
- Social unit like family, club, school etc.
- It may be individual too.
- Sampling unit must be decided before the study is taken up.

Source List:

- It is also known as *'Sampling Frame'*.
- It contains all items of Universe.
- It is to be prepared, if not available.
- It should be comprehensive, correct, reliable and appropriate.
- It should be representative of the population.

Size of sample:

- It is the number of items to be selected from Universe.
- It should be neither too large nor too small.
- Optimum sample should fulfill the characteristics of efficiency, representativeness, reliability and flexibility.
- Cost too decides the size of the sample.
- Higher variance population entails a bigger sample size.

Parameter of Interest:

- In determining the sampling design, one must consider the question of the specific population parameters which are of interest.

Budgetary Constraint:

- Cost considerations, from practical point of view, have a major impact upon decisions relating to not only the size of the sample but also to the type of sample.
- This fact can even lead to the use of a non-probability sample.

Sampling Procedure:

- Decide the type of sample and technique to be used.
- Select the procedure that gives the smallest sampling error.

CRITERIA OF SELECTING A SAMPLING PROCEDURE

In sampling analysis, two costs are involved:

- Cost of collecting data.
- Cost of incorrect inference resulting from the data. Its two causes are:

  (A) Systematic bias.

  (B) Sampling error.

(A) Systematic Bias or Non-sampling Error

- It results from sampling procedure.
- It cannot be reduced or eliminated by increasing the sample size.
- Its causes can be detected and corrected.

Usually systematic bias is the result of one or more of the following factors:

- **Inappropriate sampling frame**- It results in biased representation of the Universe leading to systematic bias.

- **Defective measuring device-** The Questionnaire or the interviewer is biased. Physical measuring device may be biased.
- **Non-respondents-** If respondents cannot be contacted and they may not respond too.
- **Indeterminacy principle-** Individuals respond differently under observed and unobserved conditions.
- **Natural bias in reporting of data-** Individuals give wrong information, for example, understate their income data for tax purpose and overstate, if asked for social status.

(B) Sampling Error

- It is the random variations in the sample estimates around true population parameter.
- It decreases with increase in size and becomes smaller in magnitude for homogenous population.
- It can be measured for a given sample size.
- Measurement of sampling error is called 'Precision of sampling plan'.
- Increasing sample size increases the precision. But increasing has a limit. As size of sample increases, cost of collecting data also increases, which enhances systematic bias.
- Hence, to increase precision, select a better sampling design, which has a smaller sampling error for a given sample size at a given cost.

Above all, while selecting a sampling procedure, one must ensure that the procedure causes a relatively small sampling error and helps to control the systematic bias in a better way.

To reduce sampling error:

- Increase sample size.
- Select a suitable sample design.
- Select a suitable method of estimation.

To reduce non-sampling error:

- Make consistency check.
- Make sample check.
- Do post-survey check.
- Do external record check.

CHARACTERISTICS OF A GOOD SAMPLE DESIGN

- True representative sample.
- Smaller sampling error.
- Viable in the context of funds available for the study.
- Systematic bias can be controlled.
- Results of sample study can be applied for the universe with a reasonable level of confidence.

METHODS OF SAMPLING

The most important point, however, is the *selection of the sample*. A sample study would give dependable conclusion only if the sample is a true representative of the population. There are several methods of collection of sample data, in short called *sampling methods*. This can be categorized into two types: Non-Probability Sampling & Probability Sampling.

Non-Probability Sampling or Non-Random Sampling

- Here there is no basis for estimating the probability.
- It is otherwise known as deliberate sampling, purposive sampling or judgment sampling.
- Sample units are purposively chosen. Samples are drawn keeping some definite purpose in view.
- This type of sampling suffers from the drawback of favoritism, nepotism and biases and does not give a representative sample of the population.
- Chances are that the results favor the investigator.
- There is always danger of bias.
- No guarantee that every element has specifiable chance of being included in the sample.
- Sampling error cannot be estimated.
- Rarely adopted in important works.
- Relatively inexpensive.
- Quicker.
- Not amenable to any statistical treatment.

PROBABILITY SAMPLING OR RANDOM SAMPLING

- Otherwise known as Chance Sampling.

- Each and every item of the universe has an equal chance of being included in the sample and each one of the possible samples, in case of finite universe, has the same probability of being selected.
- All sample choices are independent of each other.
- Results can be statistically treated.
- Standard error can be calculated.
- Sample has fairly same characteristic of the universe.
- Absolutely good random samples are not possible at all, because of human bias, which varies from individual to individual and inherent in any sampling scheme administered by human beings. But fairly good random samples can be obtained by the use of lottery system, by throwing of a dice etc.
- However the usual method of drawing random sample is done by the help of a Random Number Table. Generation of "Random Numbers" is available in a computer (e.g. Microsoft Excel).

SRSWR

- It is Simple Random Sampling With Replacement.
- Units can appear more than once.
- Replacement is done at each stage of random sampling; however such type of replacement is not necessary in case of infinite population.
- It is otherwise known as Unrestricted Random Sampling.
- It is less frequently used.

SRSWOR

- It is Simple Random Sampling Without Replacement.
- Random Sampling from a finite population is the method of sample selection which gives each possible sample combination an equal probability of being picked up and each item in the entire population to have an equal chance of being included in the sample.
- Once an item is selected for the sample, it cannot appear in the sample again.
- SRSWOR is a more precise estimator than SRSWR.
- This method has less variance or less standard error.
- Methods employed are lottery method, using random numbers (Tippets' Table)

COMPLEX RANDOM SAMPLING DESIGNS

(A) SYSTEMATIC SAMPLING

- Most practical way of sampling is to select every i^{th} item of a list.
- Only the first unit is selected randomly and then the remaining units of the sample are selected at fixed intervals.
- Can be treated to be an improvement over a simple random sample. It is spread more evenly over the entire population.
- Easier and less costly.
- Useful where there is large population.
- It is dangerous, if there is a hidden periodicity in the population, e.g. every i^{th} item is defective.

(B) STRATIFIED SAMPLING

- It is useful if the population is not homogenous.
- Population is divided into several sub-groups, those are individually more homogenous. These sub-groups or sub-populations are called *strata*.
- Since each stratum is more homogenous than the total population, more precise estimates for each stratum can be obtained.
- It results in more reliable and detailed information.
- If simple random samples are taken from each stratum, it is called *Stratified Random Sampling*.
- Number of random samples collected from each stratum is by proportional allocation. That means the sample size in each stratum varies according to the relative importance of the stratum in the population.

(C) QUOTA SAMPLING

- In stratified sampling the cost of taking random samples from individual strata is often so expensive that interviewers are simply given quota to be filled from different strata, the actual selection of items for sample being left to the interviewer's judgment.

(D) CLUSTER SAMPLING

- When the total area of interest is very big, then the entire area is divided into a number of smaller non-overlapping areas and then a number of these areas are randomly selected, called *clusters*.

- Clusters are then randomly selected for inclusion in the overall sample.
- It is cost effective, as it concentrates in survey of selected clusters.
- It is less precise than random sampling.
- It has economic advantage.
- Estimates based on cluster samples are more reliable per unit cost.
- If clusters happen to be some geographical subdivisions, it is called area sampling.

(E) MULTISTAGE SAMPLING

- It is a further development over the principle of cluster sampling.
- Suppose it is required to survey banking systems in the country

  (a) Select a large primary unit, say some states (1st stage sampling)

  (b) Select certain districts in the country (2nd stage sampling)

  (c) Select certain towns (3rd stage sampling)

  (d) Randomly select some banks (4th stage sampling)

  (e) And so on
- It is applied in big enquiries.
- It is easier to administer.
- A large number of units can be sampled.

(F) SEQUENTIAL SAMPLING

It is a somewhat complex sample design. The ultimate size of sample under this technique is not fixed in advance, but is determined according to mathematical decision rules on the basis of information yielded as survey progresses. This is usually adopted in case of acceptance sampling plan in the context of statistical quality control. When a particular lot is to be accepted or rejected on the basis of a single sample, it is known as *single sampling*; when the decision is to be taken on the basis of two samples, it is known as *double sampling* and in case the decision rests on the basis of more than two samples but number of samples is certain and decided in advance, the sampling is known as *multiple sampling*. But when the number of samples is more than two but it is neither certain nor decided in advance, this type of system is often referred to as sequential sampling. Thus, in sequential sampling, one can go on taking samples one after another as long as one desires to do so.

CLASSIFICATION OF DATA

The data thus collected needs to be arranged in a certain order for convenience. The process of arranging data into different classes and groups according to some common characteristic and similarities posed by them is called *classification.* When some statistical data is collected, there are some data, which are common in their characteristics and features in groups. So the common items are taken into one group each which subsequently facilitates comparison and proper analysis of the data. Broadly speaking, data can be classified on the following four bases: (1) *Geographical,* (2) *Chronological,* (3) *Qualitative, and* (4) *Quantitative.*

GEOGRAPHICAL CLASSIFICATION

It is based on the basis of geographical distribution and location. If, for example, someone is interested to know the living standard of rural people in a particular district, then he will collect data from persons staying in different villages and then group them according to their locations. It is also known as *spatial classification.*

CHRONOLOGICAL CLASSIFICATION

When data are classified on the basis of time, it is known as chronological classification, such as the number of air conditioners sold in different months of a year in a certain locality. In this type of classification, statistical data are arranged in sequence of time and hence are called *time series.* This type of classification is also known as temporal classification. Data on production of certain commodity in different years in a certain region, say district, state or country, when arranged chronologically falls in this category.

QUALITATIVE CLASSIFICATION

It is based on the descriptive characteristic or on the basis of attributes of statistical data and is according to non-measurable terms, such as sex, color, health condition, intelligence, degree of honesty of human beings, etc. It is always non-numeric. For example, if the workers of a factory are to be classified in respect of one attribute, say sex, we can classify them into two groups- Male workers and Female workers. When classification is done on the basis of one attribute as in the present example, it is called *dichotomous classification.* On the contrary, *manifold classification* takes into account more than one attribute. For example, workers can be classified on the basis of their sex and health condition- Infected Male, Infected Female, Uninfected Male and Uninfected Female.

QUANTITATIVE CLASSIFICATION

As the term suggests, quantitative classification refers to classification that is based on figures or in other words it is a classification which is based on such characteristics, which are capable of quantitative measurement and taking up numeric values. It is also known as *classification by variable*. For example the following table shows the number of persons in various salary groups:

Salary in Thousand of Rupees	Number of Persons
5 – 14	5
15 – 24	10
25 – 34	120
35 – 44	22
45 – 54	13
55 – 64	5
Total	175

In the above table, two elements need some explanation. (i) salary group, and (ii) number of persons. Here salary group is given in a class interval, and number of persons indicates *frequency*. So **Variable** refers to a quantitative characteristic, which varies. It is capable of taking different values in different situations. It can be measured numerically. *The measurable characteristics on which individuals differ among themselves are called a variable*. A variable can be of two types. 1) *Continuous* and 2) *Discrete*.

Continuous Variable: A continuous variable is one which is capable of assuming any fractional value within an interval. It can cover all the values in the interval. Here the data are obtained by measurement and not by counting. For example -the weights of persons in a certain locality can be 54.6 Kg, 62.8 Kg, 38.5 Kg etc. It can be a whole number or a fraction. When the continuous variables are arranged in the form of a series, they are called a *continuous series*.

Discrete Variable: On the other hand, a discrete variable takes up only whole numbers, not fractions. For example, the number of persons in salary group. It can only be expressed in whole numbers. We cannot say twenty two and half number of persons. It is not subjected to division. It is obtained by counting. When discrete variables are arranged in the form of a series, they are called a *discrete series*.

After the tedious task of collection and classification of data is over, these are then reproduced in the required form depending on the need of the investigator, which have been dealt with in the next chapter.

EXERCISES

1. What are primary and secondary data?
2. What are the differences between discrete and continuous variables?
3. Write various methods of sampling with their merits and demerits.
4. What is population and sample?
5. State advantages of sampling.
6. What is random sampling?
7. What are SRSWR and SRSWOR?
8. Distinguish between purposive sampling and random sampling.
9. Enumerate various steps in sampling design.
10. What are Sampling Error and Non-Sampling Error. How can sampling error be minimized?

REPRESENTATION OF DATA

Data is not necessarily information. Also having more data does not always produce better decisions. Hence, the data are to be summarized and presented in useful ways for taking prompt and effective decisions. Consider the following, where the grade points of 20 final year MBA students are listed.

Sl No.	Grade Point	Sl No.	Grade Point	Sl No.	Grade Point	Sl No.	Grade Point
1	19.8	6	15.0	11	15.0	16	12.5
2	17.3	7	15.0	12	12.5	17	6.0
3	17.3	8	15.0	13	15.0	18	9.5
4	19.8	9	17.3	14	15.0	19	12.5
5	17.3	10	15.0	15	15.0	20	9.5

On seeing the above data, it is not possible to arrive at any valid conclusion. These data need further arrangement. So the simplest way to present the data is in a *data array*. It arranges values in ascending or descending order. The following table arranges the grade points of students in descending order.

Sl No.	Grade Point	Sl No.	Grade Point	Sl No.	Grade Point	Sl No.	Grade Point
1	19.8	6	17.3	11	15.0	16	12.5
2	19.8	7	15.0	12	15.0	17	12.5
3	17.3	8	15.0	13	15.0	18	9.0
4	17.3	9	15.0	14	15.0	19	9.0
5	17.3	10	15.0	15	12.5	20	6.0

In this type of data representation, it is easy to find the highest and lowest grade points. It is possible to divide the data set into sections- the first 17 values pertain to upper half and the rest pertain to the lower half of grade points. It is possible to notice if any value appears more than once in the data array. It is easy to find the range, within which the students' grade points lie. Despite these advantages, the data array compels to list all the observations.

Moreover, for a large set of data, it becomes cumbersome to list all the observations. So, the data needs to be compressed and still be able to use it for interpretation and decision making.

FREQUENCY DISTRIBUTION

The most convenient way to represent the data is by using a *frequency table* or *frequency distribution*. Consider the following table, where the number of students securing various grade points are shown:

Grade Point	Number of Students (Frequency)
19.8	2
17.3	4
15.0	8
12.5	3
9.5	2
6.0	1

The table above shows the grade points (variable-x) with the number of students (frequency) securing such grade points. This frequency has been distributed over a range of variables (Grade Points). Frequency distributions are of two types: (a) Simple frequency distribution or Ungrouped frequency distribution and (b) Grouped frequency distribution. The above is an example of simple frequency distribution. Here the observations are not divided into groups or classes. Their individual values are only shown. In the grouped frequency distribution the observations are divided into groups or classes. The table below shows the same data, where the variables are in groups.

INCLUSIVE SERIES

Grade Point	Number of Students (Frequency)
1–5	6
6–10	3
11–15	11
16–20	6

The above table is called an *Inclusive Series*. This type of series is made in case of discrete variables. An item whose value is equal to the upper limit is included in that class itself. In such a case, the upper limit of one class is less than the lower limit of the next class when classes are arranged in an ascending order. In this type, overlapping of class intervals is avoided. Both upper and lower limits are included in the class intervals. For example, 5, 10, 15 and 20 are all included in the same class or group they are written. Therefore, it is known as inclusive series. The method ignores continuity in the class intervals.

EXCLUSIVE SERIES

In case of *exclusive series,* an item whose value coincides with the upper limit of a class is excluded from that class and is included in the next class that has an equivalent lower limit. In this type, a value equal to the lower limit, is included in the class interval and a value equal to the upper limit is excluded. The following table is prepared considering as exclusive series. Upper limit of one class is equal to the lower limit of the next higher class.

For any calculations or operations to be done on the data set, if given as an inclusive series, it is mandatory to first convert it to an exclusive series, to get the correct result. This can be done by applying the "Correction Factor" to the lower and upper limits. Correction factor is half of the difference between the value of lower limit of class and value of upper limit of preceding class. The above inclusive series is thus converted to an exclusive series as shown below.

Grade Point	**Number of Students (Frequency)**
0.5–5.5	0
5.5–10.5	3
10.5–15.5	3
15.5–20.5	14

This method assumes continuity in the class intervals. For example, 15.5 is not included in the class 10.5-15.5, but is included in the class 15.5-20.5 and similarly for all other classes.

Hence, a frequency distribution shows the number of observations from the data set that fall into each of the classes. If the frequencies of each class can be determined, a frequency distribution can be constructed.

RELATIVE FREQUENCY DISTRIBUTION

A more convenient way of representing the data is by *relative frequency distribution.* We can express the frequency of each value as a fraction or a percentage of the total number of observations. Consider the following, where the relative frequency of class 5.5-10.5 is obtained by dividing the frequency of that class (3) by the total frequency (20), which comes to $3 \div 20 = 0.15$. Similarly in class 15.5-20.5, the relative frequency is $14 \div 20 = 0.70$.

Grade Point	Number of Students (Frequency)	Relative Frequency
0.5 – 5.5	0	0
5.5 – 10	3	0.15
10.5 – 15.5	3	0.15
15.5 – 20.5	14	0.70
Total	20	1.00

Hence, a relative frequency distribution presents frequencies in terms of fractions or percentages. The sum of all the relative frequencies is always equal to 1.00 or 100 per cent.

Let us take up an example of how a raw data is taken and actual construction of frequency distribution is done. The ages of 50 members of a community are shown below:

83	51	66	61	82	65	54	56	92	60
65	87	68	64	51	70	75	66	74	68
44	55	78	69	98	67	82	77	79	62
41	88	76	99	84	47	60	42	66	74
91	71	83	80	68	65	51	56	73	55

In order to construct a frequency distribution, the data is arranged in ascending order first.

41	51	56	62	66	68	73	77	82	88
42	51	56	64	66	68	74	78	83	91
44	54	60	65	66	69	74	79	83	92
47	55	60	65	67	70	75	80	84	98
51	55	61	65	68	71	76	82	87	99

Next, we need to decide how many different classes to use and the range of each class should cover. The range must be divided into equal classes; that is, the width of the interval from the beginning of one class to the beginning of the next class must be the same for every class. If we choose a width of 5 years of age for each class, then the classes will be as shown in the table below:

Class in Years	Frequency
41 – 45	3
46 – 50	1
51 – 55	6

Class in Years	Frequency
56 – 60	4
61 – 65	6
66 – 70	9
71 – 75	5
76 – 80	5
81 – 85	5
86 – 90	2
91 – 95	2
96 – 100	2

In fact the number of data points and the range of data collected decide the number of classes. The more data points or the wider the range of the data, the more classes it takes to divide the data. Usually fewer than 6 or more than 15 classes are not chosen by statisticians. The rule is:

$$\text{Width of Class Interval} = \frac{\text{Next Unit Value after Largest Value}}{\text{Total Number of Class Intervals}}$$

In our current example, the next unit value after the largest value 99 is 100 and smallest value is 41. If we decide to have 6 classes, then the width of each class will be (100 – 41) ÷ 6 = 9.83 rounded to the next integer = 10. Next step is to sort the data points into classes and count the number of points in each class. Then the required frequency table can be constructed, as shown below.

Class in Years	Frequency	Relative Frequency
41 - 50	4	0.08
51 - 60	10	0.20
61 - 70	15	0.30
71 - 80	10	0.20
81 - 90	7	0.14
91 - 100	4	0.08
	Total: 50	1.00

Then it is required to prepare the exclusive series, as shown below:

Class in Years	Frequency	Relative Frequency
40.50 – 50.50	4	0.08
50.50 – 60.50	10	0.20

Class in Years	Frequency	Relative Frequency
60.50 – 70.50	15	0.30
70.50 – 80.50	10	0.20
80.50 – 90.50	7	0.14
90.50 – 100.50	4	0.08
	Total: 50	1.00

The same can also be represented in a chart also as given below.

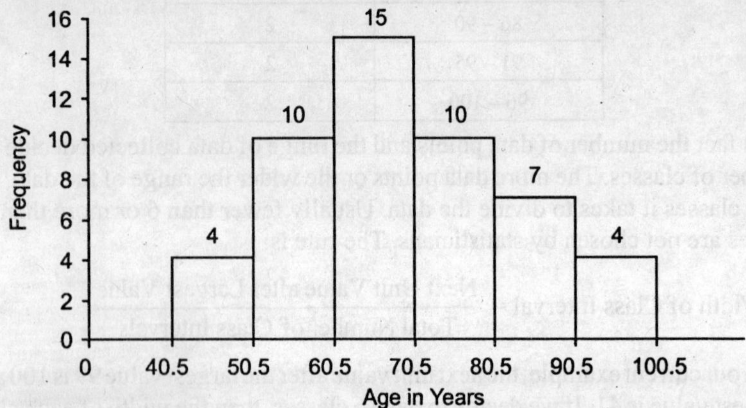

It can be seen in the above figure that the frequencies in the class width of 10 years age follow a regular progression. The number of data points begins with 4 for the first class, builds upto 10 in the second class and reaches to 15 in the third class, falls to 10, 7 and finally to 4 in the fourth, fifth and sixth class respectively. So, the larger the width of class intervals, smoother this progression will be. Nevertheless, if the classes are too wide, much information is lost so that the chart becomes almost meaningless, as shown in the figure below.

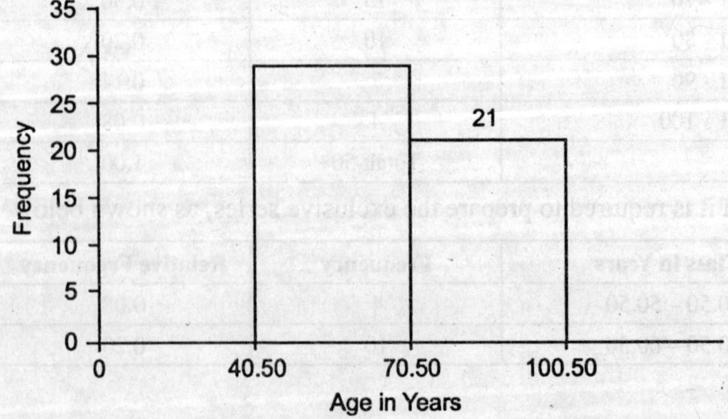

Some terms associated with grouped frequency distribution are explained below:

Class: When the data of a series are divided into certain groups and the groups are bounded by limits, each group is called a class. In the above example 41 – 50, 51 – 60, etc. are each called a class.

Class Interval: The number of class intervals should neither be too low nor too high. 6 to 15 numbers of class intervals are ideal. It is important that the class intervals should be of uniform width so that the frequency of classes can easily be compared. For business studies the class intervals should be kept in ascending order. Length of class intervals should not be very small or very large.

Class Limits: The two end limits of each class are called class limits. The smaller value of the two limits is called the *lower limit* of the class and the higher value of the same is called the *upper limit* of the class. For example- in the class 41 - 50, the lower limit is 41 and upper limit is 50.

Class Boundary: Class boundaries are the limits up to which the two limits of each class may be extended to fill up the gap which exists between classes. The lower extreme value of the two boundaries is called the *lower class boundary* and the upper extreme value of the same is called the *upper class boundary.* For example- in the class 41 - 50, the lower limit can be extended up to 40.50 and hence is the lower class boundary. The upper limit can be extended up to 50.50 and hence is the upper class boundary. Similarly, for the class 51- 60, the two class boundaries are 50.50 and 60.50 and so on.

Class Magnitude: The difference between the upper class boundary and the lower class boundary is the class magnitude. It is also known as the *width of a class* or the *size of a class*. For example, the class magnitude for class 41 - 50 is 50.5 – 40.5 = 10. The class magnitude for class 51 - 60 is 60.50 – 50.50 = 10 etc. The size of a class should not be too broad or too small.

Mid Value: The value just at the middle of a class is called its mid-value. The mid-value is also known as *mid-point* or *central value* of a class. It is the arithmetic mean of the lower class boundary and the upper class boundary of the same class. For example -the mid-value of the class 41 - 50 is and the mid-value for the class 51 - 60 is 55.50 and so on. The mid-value of a class is used to represent that class. So, the mid-value is called *class mark.*

Class Frequency: The number of observations that falls in a class is known as class frequency. In the above example the class frequency of the class 41 - 50 is 4. The sum of frequencies of all the classes of a frequency distribution is called the *total frequency.*

The unprocessed raw data are only a dry set of figures which can be arranged in tabular form, or presented in the forms of graphs or diagrams. The special feature of diagrams and graphs is that they do away with figures altogether, and present dry and uninteresting statistical facts in the shape of attractive and appealing pictures and charts. Such diagrams and pictures give a bird's eye view of the whole mass of statistical data. This is one of the most appealing ways of presenting the statistical data. While representing the data in graphical form, on horizontal axis, the values of variable are shown and on vertical axis, the frequencies of the classes are shown.

HISTOGRAM

One of the very popular ways of data representation is by *histogram*. Histogram is a series of rectangles, each proportional in width to the range of values within a class and proportional in height to the number of items falling in the class. A histogram that uses relative frequency of data points in each of the classes instead of the actual number of points is called a *relative frequency histogram*. It has the same shape as that of an absolute frequency histogram. The graph below shows the relative frequency histogram of the previously drawn absolute frequency histogram.

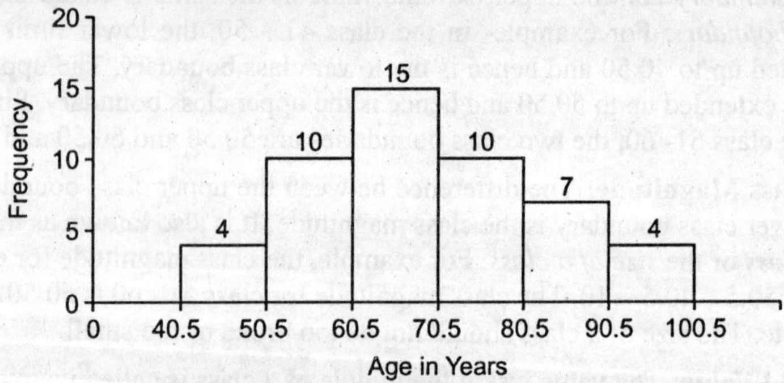

It may be noted that the sum of all relative frequencies is always 1.

FREQUENCY POLYGON

Frequency polygons are another way to represent graphically both simple and relative frequency distributions. Here, the frequencies are marked on vertical axis and the values of variables on the horizontal axis, as it is done in histograms. Then the class frequencies are marked as dots above the mid-points of all the classes, which are subsequently joined by straight lines to form a polygon. Hence, a frequency polygon is simply a line graph that connects the mid-points of all the bars in the histogram. The figure below shows the frequency polygon of the histogram drawn above. It is to be noted that classes have been added at each end of the scale of observed values. These two new classes contain zero observations but allow the polygon to reach the horizontal axis at both ends of the distribution.

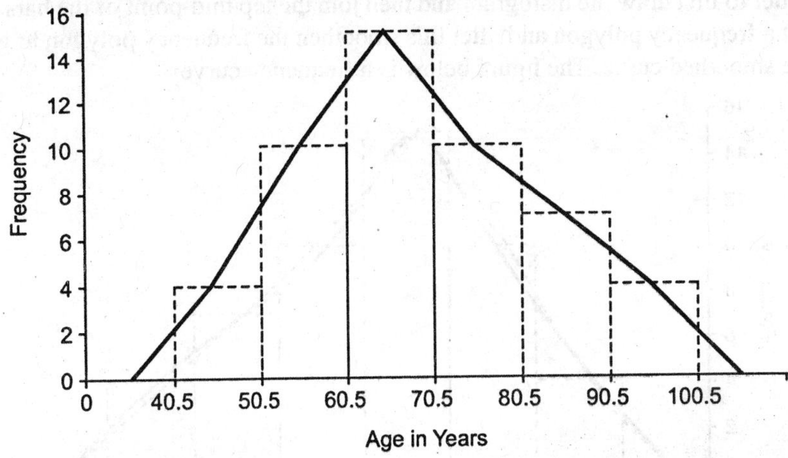

ADVANTAGES OF HISTOGRAM

1. In a histogram, the rectangles clearly show separate classes in the distribution.
2. The area of each rectangle, relative to all the other rectangles, shows the proportion of the total number of observations that occur in that class.

ADVANTAGES OF FREQUENCY POLYGON

1. Frequency polygons are simpler than histograms.
2. The outlines of the data patterns are more clearly drawn here.
3. As the number of classes increases, the polygon becomes smoother.

FREQUENCY CURVE

If we observe the frequency polygon above, it is seen that there are many angulations in the polygon at the junction points. Such angulations can be smoothed if joined by free hand. When the number of observations is very large and the group intervals are reduced, the frequency polygon tends to lose its angulations giving place to a smooth curve known as *frequency curve*. In smoothening a frequency polygon, extreme care should be exercised. The top of this smooth curve should generally be above the highest point of the frequency polygon. The smoothening should not distort the original shape of the curve. Another point to be kept in mind while smoothening a curve is that, the area under the smoothed curve should be almost the same as the area of the polygon, or it should be equal to the total of the frequencies. It is always better to first draw the histogram and then join the top mid-point of the bars to get a frequency polygon and after this smoothen the frequency polygon to get the smoothed curve. The figure below is a frequency curve:

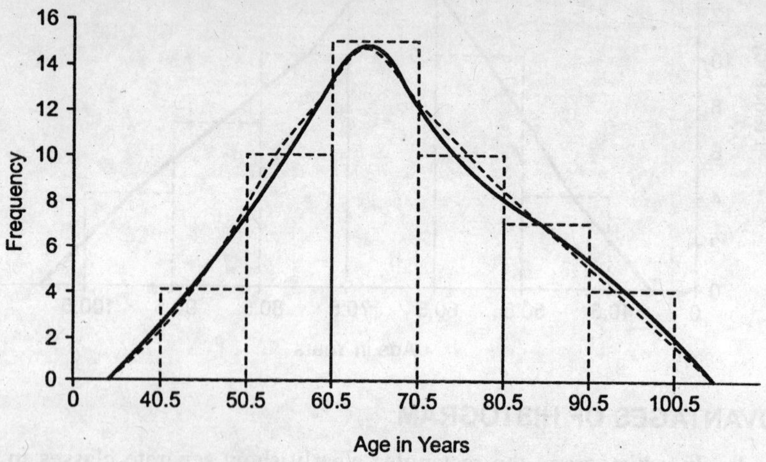

CUMULATIVE FREQUENCY CURVE or OGIVE

Sometimes it is more useful to cumulate the data and to draw what are called cumulative frequency curves or *ogives*. It enables us to see how many observations lie above or below certain values, rather than merely recording the number of items within intervals. For example, if we wish to know how many persons have age below say 80 years, we can use the curve shown below (Ogive- Less Than type) and the result is 39.

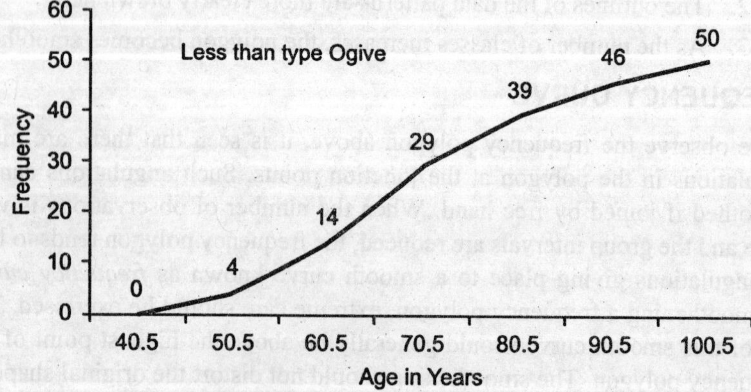

If it is required to know how many persons are above a certain age say 61 years, we can use the curve shown below (Ogive- More Than Type) and the result is 36. It may be noted that, in less than type, the upper limits of the class and in more than type, the lower limits of the class are plotted along the x-axis. We can also construct ogive of a relative frequency distribution in the same manner we draw the ogive of an absolute frequency distribution.

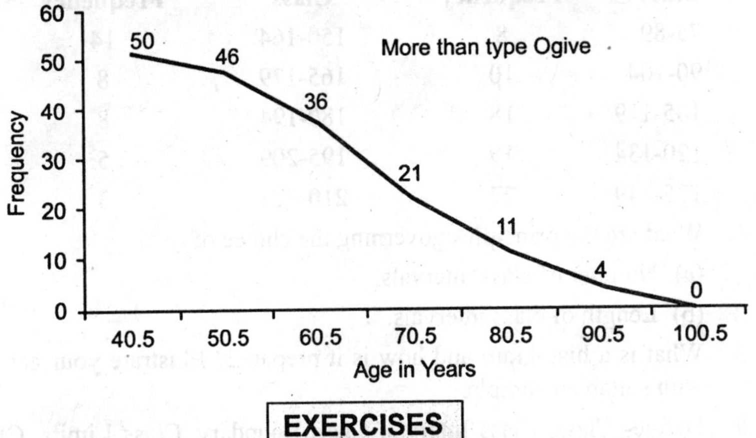

EXERCISES

1. One agricultural farm has these data representing weekly growth (in cm) of samples of newly planted corn tree.

1.1	4.9	3.8	0.8	4.1	1.1	3.8	3.1
2.1	2.3	1.8	2.3	1.8	2.3	3.8	1.3
3.8	4.4	4.6					

   (a) Arrange the data in an array from highest to lowest.

   (b) Construct a relative frequency distribution using intervals of 0.5.

   (c) Construct an ogive that will help to determine what proportion of the corn grew at more than 2.5 cm in a week.

2. A manufacturing company keeps record of the number of each item produced per month in order to examine the relative production levels. Records show the following numbers of each item produced by the company for the last month of 20 operating days:

9,898	10,050	10,030	9,720	9,908	10,098	10,588	9,873
9,966	9,929	10,123	10,508	9,910	9,995	10,238	

   Construct an ogive and answer the following:

   (a) On how many of its items did production exceed the break-even point of 10,000 units?

(b) What production level did 75% of its items exceed that month?

(c) What production level did 90% of its items exceed that month?

3. Here is a frequency distribution of 120 people who used ski-lift on a certain day for repair of street lamps. Construct a histogram for these data.

Class	Frequency	Class	Frequency
75-89	8	150-164	14
90-104	10	165-179	8
105-119	18	180-194	8
120-134	19	195-209	5
135-149	27	210-224	3

4. What are the principles governing the choice of:

   (a) Number of class intervals.

   (b) Length of class intervals.

5. What is a histogram and how is it prepared? Illustrate your answer with suitable example.

6. Define- Class, Class Interval, Class Boundary, Class Limits, Class Magnitude, Mid Value, Class Frequency.

7. What is class interval and how is its width determined in preparing the frequency distribution table for a series of ungrouped data?

8. What is an *Ogive* and how is it prepared?

9. Differentiate between frequency curve and relative frequency curve with suitable diagrams.

10. Enumerate the uses of normal frequency curve.

11. How frequency polygon and frequency curve prepared?

12. Is it necessary to prepare a grouped frequency distribution from a given data? What are the considerations that one has to bear in mind while forming a grouped frequency distribution?

13. Explain the method of constructing a histogram and a frequency polygon.

14. Here is a frequency distribution of the length of phone calls made by 170 people during a holiday weekend. Construct a histogram for these data.

Duration of call	1-7	8-14	15-21	22-28	29-35	36-42	43-49	50-56
Frequency	43	32	34	22	15	12	8	4

Construct the general shape of the histogram. Does it appear to be a pattern?

15. Before constructing a dam across a river, the engineers performed a series of tests to measure the water flow past the proposed location of the dam. The results of the testing were used to construct the following frequency distribution:

River Flow ('000 cubic metre per second)	**Frequency**
0.076 – 0.079	7
0.080 – 0.083	21
0.084 – 0.087	32
0.088 – 0.091	49
0.092 – 0.095	58
0.096 – 0.099	41
0.010 – 0.013	27
0.014 – 0.017	10
	246

(a) Use the data given in the above table to construct a "more-than" cumulative frequency distribution and ogive.

(b) Use the data given in the above table to construct a "less-than" cumulative frequency distribution and ogive.

(c) Use the ogive to estimate what proportion of the flow occurs at less than 0.0985 thousand cubic metre per second.

MEASURES OF CENTRAL TENDENCY

Statistical analyses need manipulation of huge amounts of data, where large numbers of big figures are not only confusing but also difficult to process. Hence condensation of such huge data to a meaningful representative form is absolutely necessary. In order to *reduce the complexity of data* obtained from different sources and *to make them comparable,* it is essential that data should be reduced to a single figure. If, for example, a direct comparison were made between the ages of persons belonging to two different communities consisting about 5000 numbers of persons each, it would be impossible to arrive at any conclusion. On the other hand, if ages of each community were represented by one figure, comparison would be an extremely easy affair. It is obvious that the single figure, which is used to represent a whole series (in the present case age of individual persons of one community) should neither have the lowest value in the series nor the highest value but a value somewhere between these two limits, possibly in the center, where most of the items of the series cluster. Such figures are called *Measures of Central Tendency* or *Averages.* The average represents a whole series and as such, its value always lies between the minimum and maximum values and generally it is located in the center or middle of the distribution. According to Simpson and Kafka, *"A measure of central tendency is a typical value around which other figures congregate"*, whereas according to Ya-Lun Chou, *"an average is a typical value in the sense that it is sometimes employed to represent all the individual values in a series or of a variable."*

So averages can broadly be classified into two types: (1) Mathematical Average and (2) Positional Average. Mathematical Average is further divided into Arithmetic Mean, Geometric Mean and Harmonic Mean; whereas Positional Average is further divided into Mode and Median, which are discussed below with their merits and demerits

ARITHMETIC MEAN

Arithmetic Mean of a series is the figure obtained by dividing the total values of all the items of the series by the number of items. It is denoted by $\overline{x}$.

Determination of Arithmetic Mean of Individual Series

As it has been said above, the simple arithmetic mean of a series is equal to the sum of items divided by the number of items. Let $x_1, x_2, x_3, \ldots\ldots x_n$ be the n values of variables. It can be expressed mathematically by:

$$\overline{x} = \frac{x_1 + x_2 + x_3 + \ldots\ldots\ldots\ldots x_n}{n} = \frac{\sum x}{n}$$

where,

$\overline{x}$ = Arithmetic mean

$\sum$ = Summation symbol

n = Number of items.

Example 4.1. M/s Dorenbo Furniture Company has a revolving credit agreement with the Axis National Bank. The loan showed the following ending monthly balances last year:

Jan: Rs.1,20,300 Apr: Rs.72,600 Jul: Rs.57,800 Oct:Rs.52,800

Feb: Rs.1,11,300 May: Rs.72,600 Aug: Rs.61,200 Nov:Rs.49,300

Mar: Rs.71,900 Jun: RTs.59,500 Sep: Rs.50,500 Dec:Rs.46,200

The Company is eligible for a reduced rate of interest if its average monthly balance is over Rs. 63,000. Does it qualify?

Solution: The Company will certainly be eligible for the reduced rate of interest, if the average monthly balance over the previous year is more than the qualifying amount of Rs. 63,000. Hence the arithmetic mean of all the monthly balance amounts is equal to

$$\frac{120300 + 111300 + 71900 + 72600 + 72600 + 59500 + 57800 + 61200 + 50500 + 52800 + 49300 + 46200}{12}$$

$$= \frac{826000}{12} = \text{Rs. } 68833.33$$

Hence, the Company is eligible to get the reduced rate of interest.

Determination of Arithmetic Mean of Discrete Series

In this case the values of the variables are multiplied by their respective frequencies and the products so obtained are totaled. This total is then divided by the total number of items (which is equal to the total of the frequencies). The resulting quotient is a simple arithmetic mean of the series.

If $f_1, f_2, f_3, ..., f_n$ stand for the frequencies of the values $x_1, x_2, x_3, ..., x_n$ respectively and $N (=\Sigma f)$ is the total number of items, then the arithmetic mean can be mathematically expressed as

$$\overline{x} = \frac{f_1x_1 + f_2x_2 + f_3x_3 + \ldots\ldots\ldots\ldots f_nx_n}{N} = \frac{\sum fx}{\sum f}$$

Example 4.2. Blossom orphanage school is eligible for a national social services grant as long as the average age of its children stays below 9. If the following data represent the ages of all the children currently attending the school, do they qualify for the grant?

8 5 10 9 12 12 7 8 7 13 8 10 7

12 8 13 5 5 10 8 7 13 5 9 9

Solution: The given data is arranged in a frequency table as shown below:

Age:	5	7	8	9	10	12	13
No of children:	4	4	5	3	3	3	3

Hence,

$$\text{Mean} = \frac{4\times5+4\times7+5\times8+3\times9+3\times10+3\times12+3\times13}{4+4+5+3+3+3+3} = \frac{220}{25} = 8.80$$

So, the school is eligible for the social grant.

Determination of Arithmetic Mean of Continuous Series

In case of a grouped frequency distribution, the class mid-values are taken in place of the individual values. Thus if m denotes the class mid-values, then mean will be:

$$\overline{x} = \frac{f_1m_1 + f_2m_2 + f_3m_3 + \ldots\ldots\ldots\ldots + f_nm_n}{\sum f} = \frac{\sum fm}{\sum f}$$

Example 4.3: The frequency distribution below represents the weights in kg of a sample of packages carried last month by a small aircraft company.

Class	Frequency	Class	Frequency
10.0-11.0	2	15.0–16.0	10
11.0-12.0	4	16.0–17.0	8
12.0-13.0	6	17.0–18.0	6
13.0-14.0	7	18.0–19.0	5
14.0-15.0	10	19.0–20.0	3

Compute the sample mean.

Solution: The following shows the detail calculation, which is self-explanatory.

Class	Frequency (f)	Class Mid-Value (m)	f × m
10.0-11.0	2	10.5	21
11.0-12.0	4	11.5	46
12.0-13.0	6	12.5	75
13.0-14.0	7	13.5	94.5
14.0-15.0	10	14.5	145
15.0-16.0	10	15.5	155
16.0-17.0	8	16.5	132
17.0-18.0	6	17.5	105
18.0-19.0	5	18.5	92.5
19.0-20.0	3	19.5	58.5
Total	**61**		924.5

Hence, $\text{Mean} = \dfrac{\sum fm}{\sum f} = \dfrac{924.50}{61} = 15.155 \text{ kg}$

CODING

Many times very large numbers are encountered while finding out the arithmetic mean and it becomes quite cumbersome to deal with such large numbers. In case of grouped data, to eliminate the problem of large or inconvenient mid-points, codes are assigned to them. In place of usual mid-points to carry out further calculations, small-value consecutive integers (called codes) are assigned to the mid-points. Usually zero value is assigned to the mid-point of the class at the middle or near the middle of the frequency distribution.

Accordingly negative integers are assigned to the values smaller than mid-point and positive integers to the values larger than it. So the formula for the arithmetic mean of grouped data using codes is:

$$\overline{x} = x_0 + w\frac{\sum(u \times f)}{n}$$

where,

$\overline{x}$ = mean of sample.

x_0 = mid value of class at middle position

w = width of class interval

u = code assigned to each class

f = frequency of each class

n = total number of observations

Example 4.4: The frequency distribution below represents the weights in Kg of a sample of packages carried last month by a small aircraft company.

Class	Frequency	Class	Frequency
10.0-11.0	2	15.0-16.0	10
11.0-12.0	4	16.0-17.0	8
12.0-13.0	6	17.0-18.0	6
13.0-14.0	7	18.0-19.0	5
14.0-15.0	10	19.0-20.0	3

Compute the sample mean.

Solution: The following shows the detail calculation, which is self-explanatory.

Class	Frequency (f)	Class Mid-Value (x)	Code (u)	u × f
10.0–11.0	2	10.5	-4	-8
11.0–12.0	4	11.5	-3	-12
12.0–13.0	6	12.5	-2	-12
13.0–14.0	7	13.5	-1	-7
14.0–15.0	10	14.5	0	0
15.0–16.0	10	15.5	+1	10
16.0–17.0	8	16.5	+2	16

Class	Frequency (f)	Class Mid-Value (x)	Code (u)	u × f
17.0–18.0	6	17.5	+3	18
18.0–19.0	5	18.5	+4	20
19.0–20.0	3	19.5	+5	15
Total:	**61**			**40**

$$\overline{x} = x_0 + w\frac{\sum(u \times f)}{n} = 14.5 + 1 \times \frac{40}{61} = 15.155$$

The arithmetic mean is the most popularly used measure of central tendency. There are many reasons for its popularity, which are stated below.

Merits of Arithmetic Mean

1. It is rigidly defined and its value is always definite.
2. It is based on all the observations of a series. It cannot be calculated if a single value is left out.
3. It is capable of further algebraic treatment.
4. It is easy to understand and calculate. It does not need arrangement of data.
5. It is least affected by the fluctuations of sampling.

Demerits of Arithmetic Mean

The arithmetic mean should be used with caution because of certain drawbacks.

1. Unlike median and mode it cannot be calculated merely by inspection.
2. It can be a figure, which may not exist in the series.
3. Sometimes it gives very absurd results like 5.2 number of persons or 10.8 number of machines etc.
4. Since arithmetic average is calculated from all the items of a series, sometimes the abnormal items may considerably affect this average.
5. It cannot be calculated without all the items of a series. If for example one item out of say 1000 items is left out, then mean cannot be known, whereas median and mode can be found out.

GEOMETRIC MEAN

When values of a variable change over a period of time, we need to know an average rate of change, such as an average growth rate over a period of several years. In such cases, the simple arithmetic mean is inappropriate, because it gives the wrong answers. We need to find the *geometric mean* or simply called G.M.

Geometric mean is the nth root of the product of n items of a series. Thus if the geometric mean of 4 and 6 is to be calculated it would be equal to the square root of the product of 4 and 6. Geometric mean (G.M.) of 4, 6 and 8 is cube root of the product of 4, 6, and 8 and so on.

Determination of G.M. of Individual Series:

Let $x_1, x_2, x_3, \ldots, x_n$ be *n* observations. Then G.M. is determined by

$$\text{G.M.} = \sqrt[n]{x_1 \times x_2 \times x_3 \times \cdots x_n}$$

On taking logarithm of both sides and then taking anti-logarithm the formula reduces to

$$\text{G.M.} = \text{Antilog}\left[\frac{\sum(\log x)}{n}\right]$$

Thus G.M. is the anti-log of the arithmetic average of the log of the values of a variable.

Determination of G.M. of Discrete Series

If variables $x_1, x_2, x_3, \cdots x_n$ have frequencies $f_1, f_2, f_3, \cdots f_n$ respectively, then

$$\text{G.M.} = \text{Antilog}\left[\frac{\sum(f \times \log x)}{\sum f}\right]$$

where, $\Sigma f = N$

Determination of G.M. of Continuous Series

In case of continuous series the class mid value (m) is considered and the formula is

$$\text{G.M.} = \text{Antilog}\left[\frac{\sum(f \times \log m)}{\sum f}\right]$$

where, $\Sigma f = N$

Example 4.5: The growths in bad-debt expense for Brandon Office Supply Company over the last few years are as follows: Calculate the average percentage in bad-debt expense over that time period. If this rate continues, estimate the percentage increase in bad-debt for 2012, relative to 2002.

2002	2003	2004	2005	2006	2007	2008	2009	2010
0.12	0.08	0.085	0.09	0.095	0.106	0.120	0.130	0.135

Solution: According to business mathematics rule, the growth factor is equal to $1+\frac{\text{interest rate}}{100}$

So the various interest rates are: 1.12, 1.08, 1.085, 1.09, 1.095, 1.106, 1.120, 1.130, 1.135

Hence, the average increase is 1.1066-1 = 0.1066 i.e. 10.66% per year. The estimate for bad-debt expenses in 2012 is (as per business mathematics rule) i.e. 22.45% higher than that in 2002.

Example 4.6: Mid-East Textiles has shown the following percentage increase in net worth over the last 5 years.

2001	2002	2003	2004	2005
5%	10.5%	9.0%	6.0%	7.5%

What is the average increase in the net worth over the 5-year period ?

Solution: Hence, the average increase in the net worth over the 5-year period is 7.33%.

Merits of Geometric Mean

1. The geometric mean is rigidly defined and its value is a precise figure.
2. It is capable of further algebraic treatment.
3. It is based on all observations of a series.
4. It is not affected by the extreme values.

Demerits of Geometric Mean

1. It is neither simple to calculate nor easy to understand.
2. If any value in the series is zero, then its value would also be zero.
3. Like arithmetic average, it may be a value which does not exist in the series.
4. If one of the observations is negative it cannot be calculated.

MEDIAN

As the name itself suggests, median is the value of the middle item of a series arranged in ascending or descending order of magnitude. Median is a single value from the data set that measures the central item in the data. This single item is the middlemost or most central item in the set of numbers.

Determination of Median from Individual Series

In case of individual series, if the number of items (n) is odd, then median will be the value of the $\left(\frac{N+1}{2}\right)^{th}$ item. If the number of items is even, then the median will be the arithmetic mean of the value of $\left(\frac{N}{2}\right)^{th}$ item and $\left(\frac{N}{2}+1\right)^{th}$ item. The data must be arranged in ascending or descending order.

Determination of Median from Discrete Series

Here the median always falls at the item.

Determination of Median from Continuous Series

The series must be converted to exclusive series, if initially it is an inclusive series.

Remember that in this case, median falls at $\left(\frac{N+1}{2}\right)^{th}$ item. The formula for this is:

l_1 = Lower boundary of the class where median lies

l_2 = Upper boundary of the class where median lies

f_1 = Frequency of the median class

N = Total frequency

c = Cumulative frequency preceding the median class

Example 4.7: For the following frequency distribution, determine the median class and the median value.

Class	Frequency	Class	Frequency
100-150	12	300-350	72
150-200	14	350-400	63
200-250	28	400-450	35
250-300	56	450-500	20

Solution: The given series is an exclusive series. The calculation table below is self-explanatory.

Class	Frequency	Cumulative Frequency	
100-150	12	12	
150-200	14	26	
200-250	28	54	
250-300	56	110	Median Class
300-350	**72**	**182**	
350-400	63	245	
400-450	35	280	
450-500	20	300	

Median Class will fall at 300 ÷ 2 = 150 i.e., Class 300–350

$$\text{Median} = l_1 + \left(\frac{l_2 - l_1}{f_1}\right) \times \left(\frac{N}{2} - c\right) = 300 + \left(\frac{350-300}{72}\right) \times \left(\frac{300}{2} - 110\right)$$

= 327.777

Example 4.8: For the following frequency distribution, find out (a) the estimated value of the median (b) the class containing the median.

Class	*Frequency*	*Class*	*Frequency*
11-20	*8*	*61-70*	*52*
21-30	*15*	*71-80*	*84*
31-40	*23*	*81-90*	*97*
41-50	*37*	*91-100*	*16*
51-60	*46*	*101 or more*	*5*

Solution: The given data is an inclusive series. So it is converted to an exclusive series. The table is reconstructed below.

Class	Frequency	Cumulative Frequency	
10.5-20.5	8	8	
20.5-30.5	15	23	
30.5-40.5	23	46	
40.5-50.5	37	83	
50.5-60.5	46	129	
60.5-70.5	52	181	
70.5-80.5	84	265	Median Class
80.5-90.5	97	362	
90.5-100.5	16	378	
100.5 or more	5	383	

Median Class will fall at 383 ÷ 2 = 191.5 i.e. Class 70.5 – 80.5

$$\text{Median} = = l_1 + \left(\frac{l_2 - l_1}{f_1}\right) \times \left(\frac{N}{2} - c\right) = 70.5 + \left(\frac{80.5 - 70.5}{84}\right) \times \left(\frac{383}{2} - 181\right)$$

$= 71.75$

Merits of Median

1. It is rigidly defined.
2. It is easy to calculate and easy to understand.
3. It is not affected by the values of the extreme items.
4. It can be located merely by inspection in many cases.
5. It can also be used in qualitative measures.
6. It can also be calculated even if some observational values are missing.

Demerits of Median

1. Median may not be representative of a series in many cases. This is specially so when there are wide variations between the values of different items.
2. It is not suitable for further algebraic treatment.
3. Median ignores the values of extreme items.

MODE

Mode is the most common item of a series. It is the value which occurs the largest number of times in a series. In the words of Croxton and Cowden, *"The mode of a distribution is the value at the point around which the items tend to be most heavily concentrated"*. According to A.M.Tuttle, *"Mode is the value which has the greatest frequency density in its immediate neighborhood"*.

Determining the mode from a set of data is very simple, as it is that value in the series, which occurs most number of times. Similarly, for grouped data, it is that value of the variable, which has maximum frequency. It may so happen that more than one value have the maximum frequency, in which case it becomes *bi-modal* (having two modes) or *tri-modal* (having three modes) or *multi-modal* (having many modes).

Determination of Mode of Continuous series

The following formula may be used to find out the mode from a grouped data.

Where, the class consisting of maximum frequency is called modal class.

l_1 = Lower Class Boundary of the class where maximum frequency falls.

l_2 = Upper Class Boundary of the class where maximum frequency falls.

f_0 = Frequency of class just before the modal class.

f_1 = Frequency of the modal class.

f_2 = Frequency of class just after the modal class.

Example 4.9: The ages of a sample of the students attending sophomore college during this semester are:

2017	15	20	23	41	33	21	18	20	18	33	32
2924	19	18	20	17	22	55	19	22	25	28	30
4419	20	39	20	19	21	24	22	31	33	15	28

Construct a frequency distribution with intervals 15-19, 20-24, 25-29, 30-34, and 35 and older. Then calculate the mode.

Solution: The data are arranged in the following table:

Inclusive Class	**Frequency**		**Exclusive Class**
15-19	11		14.5-19.5
20-24	**14**	**Modal Class (Maximum frequency)**	19.5-24.5
25-29	4		24.5-29.5
30-34	6		29.5-34.5
35 & Older	4		34.5 & Older

$$\text{Mode} = l_1 + (l_2 - l_1) \times \left(\frac{f_1 - f_0}{2f_1 - f_0 - f_2} \right)$$

$$= 19.5 + (24.5 - 19.5) \times [(14 - 11) \div (2 \times 14 - 11 - 4)] = 20.65$$

Example 4.10: The number of solar heating systems available to the public is quite large and their heat-storage capacities are quite varied. The distribution of heat-storage capacity of 30 systems those were recently tested is given below. Compute the mode for these data.

Days:	0.01–0.99	1–1.99	2–2.99	3–3.99	4–4.99	5–5.99	6-6.99
Frequency:	2	4	6	7	5	4	2

Solution: The data are arranged in the following table.

Days	Frequency	
0.005-0.995	2	
0.995-1.995	4	
1.995-2.995	6	
2.995-3.995	7	**Modal Class**
3.995-4.995	5	
4.995-5.995	4	
5.995-6.995	2	

Merits of Mode

1. It can be determined without much mathematical calculation. In discrete series mode can be located even by inspection.
2. It is commonly understood.
3. It is a value which always exists in the series.
4. It is not affected by the values of extreme items.
5. All the items of a series are not required for its determination.

Demerits of Mode

1. It is ill defined, indeterminate and indefinite.
2. It is not based on all the observations of a series and hence it is rarely used in any higher business or scientific purposes.
3. It is not capable of further mathematical operation.
4. It may be unrepresentative in many cases.
5. It may be impossible to get a definite value in many cases, as there may be 2, 3 or more modal values.

COMPARING MEAN, MEDIAN AND MODE

It can be seen from the above that mean is not always equal to median or mode. Thus these three (mean, median and mode) are independent of each other. They need not be equal to one another nor they are closely related.

A distribution can be symmetrical, positively skewed or negatively skewed. In a symmetrical distribution (see the figure below), mean, median and mode are identical and have the same value.

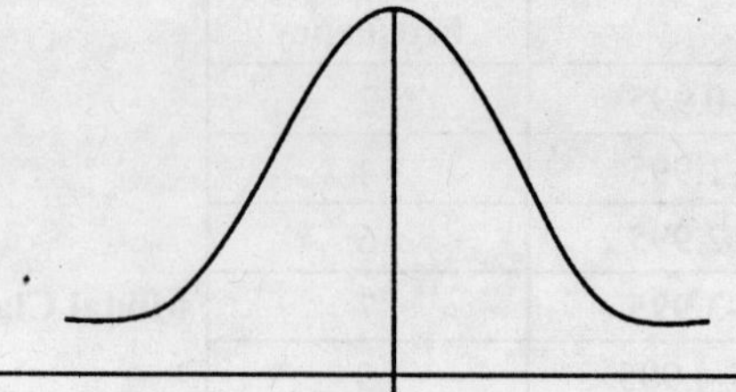

However, in actual life most distributions are not symmetrical. In case of positively skewed distribution (one skewed to the right), as shown in figure below, the mode is at the highest point of distribution, the median is to the right of that, and the mean is to the right of both the median and mode.

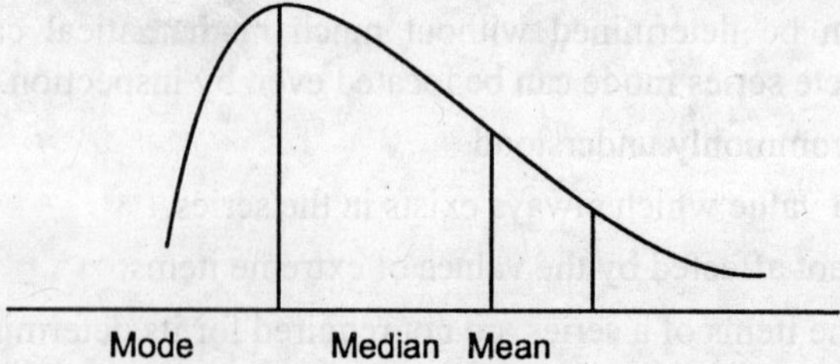

In case of negatively skewed distribution (one skewed to the left), as shown in figure below, the mode is at the highest point of distribution, the median is to the left of that, and the mean is to the left of both the median and mode. For a fairly symmetrical distribution, mathematically, the three measures are related as

MODE – MEDIAN = 2 (MEDIAN – MEAN)

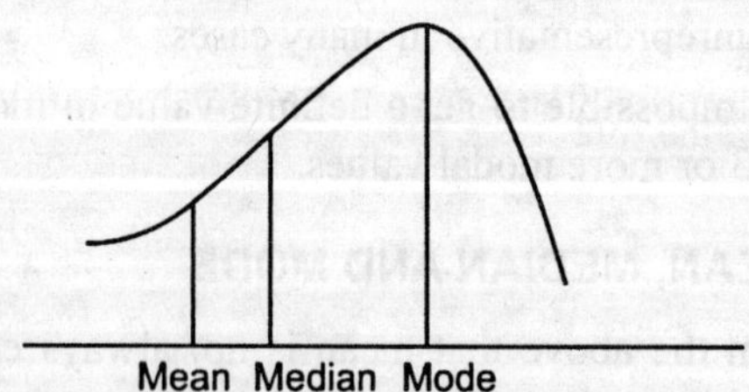

QUARTILES, DECILES & PERCENTILES

If a set of data are arranged in order of magnitude (i.e. ascending order or descending order), the middle most value (when the number of observations is odd- one middle value, or when the number of observations is even- arithmetic mean of the two middle values) that divides the set into two equal parts is the median. By extending this idea, those values which divide the set of data into four equal parts at points denoted by Q_1, Q_2 and Q_3 are otherwise called as first quartile, second quartile and third quartile respectively, the value Q_2 being equal to median. Similarly, the values that divide the set of data into 10 equal parts are called deciles and are denoted by $D_1, D_2, D_3, \cdots D_9$, the value D_5 being equal to Q_2 and median. Further, the values that divide the set of data into 100 equal parts are called percentiles and are denoted by $P_1, P_2, \ldots\ldots. P_{99}$., the P_{50} being equal to D_5, Q_2 and median, P_{25} being equal to Q_1 and P_{75} being equal to Q_3.

Computation of Quartiles, Deciles and Percentiles

For Individual Data & Discrete Data: Let the set of data is given by $x_1, x_2, \cdots x_n$. Hence we have n number of observations and suppose they are arranged in order of magnitude.

$$Q_i = i^{th} \text{ quartile} = \text{Value of } \left[\frac{i(N+1)}{4}\right]^{th} \text{ item, } i = 1, 2, 3$$

$$D_i = i^{th} \text{ decile} = \text{value of } \left[\frac{i(N+1)}{10}\right]^{th} \text{ item, } i = 1, 2, \ldots..9$$

$$P_i = i^{th} \text{ percentile} = \text{Value of } \left[\frac{i(N+1)}{100}\right]^{th} \text{ item; } i = 1, 2, 3, \ldots\ldots.99$$

For Continuous Data:

$$Q_i = i^{th} \text{ quartile} = \text{Value of } \left[\frac{i \times N}{4}\right]^{th} \text{ item, } i = 1, 2, 3$$

$$D_i = i^{th} \text{ decile} = \text{Value of } \left[\frac{i \times N}{10}\right]^{th} \text{ item, } i = 1, 2, \ldots..9$$

$$P_i = i^{th} \text{ percentile} = \text{Value of } \left[\frac{i \times N}{100}\right]^{th} \text{ item; } i = 1, 2, 3, \ldots\ldots.99$$

In case of continuous data. $Q_i = l_1 + \frac{l_2 - l_1}{f_1} \times \left[\frac{i \times N}{4} - C\right]; i = 1, 2, 3$

where, l_1 and l_2 are lower and upper class boundaries of ith quartile class.

f_i is the frequency of ith quartile class.

C is the cumulative frequency preceding to i^{th} quartile class.

N is the total frequency.

$$D_i = i^{th} \text{ decile} = l_1 + \frac{l_2 - l_1}{f_1} \times \left[\frac{i \times N}{10} - C\right];\ i = 1, 2, 3\ldots\ldots 9$$

where l_1 and l_2 are lower and upper boundaries of ith decile class.

f_i is the frequency of ith decile class.

C is the cumulative frequency preceding to i^{th} decile class.

N is the total frequency.

$$P_i = i^{th} \text{ percentile} = l_1 + \frac{l_2 - l_1}{f_1} \times \left[\frac{i \times N}{100} - C\right];\ i = 1, 2, 3\ldots\ \ldots\ldots 99$$

where l_1 and l_2 are lower and upper boundaries of ith percentile class.

f_i is the frequency of ith percentile class.

C is the cumulative frequency preceding to i^{th} percentile class.

N is the total frequency.

Example 4.16: Given the following data, calculate the quartiles, 7th decile and 45th percentile.

Age	20-25	25-30	30-35	35-40	40-45	45-50	50-55	55-60	60-65	65-70	70-75	75-80
Number	22	33	45	48	57	72	51	43	31	27	11	5

Solution: The following table is prepared. The given series is already an exclusive series.

Class Interval	Frequency	Cumulative Frequency
20-25	22	22
25-30	33	55
30-35	45	100
35-40	48	148
40-45	57	205
45-50	72	277
50-55	51	328
55-60	43	371
60-65	31	402
65-70	27	429
70-75	11	440
75-80	5	445

$N = \Sigma f = 445$

Position of first quartile $Q_1 = 1 \times N \ 4 = 445 \ 4 = 111.25^{th}$ item; it lies in the Class 35-40

Hence, Q_1 (lies in the interval 35-40) $= l_1 + \frac{l_2 - l_1}{f_1} \times \left[\frac{i \times N}{4} - C\right]$

$$= 35 + \frac{40-35}{48} \times \left[\frac{1 \times 445}{4} - 100\right] = 36.171$$

Position of 2nd quartile $Q_2 = 2 \times N \div 4 = 222.5^{th}$ item; it lies in the Class 45-50

Hence, Q_2 (lies in the interval 45-50) $= l_1 + \frac{l_2 - l_1}{f_1} \times \left[\frac{i \times N}{4} - C\right]$

$$= 45 + \frac{50-45}{72} \times \left[\frac{2 \times 445}{4} - 205\right] = 46.215$$

Position of 3rd quartile $Q_3 = 3 \times N \div 4 = 336.75^{th}$ item; it lies in the Class 55-60

Hence, Q_3 (lies in the interval 55-60) $= l_1 + \frac{l_2 - l_1}{f_1} \times \left[\frac{i \times N}{4} - C\right]$

$$= 55 + \frac{60-55}{43} \times \left[\frac{3 \times 445}{4} - 328\right] = 55.668$$

Position of 7th decile $(D_7) = 7 \times N \div 10 = 7 \ 445 \div 10 = 311.5^{th}$ item; it lies in the Class 50-55

Hence, D_7 (lies in the interval 50-55) $= l_1 + \frac{l_2 - l_1}{f_1} \times \left[\frac{i \times N}{10} - C\right]$

$$= 50 + \frac{55-50}{51} \times \left[\frac{7 \times 445}{10} - 277\right] = 53.382$$

Position of 45th percentile $(P_{45}) = 45 \times N \div 100 = 45 \times 445 \div 100 = 200.25$; it lies in the Class 40-45

Hence, P_{45} (lies in the interval 40-45) $= l_1 + \frac{l_2 - l_1}{f_1} \times \left[\frac{i \times N}{100} - C\right]$

$$= 40 + \frac{45-40}{57} \times \left[\frac{45 \times 445}{100} - 148\right] = 44.583$$

EXERCISES

1. The following data represents the ages of patients admitted to a hospital.

   85, 88, 89, 87, 75, 80, 83, 83, 66, 56, 65, 52, 43, 56, 53, 44, 40, 67, 75 and 48

   Compute the sample mean from this raw data. Construct a frequency distribution with classes 40-49, 50-59 etc. and then compute the sample mean from the frequency distribution.

2. The production manager of Hindustan Press wants to determine the average time needed to photograph one printing plate. He observed the plate makers using a stop watch and the following are his findings.

   20.4, 22.0, 20.0, 24.7, 22.2, 25.7, 23.8, 24.9, 21.3, 22.7, 25.1, 24.4, 21.2, 24.3, 22.9, 23.6, 28.2, 23.2, 24.3 and 21.0.

   An average per-plate time of less than 22.5 seconds indicates satisfactory productivity. Should the manager be concerned?

3. The table below represents the time in seconds needed to serve a sample of customers by Bradely Departmental store during a particular day.

Time (seconds):	20-29	30-39	40-49	50-59	60-69	70-79	80-89	90-99	100-109
Frequency:	6	16	21	30	26	23	12	8	5

   Compute the sample mean using coding method.

4. The owner of a stationery shop wants to shift his establishment to a new location where a larger floor space is available. But he decided to shift, if his sale during first 6 months of the year is at least 295 and the overall monthly average for the year is at least 280. The sale data for the last year is as follows:

Jan	Feb	Mar	Apr	May	Jun	Jul	Aug	Sept	Oct	Nov	Dec
235	215	195	395	310	275	300	290	275	300	375	450

   What is the owner's decision and why?

5. The Filtop Company manufacturer of electrical circuit boards, has manufactured the following units over the past 6 years.

2001	2002	2003	2004	2005	2006
12,450	13,225	14,315	15,740	16,630	17,210

Calculate the average percentage increase in units produced over this period.

6. A study was undertaken on school dropouts during 5 years and the following is the result of study showing the percentage of dropouts.

2003	**2004**	**2005**	**2006**	**2007**
4%	5%	5.6%	6%	3%

Calculate the average percentage increase/decrease in dropouts during the period.

7. Global Transport Services keeps the record on all its moving equipment. Following is the weekly mileage records(km) for its trucks.

855, 455, 765, 790, 215, 655, 590, 485, 875, 690, 1450, 560, 470, 890, 988, 559, 788, 945.

Calculate the median km a truck travelled. Also calculate the mean for the 18 trucks. Compare the results and explain which is a better measure of the central tendency of the data.

8. Following is the distribution of data on insurance claims for automobile accidents. Determine the median value.

Amount of Claim:	<250	250-499.99	500-749.99	750-999.99	1000 & above
Frequency:	50	335	1065	1775	1490

9. A librarian asked 20 different people as they left the library about how many books they checked out. Following is their responses.

1, 0, 2, 2, 3, 2, 1, 2, 0, 2, 2, 3, 1, 0, 7, 3, 5, 3

Compute the mode of the data set. Also compute the mean.

10. The ages of inhabitants of Gramin Villa Apartment has the following distribution.

Class:	47-51.9	52-56.9	57-61.9	62-66.9	67-71.9	72-76.9	77-81.9
Frequency:	5	10	12	41	38	20	8

Estimate the modal value of the distribution.

11. The numbers of apartments in 27 apartment complexes in Chandrasekharpur, Patia are given below.

90, 88, 95, 79, 98, 88, 67, 93, 87, 97, 86, 97,126, 141, 144, 140, 128, 130, 155, 185, 150, 146, 144, 162, 240, 192, 158.

Construct a frequency distribution using intervals 66-87, 88-109,···, 220-241.

Estimate the modal value. Compute the mean of the raw data. Compare the answers and comment on which of the two is the better measure of central tendency of these data and why.

12. Write brief accounts of frequency distribution and measures of central tendency within 200 words.
13. Discuss a practical situation where mode is an approximate measure of average rather than the arithmetic mean.
14. What do you understand by mode? Discuss how mode is calculated for ungrouped data and grouped data. Give two practical examples where you would recommend the use of mode.
15. Describe the different measures of central tendency with appropriate examples.

MEASURES OF DISPERSION

The variation of different items of a series from its central value is known as *dispersion*. Dispersion means scatteredness, variation or deviation of a set of observations from its central value. Two sets of data may have the same mean but still they may differ in their scatteredness. This is due to the variability in the values of the items of the series. Variability crops up in the values of observations owing to the lack of symmetry, equality and non-homogeneity in the values. The necessity for tabulating and classifying the statistical data and to condense them into a single figure called average have been discussed elaborately in the last chapter. But this cannot throw any light about how different items of a series vary from each other.

Consider the figure below. The mean of all three curves is the same, but curve A has less spread (or variability) than curve B, and curve B has less variability than curve C. If we measure only the mean of these three distributions, we will miss an important difference among the three curves. So, for any data, the mean, the median, and the mode tell us only part of what we need to know about the characteristics of the data. To better understand about the pattern of the data, we must also measure its dispersion- its spread, or variability.

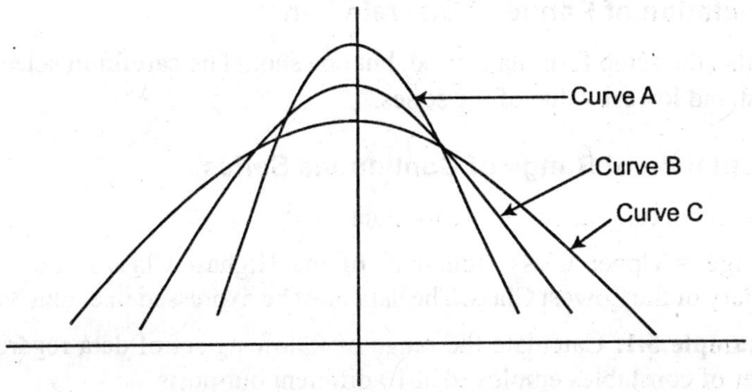

In practice, quality control experts analyze the dispersions of a product's quality levels. A drug that is average in purity but ranges from very pure to highly impure may endanger lives. Similarly financial analysts are concerned about the dispersion of a firm's earnings. Widely dispersed earnings- those varying from extremely high to low or even negative levels- indicate higher risk to stake holders and creditors than do earnings remaining relatively stable.

There are four major objectives of the measure of dispersion. They are as follows:

1. To judge the reliability of measures of central tendency.
2. To make a comparative study of the variability of two or more series.
3. To identify the causes of variability with a view to controlling it, and
4. To serve as a basis for further statistical analyses.

Again there are four kinds of measures of dispersion. They are as follows:

1. Range.
2. Quartile Deviation (or Semi-interquartile Range)
3. Mean Deviation.
4. Standard Deviation.

RANGE

Range is a measure of dispersion, which is based on the two extreme values of a series.

Range = Highest Value – Lowest Value

Computation of Range of Individual Series:

Use the above formula in case of a discrete series.

Computation of Range of Discrete Series:

Here also the same formula is used, but one should be careful in selecting the highest and lowest value of the series.

Computation of Range of Continuous Series:

In case of continuous series the formula is:

Range = Upper Class Boundary of the Highest Class – Lower Class Boundary of the Lowest Class. The data must be expressed in exclusive series.

Example 5.1: Calculate the range of following set of data regarding the number of constables employed at 10 different out-posts.

77, 82, 106, 93, 285, 115, 95, 170, 158, 220.

Solution: The highest value in the series = 285

The lowest value in the series = 77.

Range = 285 – 77 = 208.

Example 5.2: Calculate the range of the following set of data.

Percentage of defective items	7	8	9	10	11	12	14
No. of consignments	1	3	5	7	8	4	0

Solution: Here though the highest value is 14, but its frequency is 0. So the highest value = 12.

Lowest value = 7. So, Range = 12 – 7 = 5.

Example 5.3: Calculate the range of the following data set.

Systolic Pressure	71 - 75	76 - 80	81 - 85	86 - 90	91 - 95
No. of Persons	1	3	5	7	8

Here lowest class = 71 – 75.

Highest class = 91 – 95.

Lower Class Boundary of the Lowest Class = 70.5

Higher Class Boundary of the Highest Class = 95.5

So, Range = 95.5 – 70.5 = 25

Merits of Range

The merit possessed by Range is that it is easily calculated and readily understood.

Demerits of Range

1. It is affected greatly by fluctuations of sampling.
2. It is not based on all the observations of series.

QUARTILE DEVIATION (QD)

Quartile Deviation (QD) is a measure of dispersion, which is based on the two quartiles–upper and lower quartiles of a series. It is half the difference between the upper quartile (Q_3) and the lower quartile (Q_1). The difference between these two quartiles is called interquartile range. The half of this interquartile range is semi-interquartile range. So quartile deviation is also known as *semi-interquartile range*.

In the figure below Q_1, Q_2 and Q_3 divide the figure into four parts. Up to Q_1, is the first part (25%) of the distribution. From Q_1 to Q_2 is the next 25% and so on. In fact 50% of the values of a variable lie between Q_1 and Q_3. The distance between Q_1 and Q_3 i.e. $Q_3 - Q_1$ is the interquartile range (Q). When this is divided by 2, it is called semi-interquartile range.

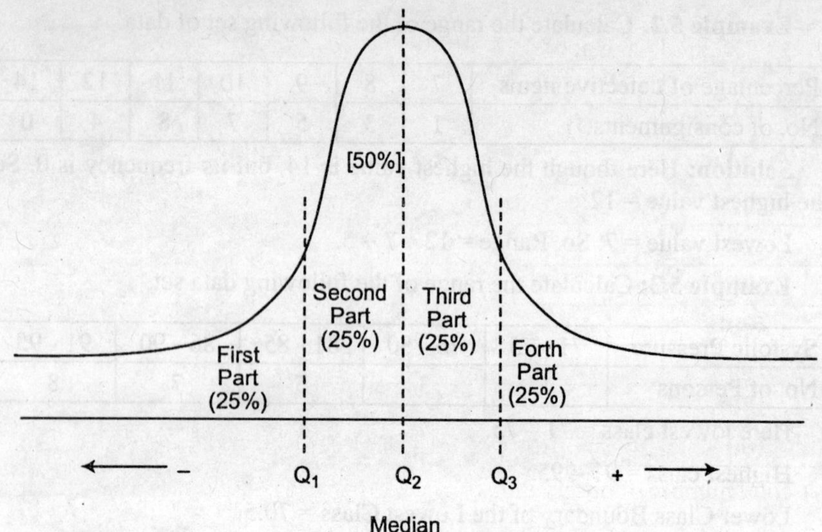

The following formula is used for finding QD. Once Q_3 and Q_1 are found out, then the quartile deviation (QD) can be found out easily.

Example 5.4: Here are student scores in GMAT. Find the quartile deviation.

124, 126, 120, 130, 120, 120, 110, 126, 118, 128, 132.

Solution: Let us arrange the data in ascending order as shown below.

110, 118, 120, 120, 120, 124, 126, 126, 128, 130, 132.

Number of items = 11, So,

Q_1 = value of $\left[\frac{N+1}{4}\right]^{th}$ item i.e. value of 12 ÷ 4 = 3rd item = 120

Q_3 = value of $\left[\frac{3(N+1)}{4}\right]^{th}$ item i.e. value of 3 12 ÷ 4 = 9th item = 128

Hence, QD = [128 – 120] ÷ 2 = 8 ÷ 2 = 4

Example 5.5: Calculate QD for the following set of data:

Average monthly monsoon rainfall	120	122	124	126	130	140	150	160
No. of Districts	1	3	5	7	10	3	1	1

Solution: The table prepared below is self-explanatory.

Rainfall (x)	**No. of Districts (f)**	**Cumulative Frequency**	**Remarks**
120	1	1	
122	3	$1 + 3 = 4$	
124	5	$4 + 5 = 9$	8th item falls here for Q_1
126	7	$9 + 7 = 16$	
130	10	$16 + 10 = 26$	24th item falls here for Q_3
140	3	$26 + 3 = 29$	
150	1	$29 + 1 = 30$	
160	1	$30 + 1 = 31$	

Total number of observations= $N = 31$, it is an odd number. So,

$Q_1 = \text{value of}\left[\dfrac{N+1}{4}\right]^{th}$ item i.e. value of $32 \div 4 = 8^{th}$ item $= 124$(shown in bold in the table)

$Q_3 = \text{value of}\left[\dfrac{3(N+1)}{4}\right]^{th}$ item i.e. value of $332 \div 4 = 24^{th}$ item $= 130$

$QD = [130 - 124] \div 2 = 6 \div 2 = 3$

Determination of QD for Continuous series:

In this case Q_1 and Q_3 are calculated using the following formulae:

$$Q_1 = l_1 + \left[\frac{N}{4} - c_1\right] \times \left[\frac{i}{f_1}\right] \text{ and } Q_3 = l_3 + \left[\frac{3N}{4} - c_3\right] \times \left[\frac{i}{f_3}\right]$$

Where l_1 = Lower Boundary of Class where Q_1 falls

l_3 = Lower Boundary of Class where Q_3 falls

c_1 = Cumulative frequency preceding the class where Q_1 falls

c_3 = Cumulative frequency preceding the class where Q_3 falls

f_1 = Frequency of Class where Q_1 falls

f_3 = Frequency of Class where Q_3 falls

i = Width of Class.

Quartile Deviation = $[Q_3 - Q_1]$ 2

Example 5.6: In one experiment the following set of observations were made regarding the number of earthworms found in different soil pH conditions. Calculate the quartile deviation.

Soil pH	5.6 – 6.0	6.1 – 6.5	6.6 – 7.0	7.1 – 7.5	7.6 – 8.0
No. Of Earthworms	3	8	9	7	1

Solution: The following table is prepared which is self-explanatory.

Soil pH (x)	**No.of Earthworms (f)**	**Cumulative Frequency**	
5.6 – 6.0	3	3	
6.1 – 6.5	**8**	3 + 8 = 11	Q_1 falls here.
6.6 – 7.0	9	11 + 9 = **20**	
7.1 – 7.5	**7**	20 + 7 = 27	Q_3 falls here
7.6 – 8.0	1	27 + 1 = 28	

Total = N = 28; i = 0.5

Q_1 = value of $\left[\frac{N}{4}\right]^{th}$ item i.e. value of 28 ÷ 4 = 7th item. So l_1 = 6.05 (Lower Class Boundary)

Q_3 = value of $\left[\frac{3N}{4}\right]^{th}$ item i.e. value of 3 28 ÷ 4 = 21st item. So l_3 = 7.05(Lower Class Boundary)

$f_1 = 8;\ c_1 = 3;\ f_3 = 7;\ c_3 = 20.$

$$\text{Hence, } Q_1 = 6.05 + \frac{6.55 - 6.05}{8} \times \left(\frac{28}{4} - 3\right) = 6.05 + \frac{0.5}{8} \times (7 - 3) = 6.30$$

$$\text{And, } Q_3 = 7.05 + \frac{7.55 - 7.05}{7} \times \left(\frac{3 \times 28}{4} - 20\right) = 7.05 + \frac{0.5}{7} \times (21 - 20) = 7.121$$

So, QD = (7.121 – 6.30) ÷ 2 = 0.821 ÷ 2 = 0.4105

Merits of Quartile Deviation

1. It is simple to calculate and easy to understand.
2. It is not affected by the values of extreme items.

Demerits of Quartile Deviation

1. It is not based on all the observations of the data as it ignores the first 25% and the last 25% of the items.
2. It is not capable of further algebraic treatment.
3. It is affected considerably by fluctuations of sampling.
4. Where there is considerable variation in the values of various items, the quartile deviation is not a suitable measure of dispersion.

MEAN DEVIATION (MD)

Mean Deviation of a set of observations of a series is the arithmetic mean of all the deviations, without their algebraic signs, taken from its central value. In other words, *mean deviation is the sum of the deviations in absolute terms from a central value divided by the number of items in the series.* But what should be the central value? The central value may be any of the averages- mean, median and mode. Generally, arithmetic mean is taken as the central value. In this chapter we discuss only mean deviation from mean. It may be noted that mean deviation from mode or median can be calculated only by substituting value of mode or median in place of mean in the mean deviation from mean.

Absolute value of x is denoted by $|x|$. It is always positive. For example, absolute value of +2 is 2 and absolute value of –2 is also 2. While finding out the absolute value of any number, the sign is omitted and only the number is taken for further calculation. Symbolically Mean Deviation from mean is denoted by δ.

$$\delta = \frac{\sum|d|}{N} \text{ or } \frac{\sum f|d|}{N}, \text{ where } N = \sum f \text{ and } d = x - \overline{x}$$

For calculation of the mean deviation, first the arithmetic mean is calculated, then deviations (d) of the individual values from the mean is calculated, these deviations are then added up ignoring their signs. Finally this sum is divided by the total number of observations

COEFFICIENT OF MEAN DEVIATION

Sometimes, it is required to find out the coefficient of Mean Deviation, which is given by:

$$\text{Coefficient of MD} = \frac{\text{Mean Deviation}}{\text{Mean}} = \frac{\delta}{\overline{x}}$$

Example 5.7: Number of workers remaining absent in first 20 days of the month are as follows. Calculate the mean deviation from this data:

7, 10, 13, 9, 10, 12, 9, 11, 8, 14, 10, 9, 11, 12, 13, 8, 7, 11, 11, 8.

Solution:

Sl. No.	No. of workers (x)	Deviations $d = x - \overline{x}$	$\lvert d\rvert = \lvert x - \overline{x}\rvert$ (Absolute Value of d)
1	7	–3.15	3.15
2	10	–0.15	0.15
3	13	+2.85	2.85
4	9	–1.15	1.15
5	10	–0.15	0.15
6	12	+1.85	1.85
7	9	–1.15	1.15
8	11	+0.85	0.85
9	8	–2.15	2.15
10	14	+3.85	3.85
11	10	–0.15	0.15
12	9	–1.15	1.15
13	11	0.85	0.85
14	12	+1.85	1.85
15	13	+2.85	2.85
16	8	–2.15	2.15
17	7	–3.15	3.15
18	11	+0.85	0.85
19	11	+0.85	0.85
20	8	–2.15	2.15

N = 20 Total = $\Sigma x = 203$ $\Sigma\lvert d\rvert = 33.3$

$$= \text{Mean} = \overline{x} = \frac{\sum x}{N} = \frac{203}{20} = 10.15$$

$$\text{Mean deviation} = \delta = \frac{\sum|d|}{n} = \frac{33.3}{20} = 1.665$$

Merits of Mean Deviation

1. Mean deviation is rigidly defined and its value is precise and definite.
2. Its calculation is not very difficult.
3. It is readily understood.
4. It is based on all the observations of the series unlike the quartile deviation.
5. It is not affected very much by the values of extreme items.

Demerits of Mean Deviation

Mean deviation ignores algebraic signs of the deviations and hence it is not capable of further mathematical treatment. It is rarely used.

STANDARD DEVIATION (SD)

As it is apparent from the above discussion, the technique of the calculation of mean deviation is mathematically illogical, because in its calculation the algebraic signs are ignored. This drawback can be removed by calculating the standard deviation of the series. One of the easiest ways of doing away with algebraic signs is to square the figures and this process is adopted in the calculation of standard deviation. So, *Standard Deviation of a set of observations of a series is the positive square root of the arithmetic mean of the squares of all the deviations from the mean.* Thus, in the calculation of standard deviation, first the arithmetic mean of all items in the series is calculated and the deviation of various items from the arithmetic mean are squared. The squared deviations are totaled and the sum is divided by the number of items. The process of omitting the algebraic signs + and – of the deviations in the mean deviation is avoided in standard deviation. Hence standard deviation is a measure of dispersion having more mathematical significance. By squaring the deviations, we make each number positive and at the same time, assign more weight to the larger deviations, that means *standard deviation gives more importance to the values of extreme items.*

Standard deviation is calculated either for the entire population or for the sample only. Population standard deviation is generally denoted by σ (sigma). The square of this standard deviation is known as Population Variance, denoted by σ^2 (sigma squared).

The formula for calculation of Population Variance is:

$$\sigma^2 = \frac{\sum(x-\mu)^2}{N} = \frac{\sum x^2}{N} - \left(\frac{\sum x}{N}\right)^2 = \frac{\sum x^2}{N} - \mu^2$$

where,

σ^2 = Population Variance and x = value of observation

μ = Population Mean

N = Total number of items in the population

Σ = sum of all values of $(x - \mu)^2$ or all the values of x^2

For discrete data the Population Variance is:

$$\sigma^2 = \frac{\sum\left[f(x-\mu)^2\right]}{N} = \frac{\sum(fx^2)}{N} - \left(\frac{\sum fx}{N}\right)^2 = \frac{\sum fx^2}{N} - \mu^2$$

The population standard deviation (σ) is simply the square root of the population variance. Because the variance is the average of the squared distances of the observations from the mean, the standard deviation is the square root of the average of the squared distances of the observations from the mean. While variance is expressed in the square of the units used in the data, the standard deviation is in the same units as those used in the data. The formula for the population standard deviation is: -

$$= \sigma = \sqrt{\sigma^2} = \sqrt{\frac{\sum\left[f(x-\mu)^2\right]}{N}} = \sqrt{\frac{\sum(fx^2)}{N} - \left(\frac{\sum fx}{N}\right)^2} = \sqrt{\frac{\sum fx^2}{N} - \mu^2}$$

Example 5.8: Percentages of impurity of 15 different sets of an alloy are shown below. Find out the variance and standard deviation by using both formulae.

0.04 0.14 0.17 0.18 0.20 0.05 0.12 0.16 0.20

0.23 0.11 0.14 0.17 0.21 0.25

Solution: The details of calculation are presented in the table below, which is self-explanatory:

i	x	x − μ	(x − μ)²	x²
1	0.04	-0.118	0.013924	0.0016
2	0.14	-0.018	0.000324	0.0196
3	0.17	0.012	0.000144	0.0289
4	0.18	0.022	0.000484	0.0324

i	x	x − μ	(x − μ)²	x²
5	0.2	0.042	0.001764	0.04
6	0.05	-0.108	0.011664	0.0025
7	0.12	-0.038	0.001444	0.0144
8	0.16	0.002	4E-06	0.0256
9	0.2	0.042	0.001764	0.04
10	0.23	0.072	0.005184	0.0529
11	0.11	-0.048	0.002304	0.0121
12	0.14	-0.018	0.000324	0.0196
13	0.17	0.012	0.000144	0.0289
14	0.21	0.052	0.002704	0.0441
15	0.25	0.092	0.008464	0.0625
Total	**2.37**	0	**0.05064**	**0.4251**

$$\text{Mean} = \mu = \frac{2.37}{15} = 0.158$$

$$\sigma^2 = \frac{\sum\left[(x-\mu)^2\right]}{N} = \frac{0.05064}{15} = 0.003376$$

By using the direct formula

$$\sigma^2 = \frac{\sum\left[x^2\right]}{N} - \left(\frac{\sum x}{N}\right)^2 = \frac{0.4215}{15} - \left(\frac{2.37}{15}\right)^2 = 0.003376$$

$$\text{So standard deviation} = \sigma = \sqrt{\sigma^2} = \sqrt{0.003376} = 0.058$$

Note that it is always zero.

Merits of Standard Deviation

1. Standard deviation is rigidly defined and its value is always definite.
2. It is based on all the observations of the data.
3. It is amenable to algebraic treatment and possesses many mathematical properties.
4. It is less affected by fluctuations of sampling.
5. The squaring of the deviations makes them positive and the difficulty about algebraic signs, which was experienced in case of mean deviation, is not found here.
6. It is a most reliable and dependable measure of dispersion.

Demerits of Standard Deviation

1. Standard deviation is not easy to calculate, nor is it easily understood.
2. It gives more weightage to extreme items and less to those which are near the mean.

Despite the drawbacks mentioned above the standard deviation is the best measure of dispersion and is recommended in all experiments. Just as mean is the best measure of central tendency, standard deviation is the best measure of dispersion.

Difference between Mean Deviation and Standard Deviation

1. Mean deviation does not take into account the algebraic signs in its calculation, which is illogical. The standard deviation takes these into account.
2. Mean deviation can be computed from mean, mode or median. The standard deviation is always calculated from the arithmetic mean.

SAMPLE STANDARD DEVIATION

To compute the sample variance and the sample standard deviation, we use the same formulae replacing μ with $\overline{x}$ and N with n-1. The formulae will be:-

$$\text{Sample Variance} = s^2 = \frac{\sum(x-\overline{x})^2}{n-1} = \frac{\sum x^2 - n\overline{x}^2}{n-1}$$

$$\text{Sample Standard Deviation} = s = \sqrt{s^2} = \sqrt{\frac{\sum(x-\overline{x})^2}{n-1}} = \sqrt{\frac{\sum x^2 - n\overline{x}^2}{n-1}}$$

where, s^2 = sample variance

s = sample standard deviation

x = value of each of n-observations

$\overline{x}$ = mean of the sample

n = number of observations in the sample

It can be shown that:

$$4\ SD = 5\ MD = 6\ QD$$

Example 5.9: Beatly Talent Ltd. is selecting a group of extras for a movie. The ages of the first 20 men to be interviewed are 50, 56, 55, 49, 52, 57, 56, 57, 56, 59, 54, 55, 61, 60, 51, 59, 62, 52, 54 & 49. The director of the movie wants men whose ages are fairly tightly grouped around 55 years. The director suggests that a standard deviation of 3 years would be acceptable. Does this group of extras qualify?

Solution: The detail calculation are shown in the table below, which is self-explanatory.

i	x	$(x-\bar{x})$	$(x-\bar{x})^2$
1	50	-5.2	27.04
2	56	0.8	0.64
3	55	-0.2	0.04
4	49	-6.2	38.44
5	52	-3.2	10.24
6	57	1.8	3.24
7	56	0.8	0.64
8	57	1.8	3.24
9	56	0.8	0.64
10	59	3.8	14.44
11	54	-1.2	1.44
12	55	-0.2	0.04
13	61	5.8	33.64
14	60	4.8	23.04
15	51	-4.2	17.64
16	59	3.8	14.44
17	62	6.8	46.24
18	52	-3.2	10.24
19	54	-1.2	1.44
20	49	-6.2	38.44
Total	1104	0	285.2

$\bar{x} = \frac{1104}{20} = 55.2$; which is close to the desired 55 years.

Sample Standard Deviation $= s = \sqrt{s^2} = \sqrt{\frac{\sum(x-\bar{x})^2}{n-1}} = \sqrt{\frac{285.2}{20-1}} = 3.874$;

This shows more variability than desired.

One may raise his eyebrows for the reason of using n-1 instead of n here. If we take many samples from a given population, find the sample variance (s^2) for each sample, and average each of these together, then this average tends not to equal the population variance , unless we use n-1 as the denominator. We shall learn the statistical significance of n-1 in later chapters.

COEFFICIENT OF VARIATION

Coefficient of variation relates the standard deviation and the mean by expressing the standard deviation as a percentage of mean. Because the standard deviation cannot be the sole basis for comparing two distributions, we cannot know the dispersion of a set of data until we know the standard deviation, the mean, and how the standard deviation compares with the mean. If we have a standard deviation of 12 and a mean of 6, the values vary by an amount twice as large as the mean itself. On the other hand, if we have a standard deviation of 12 and a mean of 6000, the variation relative to the mean is insignificant.

$$\text{Coefficient of Variation} = \frac{\text{Standard Deviation}}{\text{Mean}} \times 100$$

. It is expressed as a percentage.

Example 5.10: There are a number of possible measures of sales performance, including how consistent a salesperson is in meeting established sales goals. The data that follows represent the percentage of goal met by each of three salespersons, A, B & C over the last 5 years. Which salesperson is the most consistent ?

A :	87	67	88	91	102
B :	75	87	89	85	78
C :	103	87	117	87	122

Solution: The salesperson whose percentage of achieving goal over last 5 years has least coefficient of variation, is obviously considered to be more consistent. The calculations are shown in the table below, which is self-explanatory.

X (of A)	X^2	Y(of B)	Y^2	Z(of C)	Z^2
87	7569	75	5625	103	10609
67	4489	87	7569	87	7569
88	7744	89	7921	117	13689
91	8281	85	7225	87	7569
102	10404	78	6084	122	14884
435	38487	414	34424	516	54320

$$\text{Mean } \mu \text{ (of A)} = \frac{435}{5} = 87$$

$$\text{Mean } \mu \text{ (of B)} = \frac{414}{5} = 82.8$$

$$\text{Mean } \mu \text{ (of C)} = \frac{516}{5} = 103.2$$

$$\sigma^2 \text{ (of A)} = \frac{38487}{5} - \left(\frac{435}{5}\right)^2 = 128.4; \text{ So } \sigma = \sqrt{128.4} = 11.33$$

$$\sigma^2 \text{ (of B)} = \frac{34424}{5} - \left(\frac{414}{5}\right)^2 = 28.96; \text{ So } \sigma = \sqrt{28.96} = 5.38$$

$$\sigma^2 \text{ (of C)} = \frac{54320}{5} - \left(\frac{516}{5}\right)^2 = 213.76; \text{ So } \sigma = \sqrt{213.76} = 14.62$$

Hence, Coefficient of Variation

$$= \text{of A} = \frac{\text{Standard Deviation}}{\text{Mean}} \times 100 = \frac{11.33}{87} \times 100 = 13.02\%$$

$$= \text{of B} = \frac{\text{Standard Deviation}}{\text{Mean}} \times 100 = \frac{5.38}{82.8} \times 100 = 6.55\%$$

$$= \text{of C} = \frac{\text{Standard Deviation}}{\text{Mean}} \times 100 = \frac{14.62}{103.2} \times 100 = 14.17\%$$

So, the salesperson B is more consistent as he has least coefficient of variation among the three.

EXERCISES

1. PanNet Software Company decided to develop a new computer package for its officials, but with less hard-drive space. So it pooled 20 executives to determine the amount of available space on their PCs. The results are given below.

   6.2, 6.7, 7.8, 8.3, 9.6, 10.5, 12.4, 19.2, 29.0, 42.6, 59.7, 97.5, 135.5, 120.5, 148.5, 178.5, 200.1, 229.5 and 284.5.

   Calculate the range and inter-quartile range.

2. Here are the maximum daily temperature readings in degree Celsius during June 2001 at a place. Calculate Q_1 and Q_3.

39.5, 39.8, 40.1, 42.5, 39.6, 38.4, 42.1, 43.5, 44.8, 39.6, 39.2, 38.9, 38.7, 39.2, 40.6, 41.8, 42.2, 43.1, 44.1, 44.2, 43.8, 43.9, 43.5, 42.5, 42.8, 43.1, 43.8, 43.2, 43.00, 42.9.

3. In a survey conducted by an insurance company, here are the results of the numbers of families owning the numbers of cars.

Number of cars:	0	1	2	3	4	5
Frequency:	14	24	7	9	2	1

Calculate the variance and the standard deviation.

4. Here is the daily production rate of wooden centre table of a carpentry shop.

17, 20, 19, 27, 17, 20, 21, 22, 19, 24

The owner feels that a standard deviation of more than three tables a day indicates unacceptable production-rate variations. Should he be concerned about plant-production rate?

5. These are the total daily fares (in Rs) collected at a toll gate from cargo vehicles entering into a city for first 20 days of a month.

1054, 1189, 1265, 1345, 986, 1296, 1321, 992, 1082, 1147, 1123, 1324, 1234, 1169, 1440, 1422, 1396, 1390, 1436, 1400.

Compute the range, mean deviation and standard deviation, and state which is a better measure of dispersion and why.

6. What are the four measures of dispersion? Which is the most widely used measure of dispersion and why?

7. 10 workshops were analysed in two batches regarding their number of polishing jobs finished and the following results were obtained. Calculate the mean number of jobs completed in each batch. Which of the two batches has more consistent result? (Note:- more consistent means less coefficient of variation)

Workshop No.	1	2	3	4	5	6	7	8	9	10
No. of jobs completed in Batch-1	16	19	22	23	25	26	17	15	18	14
No. jobs completed in Batch-1I	14	16	12	15	26	18	17	29	20	13

8. What do you understand by 'Dispersion'? What is the purpose of measure of dispersion ?

9. Distinguish between mean deviation and standard deviation.

10. Describe briefly various measures of dispersion giving method of calculation both for ungrouped and grouped data.
11. Define standard deviation and mean deviation. Describe how they are measured.
12. Residual mercury of effluent sample collected from different sites of a channel was estimated in winter. The following data were obtained.

Sites	1	2	3	4	5	6	7	8	9	10
Residual Hg in winter (mg/kg soil)	1.6	2.6	1.8	2.1	1.0	1.7	0.9	2.3	2.2	2.7

The same estimate was made in summer and the following results were obtained.

Sites	1	2	3	4	5	6	7	8	9	10
Residual Hg in summer (mg/kg soil)	3.3	3.6	2.6	2.8	3.0	3.2	3.4	2.9	2.3	3.9

Make a comparison of the results obtained in two different seasons.

13. Srusti Hospital administrator surveyed the number of days 180 randomly chosen patients stayed in the hospital following an operation. The data are:-

Hospital stay in days:	1-4	5-8	9-12	13-17	18-21	22-25	26-29	30-33
Frequency:	18	90	44	20	10	10	5	6

Calculate the standard deviation and mean.

14. Owner of a departmental store received one dozen stationery item from the supplier, but insisted that she would accept the lot if the average weight is 7.5 g and standard deviation is less than 0.5 g. Here are the weights of the items (in g):

6.3, 7.2, 8.0, 7.4, 7.2, 7.5, 8.2, 7.8, 7.7, 7.7, 6.9, 7.5, 7.6, 7.4, 7.7, 8.3, 7.3, 7.3, 7.4, 7.5, 8.0, 6.3, 8.1, 7.5

What will be the owner's decision and why?

15. The amount of bajra production (in thousand tonnes) in some districts of a state are distributed as given below. Calculate the standard deviation.

Production of Bajra ('000 tonnes)	0.1-1.0	1.1-2.0	2.1-3.0	3.1-4.0	4.1-5.0	5.1-6.0	6.1- 7.0	7.1-8.0	8.1- 9.0
No. of Districts	6	8	3	4	7	3	2	3	3

SKEWNESS, MOMENTS AND KURTOSIS

SKEWNESS

A distribution is said to be skewed if its frequency curve is not symmetrical but it is stretched more on one side than to the other side. A symmetrical distribution when plotted on a graph gives a perfectly bell-shaped curve as shown in the figure. A distribution is said to be symmetrical when mean, median and mode are identical or coincide. On the contrary, in asymmetrical curves, the mean, median and mode never coincide, because it lacks symmetry. The literal meaning of "skewness" is "lack of symmetry". Skewness is opposite to symmetry and its presence tells us that a particular distribution is not symmetrical.

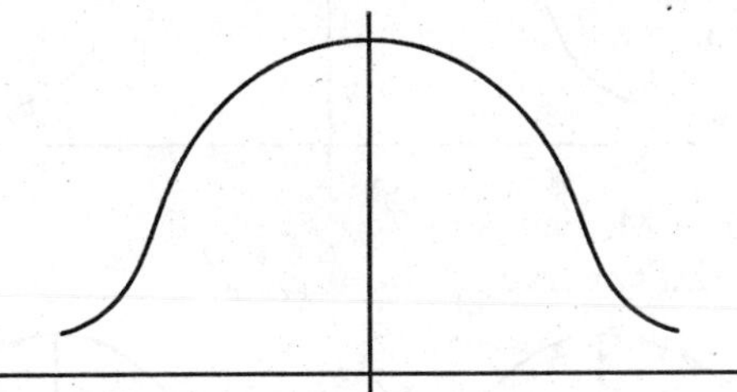

Mean = Mode = Median

Symmetrical curve

DEFINITION OF SKEWNESS

- Skewness or assymetry is the attribute of a distribution that extends further on one side of the class with the highest frequency than on the other side. – **Simpson and Kalfa**
- When a series is not symmetrical it is said to be asymmetrical or skewed. – **Croxton and Cowden**
- Skewness is the lack of symmetry. When a frequency distribution is

plotted on a chart, skewness present in the series tends to be dispersed more on one side of the mean than on the other.- **Riggleman and Frisbee**

A distribution can have three parts (i) the left tail; (ii) the middle part; and (iii) the right tail. In case of symmetrical distribution, the two tails (Right and Left tails) are of equal lengths. But in skewed distribution (or asymmetrical distribution) one tail is longer than the other. Skewed distribution may be positive or negative.

Positive skewness: If the curve of the distribution has a longer tail towards the right, it is said to be positively skewed. In this case Mean > Median > Mode. For example- Income distribution is positively skewed. Large number of population have less income.

Negative skewness: If the curve has a longer tail towards the left, it is said to be negatively skewed. In this case Mean < Median < Mode. For example- Mortality rate of countries is negatively skewed. Poor countries have less mortality. Life span of electric bulbs is also negatively skewed.

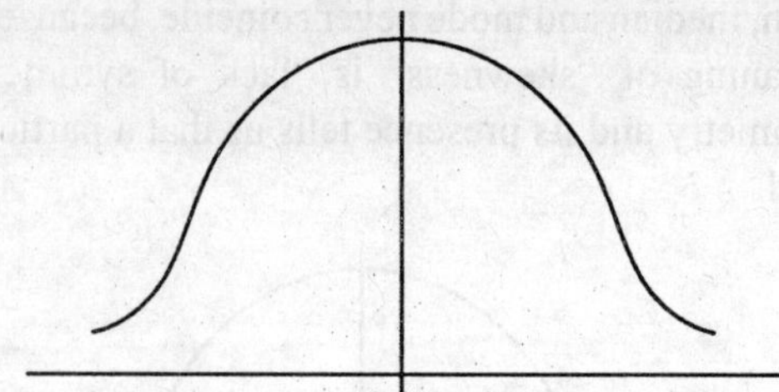

Mean = Mode = Median
Normal Probability curve

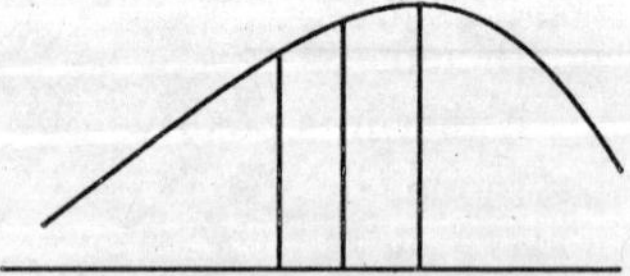

Mean Median Mode
Negatively Skewed Curve

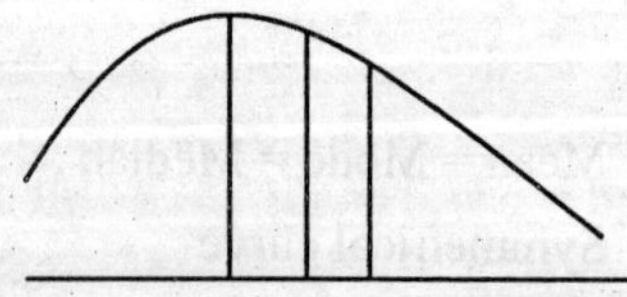

Mean Median Mode
Positively Skewed Curve

PURPOSE OF STUDY OF SKEWNESS

- To find out the nature and degree of concentration of items (or observations) of the distribution.
- To know whether the distribution is normal. Many statistical measures are based on normal distribution (i.e., bell shaped curve).

Difference between Dispersion and Skewness

Dispersion	Skewness
It judges the truthfulness of the central tendency.	It judges the difference between the central tendencies.
It shows the degree of variability.	It shows whether the concentration is in higher or lower values.
It shows the spreading of individual values about the central value. i.e., mean.	It shows the departure from symmetry, i.e., directions of variations.
It is a type of average or deviation average of the second order.	It is not an average but is measured by use of the mean, median and mode.

MEASURES OF SKEWNESS

The measures of asymmetry are usually called measures of skewness.

- Skewness is used to find out the extent of asymmetry (i.e., departure from symmetry and direction in series).
- Measures of skewness indicate not only the extent of skewness (in numerical expressions), but also the direction, i.e., the manner in which the deviations are distributed. These measures can be absolute or relative.

ABSOLUTE MEASURES

The absolute measures are otherwise known as measures of skewness, whereas the relative measures are known as the coefficient of skewness. Absolute measure signifies the extent of asymmetry, whether it is positive or negative and is based on difference between mean and mode.

Absolute Skewness = Mean – Mode

In a symmetrical distribution absolute skewness will be zero because in this case Mean = Mode.

In a positively skewed frequency distribution, absolute skewness will be positive, because in this case "Mean is greater than Mode" and in a negatively skewed frequency distribution absolute skewness will be negative, because in this case "Mean is less than Mode".

Absolute measure of skewness is expressed in the units of the original data and therefore cannot be used for comparison of skewness in two different distributions if they are in different units and hence it is not a very useful measure.

Thus for comparison purpose, we use relative measure of skewness known as **coefficient of skewness.**

RELATIVE MEASURES

There are four important relative measures of skewness.

- Karl Pearson's coefficient of skewness.
- Bowley's coefficient of skewness.
- Kelly's coefficient of skewness.
- Measures of coefficient based on moments.

KARL PEARSON'S COEFFICIENT OF SKEWNESS

This method is based on the fact that, in an asymmetrical distribution mean and mode pull apart from one another and greater the distance between the two, the greater is the degree of skewness. Karl Pearson's coefficient of skewness enables us to find out the direction as well as **extent of skewness.** It is denoted by S_k

$$S_k = \frac{\text{Mean} - \text{Mode}}{\text{Standard Deviation}}$$

If the mode is ill defined, then coefficient of skewness is determined by the following formula:

$$S_k = \frac{3(\text{Mean} - \text{Median})}{\text{Standard Deviation}}$$

Properties of Karl Pearson's coefficient of skewness

- When its value is negative, the distribution is negatively skewed.
- When its value is positive, the distribution is positively skewed.
- When its value is zero, there is no skewness, i.e. the distribution is symmetrical.
- It's value usually lies between ±3, but rarely reaches these limits.

Example 6.1: Calculate Karl Pearson's coefficient of skewness for the following data of number of hours, nine machines have been put to run.

25, 15, 23, 40, 27, 25, 23, 25, 20.

Solution: Calculation of Mean and Standard Deviation

Sl.No	Machine running hours (x)	$(x-\bar{x})$	$(x-\bar{x})^2$
1	25	0.222	0.049
2	15	-9.778	95.605
3	23	-1.778	3.160
4	40	15.222	231.716
5	27	2.222	4.938
6	25	0.222	0.049
7	23	-1.778	3.160
8	25	0.222	0.049
9	20	-4.778	22.827
Total:	223		361.556

$\bar{x}$ = Mean = 223 / 9 = 24.778

$$\sigma = \sqrt{\frac{(x-\bar{x})^2}{n}} = \sqrt{\frac{361.556}{9}} = 6.338$$

Mode = 25, as it occurs maximum number of times.

$$\text{Coefficient of Skewness } S_k = \frac{\text{Mean} - \text{Mode}}{\text{Standard Deviation}} = \frac{24.778-25}{6.338} = -0.035$$

It indicates that the distribution is negatively skewed.

Example 6.2: Consider the following distribution of blood test for fasting sugar of 100 persons in two pathology laboratories.

	Laboratory A	Laboratory B
Mean	98	94
Median	90	86
Standard deviation	8	8

Both the results have the same degree of skewness. True/ False?

Solution: Since mode is not known,

Karl Pearson's coefficient of skewness is:

$$S_k = \frac{3(\text{Mean} - \text{Median})}{\text{Standard Deviation}}$$

Skewness for the Laboratories A and B.

Laboratory A: Coefficient of skewness: S_{kA} = 3(98-90) ÷ 8 = 3

Laboratory B: Coefficient of skewness: S_{kB} = 3(94-86) ÷ 8 = 3

Since S_{kA} and S_{kB} are equal, the statement that both the laboratories have the same degree of skewness is true.

Example 6.3: 100 patients tested their blood for total cholesterol from two pathology laboratories A and B. The following results were obtained in mg/dl.

Laboratory A: Mean= 45.43; Mode= 50.68; SD= 13.2

Laboratory B: Mean= 46.43; Mode= 45.52 ; SD=13.2

Determine the results of laboratory which is more skewed.

Solution:

Laboratory A: $S_{kA} = \frac{\text{Mean} - \text{Mode}}{\sigma}$ = (45.43-50.68) 13.2 = -0.398

Laboratory B: $S_{kB} = \frac{\text{Mean} - \text{Mode}}{\sigma}$ = (46.43-45.52) 13.2 = - 0.069

Thus we see that = 0.398 is greater than so the result in pathology laboratory A is more skewed.

Example 6.4: From a moderately skewed distribution of retail prices of men's shoe, it is found that the mean price is Rs 18 and the median price is Rs. 15. If the coefficient of variation is 20%, find Karl- Pearson's coefficient of skewness of the distribution.

Solution: It is given that:

Mean = 18 and Median = 15. To find the coefficient of skewness, we need to find out the standard deviation.

Coefficient of variation = CV = $\frac{\text{Standard Deviation}}{\text{Mean}} \times 100$

Hence, Standard Deviation = CV Mean ÷ 100 = 20 18 100 = 3.6

$S_k = \frac{3(\text{Mean} - \text{Median})}{\text{Standard Deviation}}$ = 3(18 – 15) 3.6 = 2.5

Example 6.5: From the following data find out Karl Pearson's coefficient of skewness :

No of questions left unattended	0	1	2	3	4	5	6	7
No of Students	11	16	28	18	7	3	2	0

Solution: x: No of questions ; f: No of students (frequency).

X	f	fx	fx²
0	11	0	0
1	16	16	16
2	28	56	112
3	18	54	162
4	7	28	112
5	3	15	75
6	2	12	72
7	0	0	0
Total:	85	181	549

$N = \Sigma f = 85$

Mean $\frac{\sum fx}{N} = 181 \div 85 = 2.129$

$$\sigma = \sqrt{\frac{\sum(fx^2)}{N} - \left(\frac{\sum fx}{N}\right)^2} = \sqrt{\frac{549}{85} - \left(\frac{181}{85}\right)^2} = 1.387$$

Mode = 2 as f_{max} = 28 occurs at x = 2

S_k = (Mean – Mode) ÷ σ= (2.129-2) 1.387 = 0.093

BOWLEY'S COEFFICIENT OF SKEWNESS

Prof. A.L. Bowley's coefficient of skewness is based on the quartiles and is given by

$$S_k = \frac{(Q_3 - M) - (M - Q_1)}{(Q_3 - M) + (M - Q_1)} = \frac{Q_3 + Q_1 - 2\text{ Median}}{Q_3 - Q_1}$$

where Q_3 is 3rd quartile and Q_1 is 1st is quartile.

Bowley's coefficient of skewness ranges from -1 to +1.

i.e., $-1 \le S_k$(Bowley's) ≤ 1.

Example 6.6: A distribution had Q_1 = 43.2; median = 45 and Q_3 = 46.8 Calculate the coefficient of skewness.

Solution: Here Q_3 = 46.8 ; Q_1 = 43.2; Median = 45

$$S_k = \frac{Q_3 + Q_1 - 2\text{ Median}}{Q_3 - Q_1}$$

= (46.8 + 43.2-2 45) (46.8 - 43.2) = 0

Hence the distribution is perfectly symmetrical.

KELLY'S MEASURE OF SKEWNESS

Bowley's coefficient of skewness uses only first and third quartile values, it ignores 50% of data towards either extreme. This difficulty can partially be removed by taking two deciles or percentiles equidistant from the median value, as suggested by Kelly. Kelly has suggested the following formula for measuring skewness upon 10th and 90th percentiles. Its value also lies within.

$$S_k = \frac{(P_{90} - M) - (M - P_{10})}{(P_{90} - M) + (M - P_{10})} = \frac{P_{10} + P_{90} - 2\text{ Median}}{P_{90} - P_{10}}$$; where P_{10} and P_{90} are 10th and 90th percentiles respectively.

Similarly, if D_1 and D_9 are 1st and 9th deciles respectively,

$$S_k = \frac{(D_9 - M) - (M - D_1)}{(D_9 - M) + (M - D_1)} = \frac{D_1 + D_9 - 2\text{ Median}}{D_9 - D_1}$$

Note: As this method is primarily of theoretical importance only, it is seldom used in practice.

Example 6.7: Calculate percentile coefficient of skewness from the following positional measures given below:

P_{90} = 105; P_{10} = 60.35; P_{50} = 80.65

Solution:

$$S_k = \frac{P_{10} + P_{90} - 2\text{ Median}}{P_{90} - P_{10}}$$ = (60.35+105-2 80.65) (105-60.35) = 0.090

Hence the distribution is positively skewed.

COEFFICIENT OF SKEWNESS BASED ON MOMENTS

The turning or rotating effect of a force is termed as "moment". In statistics, it is used to describe the eccentricity of a frequency distribution. Using moments we can measure the central tendency of set of observations, their scatter, asymmetry and also the peakedness of the curve. Deviations of items are taken from the arithmetic mean of the distribution. The arithmetic mean of various powers of the deviations will give the required moments of the distribution. The moments about the actual mean is denoted by the Greek letter μ(mu).

The rth order moment about any point á' (r[th] non-central moment) is given by:

$$\mu_{ra} = \frac{\sum f_i (x_i - a)^r}{\sum f_i}$$

If a is taken as 0, the rth moment about origin (r[th] raw moment) is given by:

$$\mu_r' = \frac{\sum f_i (x_i)^r}{\sum f_i}$$

If a is taken as AM, the rth moment about mean (r[th] central moment) is given by:

$$\mu_r = \frac{\sum f_i (x_i - \overline{x})^r}{\sum f_i}$$

Where, $\overline{x}$ is the mean of items $x_1, x_2 \cdots, x_n$ and

$$\overline{x} = \frac{\sum f_i x_i}{\sum f_i}$$

The first four raw and central moments are given by the following formulae:

	Raw Moment	Central Moment
First moment	$\mu_1' = \frac{\sum f_i x_i}{\sum f_i} = \overline{x}(AM)$	$\mu_1 = \frac{\sum f_i (x_i - \overline{x})}{\sum f_i} = 0$
Second moment	$\mu_2' = \frac{\sum f_i (x_i)^2}{\sum f_i}$	$\mu_2 = \frac{\sum f_i (x_i - \overline{x})^2}{\sum f_i} = \text{Variance}(\sigma^2)$
Third moment	$\mu_3' = \frac{\sum f_i (x_i)^3}{\sum f_i}$	$\mu_3 = \frac{\sum f_i (x_i - \overline{x})^3}{\sum f_i}$
Fourth moment	$\mu_4' = \frac{\sum f_i (x_i)^4}{\sum f_i}$	$ì_4 = \frac{\sum f_i (x_i - \overline{x})^4}{\sum f_i}$

USE OF MOMENTS

1. The first moment (μ_1) of a frequency distribution is always zero, i.e., $\mu_1 = 0$. It measures the mean of the distribution, i.e., $\mu_1 = 0$.
2. The second moment (μ_2) of a frequency distribution about the mean is the variance of the distribution, i.e., $\mu_2 = \sigma^2$. It measures the variance i.e., the spread of the different terms in the distribution.
3. The third moment (μ_3) gives an idea about the degree of skewness present in the series.
4. The fourth moment (μ_4) throws light on the height of the frequency distribution, i.e., whether it is more peaked or more flat topped than the normal curve. It measures Kurtosis.
5. Co-efficient of skewness is given by β_1, where $\beta_1 = \dfrac{\mu_3^2}{\mu_2^3}$
6. Kurtosis is measured by β_2, where $\beta_2 = \dfrac{\mu_4}{\mu_2^2} = \dfrac{\mu_4}{\sigma^4}$

KURTOSIS

According to C.M. Mayer, "Kurtosis refers to the degree of peakedness of the hump of the distribution". Kurtosis enables us to have an idea about the shape and nature of the hump (middle part - bell portion) frequency distribution. It is concerned with the flatness or peakedness of the frequency curve.

Karl Pearson called it a "Measures of Convexity" of the curve. He introduced three broad patterns of peakedness, which are illustrated in the following diagrams:

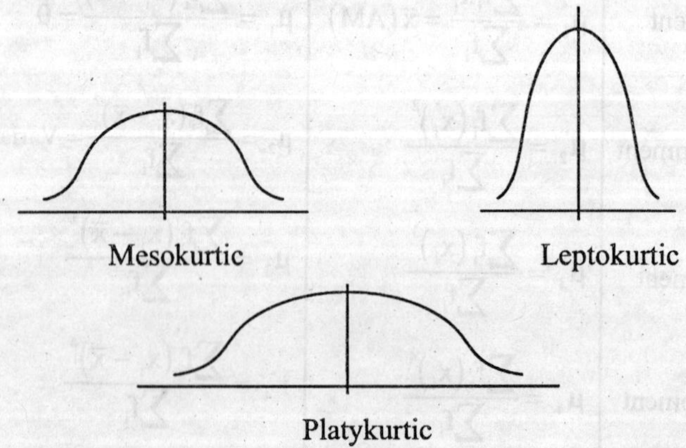

The curve which is neither flat nor peaked is known as normal curve or Mesokurtic. A curve which is more peaked than the normal curve is known as Leptokurtic and the curve which is flatter than the normal curve is called the Platykurtic.

Measures of Kurtosis: As a measure of kurtosis, Karl Pearson gave the coefficient of kurtosis as coefficient of beta two (β_2) and its next derivative as r_2.

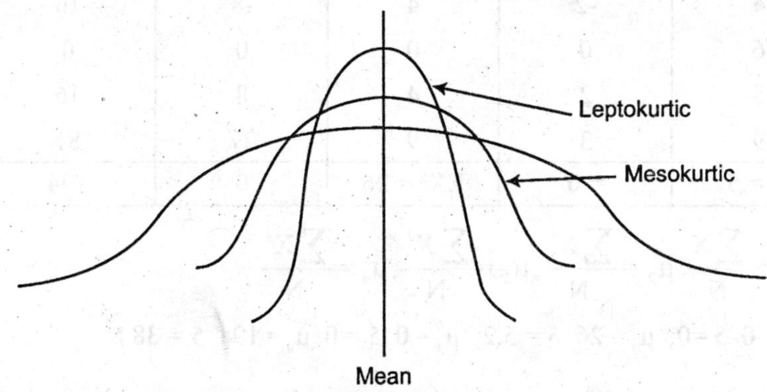

The measures are defined as:

$$\beta_2 = \frac{\mu_4}{\mu_2^2} = \frac{\mu_4}{\sigma^4}$$

$$r_2 = \beta_2 - 3 = \frac{\mu_4}{\sigma^4} - 3 = \frac{\mu_4 - 3\sigma^4}{\sigma^4}$$

I. When the value of $\beta_2 > 3$, it is a leptokurtic.

II. When the value of $\beta_2 < 3$, it is a platykurtic.

III. When the value of $\beta_2 = 3$, it is a mesokurtic.

Example 6.8: The number of bacteria in 1ml of blood from 5 persons are 3, 4, 6, 8, 9. Calculate the first, second, third and fourth moments about the mean. Also find the skewness and kurtosis.

Solution:

Table: Calculation of moments

x	$x-\bar{x}=X$	$(x-\bar{x})^2=X^2$	$(x-\bar{x})^3=X^3$	$(x-\bar{x})^4=X^4$
3	-3	9	-27	81
4	-2	4	-8	16
6	0	0	0	0
8	2	4	8	16
9	3	9	27	81
N = 5	= 0	$\Sigma X^2 = 26$	0	194

$$\mu_1 = \frac{\sum X}{N}; \mu_2 = \frac{\sum X^2}{N}; \mu_3 = \frac{\sum X^3}{N}; \mu_4 = \frac{\sum X^4}{N}$$

$\mu_1 = 0\ \ 5 = 0$; $\mu_2 = 26\ \ 5 = 5.2$; $\mu_3 = 0\ \ 5 = 0$ $\mu_4 = 194\ \ 5 = 38.8$

$$\text{Skewness } (\beta_1) = \frac{\mu_3^2}{\mu_2^3} = 0 \div (5.2)^3 = 0$$

$$\text{Kurtosis } (\beta_2) = \frac{\mu_4}{\mu_2^2} = 38.8 \div (5.2)^2 = 1.435$$

As the kurtosis is less than 3, i.e. 1.435, the distribution is platykurtic.

Example 6.9: The first four central moments of a distribution are 0, 1.4, 0.6, 5.88. Test the skewness and kurtosis of the distribution.

Solution:

$$\text{Skewness } (\beta_1) = \frac{\mu_3^2}{\mu_2^3} = \frac{(0.6)^2}{(1.4)^3} = 0.364$$

The distribution is not perfectly symmetrical as $\beta_1 = +0.131 + 0.364 \neq 0$

$$\text{Kurtosis } (\beta_2) = \frac{\mu_4}{\mu_2^2} = \frac{5.88}{(1.4)^2} = 3$$

The distribution is mesokurtic as $\beta_2 = 3$

Note: In case of calculation of moments from the group data generated by a continuous variable and if the number of observations is very large, the distribution is not very symmetrical and the frequencies taper off to zero in both ends. Hence **Shepperd's Correction** is applied to second and fourth moments only. First and third moments are not affected and they need no correction. (d is the size of the class interval)

EXERCISES

1. In a distribution, the difference between two quartiles is 20, their sum is 60 and median is 40. Find the coefficient of skewness.
2. The first four central moments of a distribution of blood samples of 80 patients for their lipid profile are 0, 1.9, 2.4, 12.42. Test the skewness and kurtosis of the distribution.
3. In a distribution, the difference between two quartiles is 3.24 and their sum is 62.25 and median is 21.7. Find the coefficient of skewness.
4. Calculate Karl Pearson's coefficient of skewness for the following data on the number of red flowers on a plant 11, 16, 13 and 20.
5. From the information given below, calculate Karl Pearson's coefficient of skewness and quartile coefficient of skewness.

Measure	**Firm A**	**Firm B**
Mean	155	142
Median	140	150
Standard deviation	32	52
Third quartile	186	140
First quartile	59	76

6. For a group of 8 monkeys $\Sigma X = 254$, $\Sigma X^2 = 22480$ and Mode = 46.4. Find their Karl Pearson's coefficient of skewness.
7. Calculate any measure of skewness for the following:

No. of flowers on a plant	0	1	2	3	4	5
No. of plants	8	2	4	5	1	3

8. The data on the number of peas in a pod is 2, 4, 2, 3, 6. Find its (i) first four moments about the mean ; (ii) skewness and (iii) kurtosis.
9. Find the measures of skewness for the following data of white flowers on a plant: 15, 4, 8, 2 and 6.
10. From a moderately skewed distribution of prices of a commodity, it is found that the mean price is Rs 15 and the median price is Rs. 20. If the coefficient of variation is 15%, find the Pearson coefficient of skewness of the distribution.

PROBABILITY–BASIC CONCEPTS

The theory of Probability dates back to the 17th century through the works of Galileo, Huygens, Pascal, Ferment and Bernoulli while trying to solve the problems faced by gamblers. The theory of errors of observations put forward by De Moivre, Laplace and Gauss pioneered further development on theory of probability. Subsequently, this theory got enriched by the fundamental contributions of Russian mathematicians like Chebyshev, Markov, Lyapunov and above all Kolmogorov. Today the theory of probability is a part of our everyday life.

In personal and managerial decisions, we face uncertainties and use probability theory whether or not we admit its use. The insurance industry uses probability theory to calculate premium rates. Drug manufacturers use this theory in testing effectiveness of the drug being manufactured. There are many more practical areas where theory of probability plays a dominant role in decision making processes.

DEFINITION

The term "*probability*" is difficult to define, especially in precise terms. In common parlance the term "probability" refers to the *chance* of happening or not happening of an event. The moment we use the word "chance", we indicate that there is an element of uncertainty about the statement that has been made or is being made. Consider a statement- "the chances that the new offspring will be a dwarf one from a tall father and a dwarf mother". There is an element of uncertainty associated with this statement. We have not assigned any *numerical value* to this statement. If we could assign, the statement would become more precise. *The theory of probability provides a numerical measure of the element of uncertainty. It enables us to take decision under conditions of uncertainty with a calculated risk.*

MEASURE OF PROBABILITY

The value of probability ranges from 0 to 1. If an event is sure to happen, then its probability would be 1 and if it is sure that the event would not take place, then the probability of happening is 0. Ordinarily in physical science the probability of the happening of an event is rarely 1 or 0. The reason is that in physical studies we deal with situations where there is always an element of uncertainty about the happening or not happening of an event. For this reason the probability of the events is somewhere between 0 and 1.

APPROACH TO STUDY PROBABILITY

There are different approaches to study probability. But whatever approach is undertaken, each of them must satisfy the basic requirements and those are (i) Sum total of all probabilities must be equal to 1 and (ii) probability must remain in the range 0 to 1, both inclusive. Mathematically, (i) $\sum p_i = 1$ and (ii) $0 \le p_i \le 1$. The basic approaches to probability are (i) Classical Approach, (ii) Relative Frequency Approach and (iii) Subjective Approach (iv) Axiometric Approach. Let us discuss these one by one.

CLASSICAL APPROACH TO PROBABILITY

This is otherwise known as Priori Probability, where orderly decision problems are dealt with. For example- card games, dice games, coin tossing and like. Here, each possible outcome is equally likely. Classical approach to probability cannot be applied where probabilistic situations are not equally likely. So the drawback here is, it cannot be applied to less orderly decision problems, which are mostly encountered in management. Very unlikely situations, though can conceivably happen, are excluded from this approach. For example- probability that I will go on a business trip to Greenland, President of USA will have dinner with me in my village, coin landing on its edge etc.

If there are n exhaustive, equally likely and mutually exclusive outcomes and m of them are favourable to the event, then the probability of happening of A is given by,

$$P(A) = \frac{\text{Favourable number of outcomes}}{\text{Exhaustive number of outcomes}} = \frac{m}{n}$$

$$P(\text{not happening of A}) = P(A^c) = \frac{n-m}{n} = 1 - P(A)$$

And $P(A) + P(A^c) = 1$

If the outcomes are not equally likely and the number of outcomes is infinite, this approach fails.

Consider another example on the survivality, mortality and abnormality of certain animal when administered with a drug, where say 12 number of animals are used in the experiment. Out of 12 number of animals, let us say, only one survived and 11 died. From these results, we can safely say, from our common mathematical knowledge, that the survival rate is $(1 \div 12) \times 100 = 8.3$ %. The mortality rate is $(11 \div 12) \times 100 = 91.7$ %. That means the chance of the animals remain alive is 8.3% or 0.083 and the chance of the animals being dead is 91.7 % or 0.917.

In the above example, let us say the chance of being alive = p and the chance of being dead = q

So, $p = 0.083$ and $q = 0.917$

$p + q = 0.083 + 0.917 = 1$

In terms of probability, p is called the probability of success and q is called probability of failure. The total probability p + q is always 1.

Now, we can put this chance occurrence in a more generalized manner in the following way:

If an event can happen in 'a' ways and fail to happen in 'b' ways, and all these ways are equally likely (means both success and failure have equal chance of happening), then

The probability or chance of its happening is $\dfrac{a}{a+b}$

The probability or chance of its not happening is $\dfrac{b}{a+b}$

The probability of happening of an event is generally denoted by 'p' and the probability of it's failing to happen by 'q'.

The probability of happening of an event is known as *success* and the probability of its failing to happen as *failure*.

Thus, probability of happening of an event

$$= p = \frac{\text{Total number of cases favourable to the event}}{\text{Total number of possible cases for the event (Exhaustive Cases)}} = \frac{a}{a+b}$$

Probability of non-happening of an event

$$= q = \frac{\text{Total number of cases not favourable to the event}}{\text{Total number of possible cases for the event (Exhaustive Cases)}} = \frac{b}{a+b}$$

$$p + q = \frac{a}{a+b} + \frac{b}{a+b} = 1$$

Now it can be concluded that if p is the probability of happening of an event, the probability of it's failing to happen is 1 – p.

Similarly, if q is the probability of failing to happen an event, the probability of its happening is 1 – q.

If p = 1, then q = 0, It is a case of sure success. i.e. p = 1 for occurrence of certain events.

If q = 1, then p = 0, It is a case of sure failure. i.e. p = 0 for occurrence of impossible events.

From the above discussion we have the following conclusions:

(i) p

(ii) $0 \leq p \leq 1$

RELATIVE FREQUENCY APPROACH

This method uses the relative frequencies of past occurrences. Percentage of happening of an event past determines the probability that it will happen in future. Suppose in a village it is recorded in the past 100 years that 8 new born infants had died immediately after birth. Then, the probability that a still-born baby take birth in the village is $\frac{8}{100} = 0.08$.

Another important characteristic of this approach is the relative frequency becomes stable, as the number of times the experiment is conducted increases. For example: Let 2 Heads occur in tossing a fair coin 10 times. So the probability of Head from this experiment is P(H) = 0.2. Again for second time, same fare coin is tossed for 10 times and let us say 4 Heads occur. After the second experiment, total 6 Heads occur in a total of 20 tosses, hence P(H) = $\frac{6}{20} = 0.30$. Let again 7 Heads occur in third time experiment, hence P(H) after this will be = $\frac{13}{30} = 0.433$. This way, the probability figure goes on increasing and approaches to a stable figure of 0.5, as the number of experiments increases. Gradually it gains accuracy as we increase the number of observations. But increasing the number of observations also adds to the cost. Another difficulty is it is often used without evaluating a sufficient number of outcomes.

Thus, if S_n is the number of occurrences of an event A in n independent repetitions of a random experiment, the relative frequency $\frac{S_n}{n}$ hovers around a constant with increasing n. This makes us to conclude that certain regularities govern the phenomenon beyond human control, thus making the relative frequency constant. Unfortunately this approach is criticized for not having any mathematical character, rather descriptive in nature.

SUBJECTIVE APPROACH

Most higher-level managerial decisions are concerned with specific, unique situations, rather than with a long series of identical situations or sufficient past records. For example: Suppose a decision is to be taken as to whether a nuclear power plant to be constructed at a moderately seismic zone or not. Here, past record of any nuclear accident at the site is certainly not available and that does not mean that decision cannot be taken. Rather, best judgement in trying to determine the subjective probability of a nuclear accident must be taken. It depends on a person's degree of belief.

Subjective approach is based on the beliefs of a person making the probability assessment. It can be defined as the probability assigned to an event by an individual, based on available evidence. If no evidence is available, it requires an educated guess.

AXIOMETRIC APPROACH

Before we go into details of probability it is necessary that we familiarize ourselves with certain terms that are used in this context.

Random Experiment

It is an experiment, which if conducted repeatedly under homogeneous conditions does not give the same result. The result may be any one of the various possible "*outcomes*". For example, in case of throwing a dice, the possible outcomes are any of 1, 2, 3, 4, 5 or 6.

Trial and Event

The performance of a random experiment is called a *trial* and the outcome is called an *event*. For example, the throwing of a dice is a trial and the result (falling with nos. 1, 2, 3, 4, 5, or 6) is an event. There are several types of events.

(i) ***Simple Events***: Occurrence or non-occurrence of a single event. Drawing of a particular card from a pack is a simple event. A woman giving birth to a child male or female is a case of simple event.

(ii) ***Mixed or Compound or Joint Events***: Occurrence of two or more simple events simultaneously is called *mixed event*. For example, if a bag contains 6 black balls and 4 white balls and we make a draw of 2 balls at random, then the events that 'both are black' or 'one is black' and 'one is white' are compound events. Again compound events may be of two types:

(a) *Independent Events*: An event is said to be independent if it's happening is not affected by the happening of other events and if it does not affect the happening of other events. For example, a coin is thrown twice. The results of the two tosses are independent.

(b) *Dependent Events*: If the occurrence of one event influences the occurrence of the other, then the second event is said to be dependent on the first. For example, a card is drawn from a pack and is not replaced; the subsequent draws would definitely be dependent on the first draw.

Exhaustive Cases

All possible outcomes of a random experiment is known as exhaustive cases. In a throw of single die (the plural is dice) the exhaustive cases are 6 as any one of the numbers (1 to 6) may come up.

Favorable Cases

The numbers of outcomes which result in the happening of a desired event are called favorable cases. Thus in a single throw of a die the number of favorable cases of getting an even number are three- 2, 4 and 6.

Mutually Exclusive Cases

Two or more cases are said to be mutually exclusive if the happening of any one of them excludes the happening of all others in a single experiment. In tossing a coin, either head or tail will come up. Here *one and only one* of the events can take place excluding all others.

Equally Likely Event

If the likelihood of the occurrence of every possible outcome is the same, it is called equally likely event. The birth of a baby, the male and female has an equal chance.

Sample Space

The set of all possible outcomes of a random experiment is known as the sample space and is denoted by S. In other words, sample space is the set of all exhaustive cases of the random experiment. One can write n(S) to denote the number of elements n in sample space S called sample points. For example, a cow can give birth to either a male calf (M) or a female calf (F). So the sample space is {M, F} in this context. In case of throwing of a die, the possible outcomes are any one of 1, 2, 3, 4, 5 or 6. So, here the sample space is {1, 2, 3, 4, 5, 6}.

Sample Points

Components of sample space are known as sample points. In the above example, if the cow gives birth to a male calf, then the sample point would be {M}.

Null Set (Φ)

A set having no elements in it is called a null set or an empty set and is denoted by Φ. Hence, n(Φ) = 0. For example, probability of getting a 7 in throwing a die is 0, because the exhaustive cases in throwing a dice do not contain the sample point as 7 in it. So probability based on null set is zero.

A prior knowledge of permutation and combination is absolutely necessary while studying probability.

Permutations and Combinations

In simple language, the word *permutation* means *arrangement with considering order* and *combination means arrangement without order* or *group selection.* It will be clearly understood if an example is taken. Consider three letters A, B and C. So, the permutations (arrangement) of these letters taken two at a time will be AB, BC, CA, BA, CB and AC, i.e. 6 in all. The combination (selection) of these letters taken two at a time will be AB, BC and CA, i.e. 3 in all. It should be noted that in all combinations, the order of the elements is immaterial i.e. AB and BA form the same combination but these are different arrangements. Similarly, in case of 4 letters A, B, C, and D, the total number of combinations taking three at a time is: ABC, ABD, ACD, BCD, i.e. 4 in all. However, each of these combinations with considering order gives six different arrangements (permutations), i.e. ABC, ACB, BAC, BCA, CAB, CBA.

Hence the total number of permutations (arrangements) of 4 letters taking 3 at a time is

$4 \times 6 = 24$. i.e. $4 \times 3 \times 2 = 24$

Permutations of n different objects taken r at a time is denoted by ${}^{n}P_{r}$.

$^nP_r = n\,(n-1)\,(n-2)\cdots(n-r+3)\,(n-r+2)\,(n-r+1)$

e.g. $^3P_2 = 3.(3-1) = 3.2 = 6$

$^4P_3 = 4.(4-1)\,.\,(4-2) = 4.3.2 = 24$

Now, total number of permutations of n different objects, taken all at a time is given by

(Take r = n):

$^nP_n = n\,.\,(n-1)\,.\,(n-2)\,.\,(n-3)\cdots(n-n+1)$

$= n\,.\,(n-1)\,.\,(n-2)\,.\,(n-3)\cdots 1 = n!$

(Pronounced as Factorial n)

So, $4! = 4.3.2.1 = 24$

$6! = 6.5.4.3.2.1 = 720$

Combinations of n different objects taken r at a time is given by:

$$^nC_r = \frac{n!}{r!\times(n-r)!} = \frac{P^n_r}{r!} \text{ as } {}^np_r = \frac{n!}{(n-r)!}$$

There is a nice diagrammatic way to illustrate the probability concepts. A pictorial representation called *Venn diagram* is used, after 19th century English mathematician *John Venn*. In these diagrams, the entire sample space is represented by a rectangle and the events are represented by parts of the rectangle.

If two events are mutually exclusive, their parts will not overlap each other, as shown in the diagram below. In case of mutually exclusive events, either A or B happens at any time. Both cannot happen simultaneously. Here A and B cannot intersect. For example, in tossing of a coin, either Head will occur or Tail will occur and both cannot occur. This is called Disjoint events. Here, Φ(Null Set).

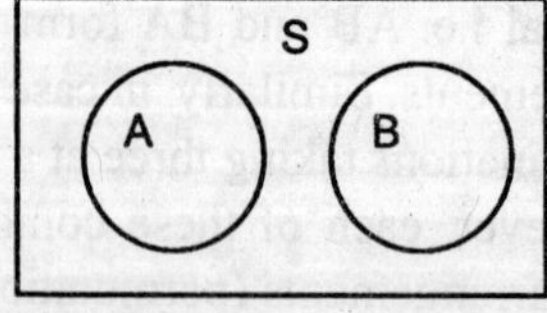

If two events are *not* mutually exclusive, their parts *will* overlap. An event which represents the happening of *at least one* of the events A and B (i.e. either A occurs or B occurs or both A and B occur). This is called union of sets, symbolically represented by $A \cup B$.

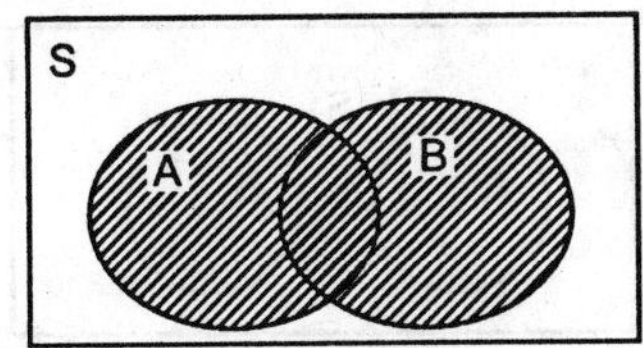

There is another event called simultaneous happening of events. For example, drawing a diamond ace card from a pack of cards. Here the card drawn is simultaneously an ace and also a diamond. This is called "intersection of sets", symbolically represented by $A \cap B$.

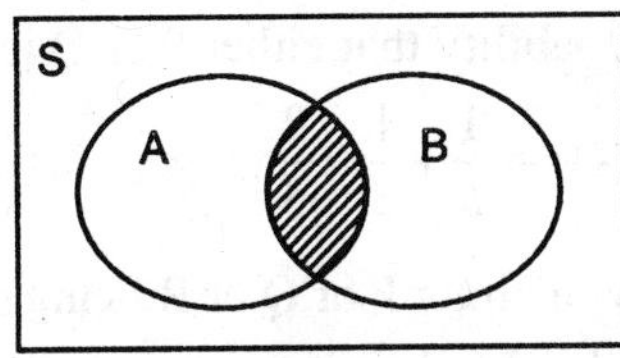

An event A and its complementary together form the sample space S. That means $S = A \cup A^c$.

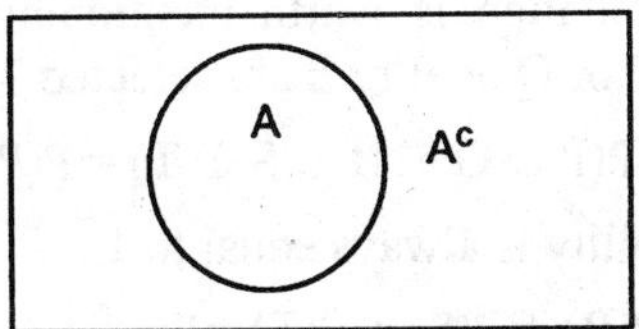

Example 7.1: In how many ways 4 students can be chosen for taking leadership out of 10 students in a class?

Solution: The number of ways 4 students can be selected out of 10 is $^{10}C_4$.

$$^{10}C_4 = \frac{10!}{4! \times (10-4)!} = \frac{10\times9\times8\times7\times6\times5\times4\times3\times2\times1}{(4\times3\times2\times1)\times(6\times5\times4\times3\times2\times1)} = 210$$

Example 7.2: From a bottle containing 20 groundnut seeds, 6 seeds are to be taken out for some experiment. What is the number of exhaustive cases?

Solution: The exhaustive cases are $^{20}C_6$ number of ways = 38760.

ADDITION LAW OF PROBABILITY (Mutually Exclusive Events)

We may be interested to know that one thing or another will occur. In case of mutually exclusive events

$$P(A \text{ or } B) = P(A \cup B) = \text{Probability of either A or B} = P(A) + P(B)$$

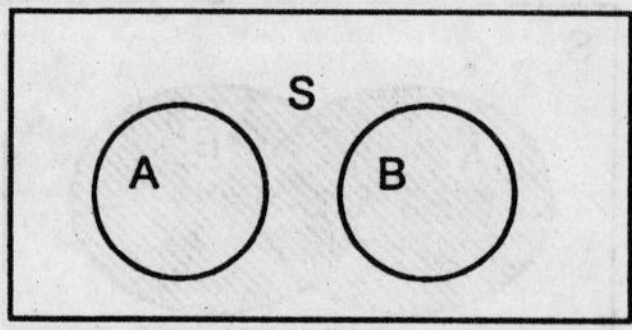

Take an example that, 5 equally good students (P, Q, R, S & T) applied for a job in a company and only one is to be selected. Hence each is having a probability of $\frac{1}{5}$ of getting selected. So, P(P) = P(Q) = P(R) = P(S) = P(T) $= \frac{1}{5}$. But what is the probability that either P or Q is selected $P(P \cup Q)$? The answer will be $P(P) + P(Q) = \frac{1}{5} + \frac{1}{5} = \frac{2}{5}$.

Similarly, probability of either P or Q or R being selected is $P(P \cup Q \cup R)$ $= P(P) + P(Q) + P(R) = \frac{1}{5} + \frac{1}{5} + \frac{1}{5} = \frac{3}{5}$

One important calculation is worth mentioning here that, what is the probability that either P or Q or R or S are selected ?

The answer will be $P(P \cup Q \cup R \cup S \cup T) = P(P) + P(Q) + P(R) + P(S)$.

But the total probability is always equal to 1.

So, P(P) + P(Q) + P(R) + P(S) + P(T) =1.

Hence, the required probability that either P or Q or R or S are selected is $1 - P(T) = 1 - \frac{1}{5} = \frac{4}{5}$.

Example 7.3: A box contains 6 white puppies, 4 black puppies and 2 gray puppies. A puppy is taken out. What is the probability that it is either a black or gray puppy.

Solution: Total number of puppies = 6 + 4 + 2 = 12 i.e. exhaustive cases = 12

Probability that a black puppy is taken = $P(B) = \frac{4}{12} = \frac{1}{3}$

Probability that a gray puppy is taken = $P(G) = \frac{2}{12} = \frac{1}{6}$

Probability that it is black or gray = P(B or G) = P(B U G)

Now applying the additive law of probability

$$P(B \cup G) = P(B) + P(G) = \frac{1}{3} + \frac{1}{6} = \frac{1}{2}$$

ADDITION LAW OF PROBABILITY (Not Mutually Exclusive Events)

In case of *not* mutually exclusive events,

$$P(A \text{ or } B) = P(A \cup B) = P(A) + P(B) - P(AB)$$

Here, P(A) is probability of A, P(B) is probability of B and $P(A \cap B)$ is probability of A and B occurring simultaneously. In the diagram below, the double shaded portion is the region of simultaneous happening of both A and B, which is to be deducted because of double counting in P(A) + P(B).

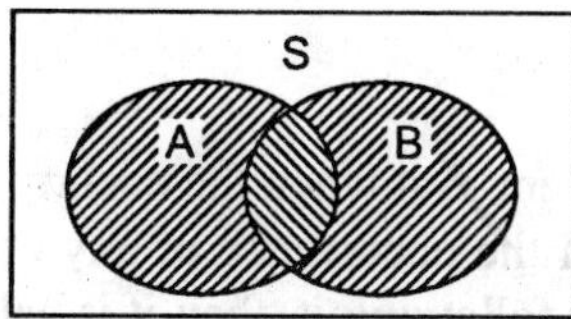

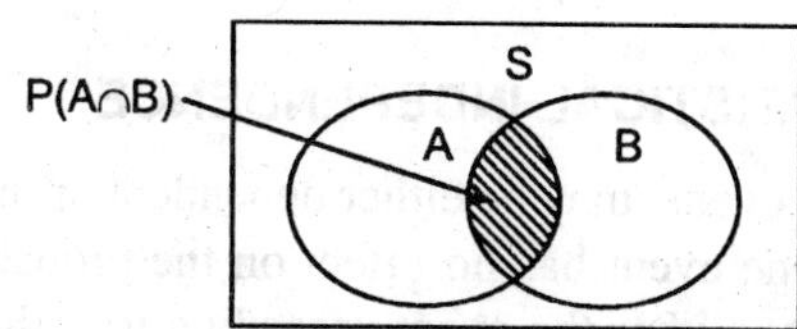

Take an example that, a card is drawn from a pack of cards so that it is either a King or a Heart. In a pack of 52 cards, there are 4 Kings and 13 Hearts. But King and Heart are not mutually exclusive, because there is 1 card being both King and Heart(simultaneous happening). Probability of drawing of king P(King) is 4/52; probability of drawing of heart P(Hear) is 13/52 and probability of drawing of both King and Heart simultaneously $P(\text{King} \cap \text{Heart})$ is 1/52. Hence the required probability of drawing of either King or heart will be,

$$P(\text{King} \cup \text{Heart}) = P(\text{King}) + P(\text{Heart}) - P(\text{King} \cap \text{Heart}) = \frac{4}{52} + \frac{13}{52} - \frac{1}{52} = \frac{16}{52}$$

Let us take another example. The labour union of a company selects 5 persons, out of which only one will represent in the management decision making process. The ages of the five persons are as under:

1. Female aged 30
2. Male aged 38
3. Male aged 42
4. Male aged 50
5. Female aged 55

Now the question is what is the probability that the person selected is either a male or age over 40? Here the probability of male is $P(\text{Male}) = \frac{3}{5}$.

Probability of age over 40 is $P(\text{age over } 40) = \frac{3}{5}$. Probability of both male and age over 40 is $P(\text{Male} \cap \text{age over} 40) = \frac{2}{5}$.

Hence, P(Male or age over 40) = P(Male) + P(age over 40) – P(Male $\cap$ age over 40) $= \frac{3}{5} + \frac{3}{5} - \frac{2}{5} = \frac{4}{5}$

On visual inspection of data, there are 4 persons either male or age over 40.

Similarly, for three events A, B and C (not mutually exclusive),

$$P(A \cup B \cup C) = PA) + P(B) + P(C) - P(A \cap B) - P(B \cap C) - P(C \cap A) + P(A \cap B \cap C)$$

STATISTICAL INDEPENDENCE

The events may be either dependent or independent. But when the occurrence of one event has no effect on the probability of the occurrence of any other event, either the event preceding it or the event following it, then it is called *statistical independence.* For example, tossing a coin. In an unbiased coin, P(Head) = P(Tail) = 0.5 always, no matter how many times the coin is tossed. Probability in each toss has no relationship with any of the tosses either preceding it or following it. The outcomes of several tosses of this coin are statistically independent events.

JOINT PROBABILITY UNDER STATISTICAL INDEPENDENCE

When two or more independent events occur simultaneously or in succession, then their joint probability is product of their individual probabilities.

$$P(A \cap B) = P(A) \times P(B)$$

For example, probability of two Heads in tossing of two unbiased coins is $P(H_1H_2) = 0.5 \times 0.5 = 0.25$. Similarly, in three successive tosses of a coin the probability of three Heads is $P(H_1H_2H_3) = 0.5 \times 0.5 \times 0.5 = 0.125$.

Suppose the coin is biased with probability of Head 0.8 and probability of tail 0.2. Then joint probability of two Heads will be 0.8 0.8 = 0.64 and so on.

Example 7.4: What is the probability of getting Head, Tail and Head in that order on three successive tosses of a fair coin?

Solution: $P(H_1T_2H_3) = 0.5 \times 0.5 \times 0.5 = 0.125$

Example 7.5: What is the probability of getting at least two heads on three tosses of a fair coin?

Solution: The various possibilities are:

$H_1H_2H_3$, $H_1H_2T_3$, $H_1T_2H_3$ and $T_1H_2H_3$

As seen earlier, each of these has individual probability of $0.5 \times 0.5 \times 0.5 = 0.125$

Hence, the total probability will be $0.125 + 0.125 + 0.125 + 0.125 = 0.5$

Example 7.6: What is the probability of at least one Head on three tosses of a fair coin?

Solution: At least one Head means, it may be one Head or two Heads or three Heads. So the various possibilities are as under:

One Head: $H_1T_2T_3$, $T_1H_2T_3$, $T_1T_2H_3$. Probability = $0.125 + 0.125 + 0.125 = 0.375$

Two Heads: $H_1H_2T_3$, $H_1T_2H_3$, $T_1H_2H_3$. Probability = $0.125 + 0.125 + 0.125 = 0.375$

Three Heads: $H_1H_2H_3$. Probability = 0.125

Total Probability = $0.375 + 0.375 + 0.125 = 0.875$

In another way also the problem can be solved. As total probability is always 1, the probability of at least one head is equal to 1 Probability of no Heads. Probability of no Heads is $P(T_1T_2T_3) = 0.125$. So, the required probability of at least one Head = $1 - 0.125 = 0.875$.

CONDITIONAL PROBABILITY UNDER STATISTICAL INDEPENDENCE

Conditional probability is the probability that a second event (B) will occur if a first event (A) has already occurred, which is denoted by $P(B|A)$.

But for statistically independent events, the probabilities are no way affected by occurrence of each other and hence $P(B|A) = P(B)$. For example, what is the probability that a second toss will be head, given that the first one was head? Symbolically, $P(H_2|H_1) = ?$ As the probabilities are independent of each other, i.e. the result of the first toss has no effect on the result of the second toss, the required probability, $P(H_2|H_1) = P(H_2) = 0.5$.

PROBABILITY UNDER CONDITIONS OF STATISTICAL DEPENDENCE

Statistical dependence exists when the probability of some event is dependent on or affected by the occurrence of some other event. To understand this concept, let us take an example. Suppose a bag contains 10 balls of which 3 are red and dotted, 1 is red and striped, 2 are blue and dotted, 4 are blue and striped. This can be tabulated in the following manner, called as Frequency Table.

	Dotted (D)	Striped (S)	Total
Red(R)	3	1	4
Blue(B)	2	4	6
Total	5	5	10

Now let us prepare the Probability Table in the following manner.

	Dotted(D)	Striped(S)	Total
Red(R)	$P(R\cap D)=\frac{3}{10}$	$P(R\cap S)=\frac{1}{10}$	$P(R)=\frac{4}{10}$
Blue(B)	$P(B\cap D)=\frac{2}{10}$	$P(B\cap S)=\frac{4}{10}$	$P(B)=\frac{6}{10}$
	$P(D)=\frac{5}{10}$	$P(S)=\frac{5}{10}$	Total Probability = 1

$P(R \cap D)$, $P(R\cap S)$, $P(B\cap D)$ and $P(B\cap S)$ are each called **Joint Probability**. $P(R) = \frac{4}{10}$, $P(B) = \frac{6}{10}$, $P(D) = \frac{5}{10}$ and $P(S) = \frac{5}{10}$ are each called **Marginal Probability**, as those are written on the margin of the table.

Suppose, a red ball is drawn from the box. Now the question is what is the probability that it is dotted or symbolically, $P(D|R)$ = ? In other way what is the probability that this ball is dotted, given the condition that it is drawn from red category ?

To answer this question, the blue balls are to be totally ignored and we have to consider only red balls, because draw of dotted ball is only from the red category. Here, total red balls are 4 and out of these 4 red balls 3 are dotted. Hence probability of dotted ball, given that it has come from the red category is, $P(D|R) = \frac{3}{4}$.

Now, we have to use this reasoning to develop the formula for conditional probability under statistical dependence. We will have to divide the probabilities of red and dotted balls. That means, probability of red ball is $P(R) = \frac{4}{10}$ and probability of red and dotted ball is, $P(RD) = \frac{3}{10}$.

So, $P(D|R) = \frac{P(R\cap D)}{P(R)}$. Using this formula, the required answer to our question will be, $\frac{3 \div 10}{4 \div 10} = \frac{3}{4}$

So the general formula is,

$$P(A|B) = \frac{P(A \cap B)}{P(B)} \text{ and } P(B|A) = \frac{P(B \cap A)}{P(A)} = \frac{P(A \cap B)}{P(A)}$$

In other words, $P(A \cap B) = P(A) \times P(B|A) = P(B) \times P(A|B)$. This is known as **Multiplication Law of Probability.**

$$\text{Conditional Probability} = \frac{\text{Joint Probability}}{\text{Marginal Probability}}$$

Example 7.7: In the example above, find the probabilities of drawing a

(a) dotted ball given that a blue ball has been drawn?
(b) stripped ball given that a blue ball has been drawn?
(c) blue ball given that a dotted ball has been drawn?
(d) red ball given that a dotted ball has been drawn?
(e) red ball given that a striped ball has been drawn?
(f) blue ball given that a striped ball has been drawn?

Solution:

(a) $P(D|B) = \frac{P(D \cap B)}{P(B)} = \frac{2 \div 10}{6 \div 10} = \frac{2}{6} = \frac{1}{3}$

(b) $P(S|B) = \frac{P(S \cap B)}{P(B)} = \frac{4 \div 10}{6 \div 10} = \frac{4}{6} = \frac{2}{3}$

(c) $P(B|D) = \frac{P(B \cap D)}{P(D)} = \frac{2 \div 10}{5 \div 10} = \frac{2}{5}$

(d) $P(R|D) = \frac{P(R \cap D)}{P(D)} = \frac{3 \div 10}{5 \div 10} = \frac{3}{5}$

(e) $P(R|S) = \frac{P(R \cap S)}{P(S)} = \frac{1 \div 10}{5 \div 10} = \frac{1}{5}$

(f) $P(B|S) = \frac{P(B \cap S)}{P(S)} = \frac{4 \div 10}{5 \div 10} = \frac{4}{5}$

In the above it is seen that probability of B under the condition that A has been drawn is

$$P(B|A) = \frac{P(A \cap B)}{P(A)} = \frac{\text{Joint Probability of A and B}}{\text{Marginal Probability of A}}$$

Or,

$$P(A|B) = \frac{P(A \cap B)}{P(B)} = \frac{\text{Joint Probability of A and B}}{\text{Marginal Probability of B}}$$

Hence, Joint Probability of A&B is $= P(A \cap B) = P(B|A) \times P(A) = P(A|B) \times P(B)$

Example 7.8: According to a survey, the probability that a family owns a car if its monthly income is more than Rs. 50,000.00 is 0.75. Of the households surveyed, 60% had incomes over Rs. 50,000.00 per month and 52% had a car. What is the probability that a family has a car and income over Rs. 50,000.00 per month?

Solution: Let probability of owning a car is P(C)

Let probability of having income over Rs.50,000.00 (affluent) is P(A)

So, probability of owning a car with the condition of being affluent is $P(C|A) = 0.75$

Given that $P(A) = 0.6$

$$\text{But, } P(C|A) = \frac{P(C \cap A)}{P(A)} = 0.75$$

So, $P(C \cap A) = P(A) \times 0.75 = 0.6 \times 0.75 = 0.45$

Example 7.9: A departmental store has been the target of many shoplifters during the past month, but owing to increased security precautions, 250 shoplifters have been caught. Each shoplifter's sex is noted; also noted is whether the perpetrator was a first-time or repeat offender. The data are summarized below in the table below:

Sex	1st Time Offender	Repeat Offender	Total
Male	60	70	130
Female	44	76	120
Total	104	146	250

Assuming that an apprehended shoplifter is chosen at random, find the probability that:

(a) the shoplifter is a male.

(b) the shoplifter is a first-time offender, given that the shoplifter is a male.

(c) the shoplifter is a female, given that the shoplifter is a repeat offender.

(d) the shoplifter is a female given that the shoplifter is first-time offender.

(e) the shoplifter is both a male and a repeat offender.

Solution: Let the symbols are: Male = P(M); Female = P(F); first-time offender = P(T); repeat offender = P(R)

Given that,

$P(T) = 104/250 = 0.416; P(R) = 146 \div 250 = 0.584$

$P(M) = 130/250 = 0.52; P(F) = 120 \div 250 = 0.48$

(a) the probability that the shoplifter is male = 0.52

(b) the probability that the shoplifter is a first-time offender, given that the shoplifter is a male $= P(T|M) = \dfrac{P(T \cap M)}{P(M)} = \dfrac{60 \div 250}{130 \div 250} = 0.4615$

(c) the probability that the shoplifter is female, given that the shoplifter is a repeat offender $= P(F|R) = \dfrac{P(F \cap R)}{P(R)} = \dfrac{76 \div 250}{146 \div 250} = 0.5205$

(d) the probability that the shoplifter is female given that the shoplifter is a first-time offender $= P(F|T) = \dfrac{P(F \cap T)}{P(T)} = \dfrac{44 \div 250}{104 \div 250} = 0.4230$

(e) the probability that the shoplifter is both male and a repeat offender

$$= P(MR) = \frac{70}{250} = 0.28$$

BAYES' THEOREM

The concept of conditional probability has been discussed above, which evaluates the probability of occurrence of one event with the condition that another has already occurred. In other words, it relates to the prediction of certain events considering the information about the occurrence of another earlier event. So the determination of probability of happening of an event can be based on the probability of happening of an earlier event. In practical life, we say that the probability that a particular effect is due to a specific cause. Thus the actual probability of happening of an event without knowing any prior probabilities can be revised if any prior probability is known. The procedure of revising these probabilities is based on *Bayes' theorem.* An English clergy man with an interest in mathematics Rev. Thomas Bayes (1702-1763) propounded this theory in 1763.

The so called "Bayesian" approach addresses to the question of determining the probability of some event A, given that another event B, has already (or will be) occurred i.e. determining the value of P(A|B). In this approach the sample information can be used to determine the probability of an event. For example, a sample experiment on the effect of certain drug administered on 50 albino rats revealed that only 2 of them died. This prior information (*prior probability*) can be used as an estimate on the marketability of the drug i.e. probability of the drug to be effective when launched in the market (*posterior probability*).

Let A_1 and A_2 = the set of events which are mutually exclusive and exhaustive.

B = a simple event which occurs both in A_1 and A_2.

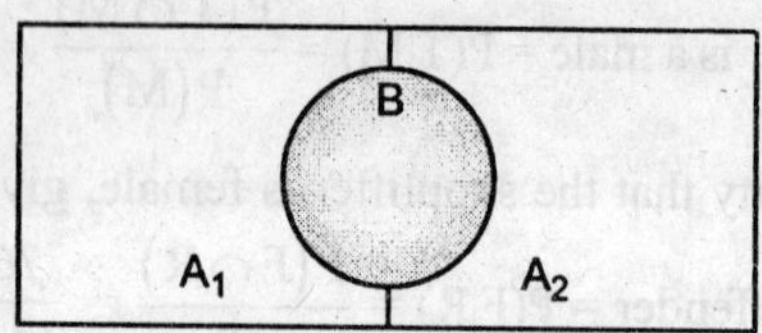

Observing the Venn diagram above, we have the part of B within A_1 represents the area 'A_1 & B' and the part of B within A_2 represents the area 'A_2 & B'.

Thus the probability of event A_1, given the event B has occurred, is

$$P(A_1|B) = \frac{P(A_1 \cap B)}{P(B)};\ P(B) \neq 0$$

Similarly, the probability of event A_2, given that B has occurred, is,

$$P(A_2|B) = \frac{P(A_2 \cap B)}{P(B)};\ P(B) \neq 0$$

Where, $P(B) = P(A_1 \cap B) + P(A_2 \cap B)$

$P(A_1 \cap B) = P(A_1) \times P(B|A_1)$

$P(A_2 \cap B) = P(A_2) \times P(B|A_2)$

Considering the above, we state the Bayes theorem for n events.

If $A_1, A_2, A_3, \ldots A_n$ are n mutually disjoint events with $P(A_i) \neq 0$ (i = 1,

2, 3 ··· n) then for any arbitrary event B which is a subset of $\bigcup_{i=1}^{n} A_i$ such that P(B) > 0, we have

$$P(A_i|B) = \frac{P(A_i \cap B)}{\sum_{i=1}^{n}[(P_i) \times P(B|A_i)]} = \frac{P(A_i) \times P(B|A_i)}{P(B)}$$

The most widely used application of such theory is the areas involving estimation of unknown probabilities and making decisions on the basis of new (sample) information. Probabilities before revision by Bayes' rule are called a *prior probabilities* or simply *priori*, because they are determined before the sample information is taken into consideration. A probability which has undergone revision in the light of sample information (using Bayes' rule), is called a *posterior probability*, because it represents a probability calculated after this information is taken into account. Thus, *a priori* probability, which is unconditional probability, becomes a posterior probability, which is a conditional probability, by using Bayes' rule.

Based on Bayesian statistics, the table below is formed for a clear understanding of the theory.

	Happening of X	**Not happening of X**	**Total**	**P(X|Event)**	**P($\bar{X}$| Event)**
Event A:	P(AX)	$P(A \cap \bar{X})$	P(A)	$\frac{P(X \cap A)}{P(A)}$	$\frac{P(\bar{X} \cap A)}{P(A)}$
Event B:	P(BX)	$P(B \cap \bar{X})$	P(B)	$\frac{P(X \cap B)}{P(B)}$	$\frac{P(\bar{X} \cap B)}{P(B)}$
Total:	P(X)	$P(\bar{X})$	100		1.00

P(A|X) or P(A|$\bar{X}$)

$$\frac{P(A \cap X)}{P(X)} \qquad \frac{P(A \cap \bar{X})}{P(\bar{X})}$$

P(B|X) or P(B|$\bar{X}$)

$$\frac{P(B \cap X)}{P(X)} \qquad \frac{P(B \cap \bar{X})}{P(\bar{X})}$$

Example 7.10: The probabilities of three events, A, B and C occurring are P(A) = 0.35; P(B) = 0.45; P(C) = 0.2. Assuming that A, B and C have occurred, the probabilities of another event X, occurring are P(X|A) = 0.8; P(X|B) = 0.65 and P(X|C) = 0.3. Find P(A|X), P(B|X) and P(C|X).

Solution: The given data are summarized in the following table:

Event	X	Not X	Total	P(X|Event)
A			0.35	0.80
B			0.45	0.65
C			0.20	0.30
Total			1.00	

P(A|X) = ?

P(B|X) = ?

P(C|X) = ?

Hence, $P(X \cap A) = P(X|A) \times P(A) = 0.80 \times 0.35 = 0.28 = P(A \cap X)$

$(X \cap B) = P(X|B) \times P(B) = 0.65 \times 0.45 = 0.2925 = P(B \cap X)$

$P(X \cap C) = P(X|C) \times P(C) = 0.30 \times 0.20 = 0.06 = P(C \cap X)$

The calculated results are tabulated below:

Event	X	Not X	Total	P(X|Event)
A	0.28	0.35 – 0.28 = 0.07	0.35	0.80
B	0.2925	0.45 – 0.2925 = 0.1575	0.45	0.65
C	0.06	0.20 – 0.06 = 0.14	0.20	0.30
Total	0.6325	0.3675	1.00	

So, $P(A|X) = \frac{P(A \cap X)}{P(X)} = \frac{0.28}{0.6325} = 0.4427$

$$P(B|X) = \frac{P(B \cap X)}{P(X)} = \frac{0.2925}{0.6325} = 0.4624$$

$$P(C|X) = \frac{P(C \cap X)}{P(X)} = \frac{0.06}{0.6325} = 0.0948$$

Example 7.11: Past record indicates that 2% of the management students passing from Synergy Institute and 5% of the students passing from Ducktech Institute are weak in statistics. 30 students from Synergy and 50 students from Ducktech appeared in an interview for the post of statistical investigator.

What is the probability that the candidate being selected is from Synergy or Ducktech Institute?

Solution: From the data set we have, 100 – 2 = 98% students from Synergy and 100 – 5 = 95% students from Ducktech Institute are strong in statistics.

Let A_1 = the event of selecting a student from Synergy

A_2 = the event of selecting a student from Ducktech

B = the event of selecting a student strong in statistics either from Synergy or Ducktech

Total students = 80

Hence, $P(A_1) = 30 \div 80 = 0.375$

$P(A_2) = 50 \div 80 = 0.625$

From the additional information

$P(B/A_1) = 98\% = 0.98$

$P(B/A_2) = 95\% = 0.95$

The following table is prepared, which is self-explanatory:

Events	Prior Probability $P(A_i)$	Conditional Probability of event B, given event A_i $P(B\|A_i)$	Joint Probability $P(A_i \& B)$ *(2) × (3)*	Posterior (revised) Probability $P(A_i\|B)$ = *(4)* ÷ P(B)
1	2	3	4	5
A_1	$P(A_1)$ = 0.375	$P(B\|A_1)$ = 0.98	$P(A_1) \times P(B\|A_1)$ = 0.375 × 0.98 = 0.3675	$P(A_1\|B)$ = 0.3675 ÷ 0.96125 = 0.3823
A_2	$P(A_2)$ = 0.625	$P(B\|A_2)$ = 0.95	$P(A_2) \times P(B\|A_2)$ = 0.625 × 0.95 = 0.59375	$P(A_2\|B)$ = 0.59375 ÷ 0.96125 = 0.6176
Total	1.00		**P(B)** = 0.3675 + 0.59375 = **0.96125**	1.000

Had we not obtained any additional information, we would have certainly inclined to draw the conclusion that the student will be from Synergy, because its students are stronger (98%) than the Ducktech (95%). But with the additional information, we may give a better answer. We may now say that the probability of a candidate being selected will be from the Ducktech Institute (0.618), though its students are only 95% strong in comparison to the Synergy, whose students are 98% strong.

TREE APPROACH OF BAYES THEOREM

Considering the above problem, the probability tree can be drawn in the following manner, which is self-explanatory. If B denotes strong student, then B^c denotes weak student.

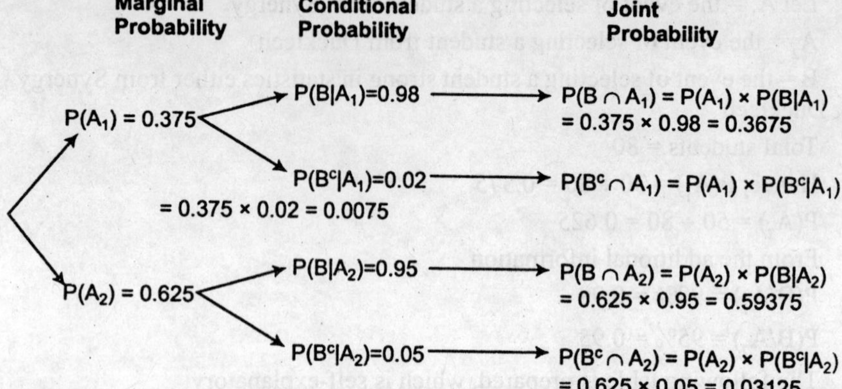

It can be seen that the probability of strong student being chosen is P(B) $= P(B \cap A_1) + P(B \cap A_2) = 0.3675 + 0.59375 = 0.96125$ (He can be chosen either from Ducktech or from Synergy)

Similarly, the probability of weak student being chosen is $P(B^c)$ $= P(B^c \cap A_1) + P(B^c \cap A_2) = 0.0075 + 0.03125 = 0.03875$ (He can be chosen either from Ducktech or from Synergy)

From the above probability tree diagram, any conditional probability can be found. For example:

(i) If a student is randomly chosen and found to be strong, the probability that he belongs to Ducktech is $P(A_1|B) = \dfrac{P(A_1 \cap B)}{P(B)} = \dfrac{0.3675}{0.96125}$ $= 0.3823$

(ii) If a student is randomly chosen and found to be strong, the probability that he belongs to Synergy is $P(A_2|B) = \dfrac{P(A_2 \cap B)}{P(B)} = \dfrac{0.59375}{0.96125}$ $= 0.6176$

(iii) If a student is randomly chosen and found to be weak, the probability that he belongs to Ducktech is $P(A_1|B^c) = \dfrac{P(A_1 \cap B^c)}{P(B^c)} = \dfrac{0.0075}{0.03875}$ $= 0.1935$

(iii) If a student is randomly chosen and found to be weak, the probability that he belongs to Synergy is $P(A_2|B^c) = \frac{P(A_2 \cap B^c)}{P(B^c)} = \frac{0.03125}{0.03875} =$ 0.8065

MORE ILLUSTRATIVE EXAMPLES

Example 7.12: Two unbiased dice are thrown. Find the probability that: (i) the first dice shows 6 (ii) Total of the numbers on the dice is 9 (iii) total of the numbers on the dice is greater than 9 (iv) both the dice show the same number (v) total of the numbers on the dice is any number from 2 to 12, both inclusive (vi) total of the numbers on the dice is 13.

Solution: A dice has six faces. In the random throw of the two dice, each of the six faces of one dice can be associated with each of the six faces of the other dice. So there will be 6 × 6 = 36 different combinations (exhaustive cases) in throwing of two dice, as shown below

(1,1), (1,2), (1,3), (1,4), (1,5), (1,6)
(2,1), (2,2), (2,3), (2,4), (2,5), (2,6)
(3,1), (3,2), (3,3), (3,4), (3,5), (3,6)
(4,1), (4,2), (4,3), (4,4), (4,5), (4,6)
(5,1), (5,2), (5,3), (5,4), (5,5), (5,6)
(6,1), (6,2), (6,3), (6,4), (6,5), (6,6)

(i) Favourable cases that the first dice will have 6 are (6,1), (6,2), (6,3), (6,4), (6,5), (6,6); i.e. total 6 cases.

$$\text{So } P = \frac{\text{Favourable cases}}{\text{Exhaustive cases}} = \frac{6}{36} = \frac{1}{6}$$

(ii) Cases favourable to get 9 will be (3,6), (4,5), (5,4), (6,3); i.e. total 4 cases.

$$\text{Hence } P = \frac{\text{Favourable cases}}{\text{Exhaustive cases}} = \frac{4}{36} = \frac{1}{9}$$

(iii) Cases favourable to get more than 9 will be (4,6), (5,5), (6,4); i.e. total 3 cases.

$$\text{Hence } P = \frac{\text{Favourable cases}}{\text{Exhaustive cases}} = \frac{3}{36} = \frac{1}{12}$$

(iv) Favourable cases that both dice show same number will be (1,1), (2,2), (3,3), (4,4), (5,5) and (6,6); total 6 cases.

$$\text{So P} = \frac{\text{Favourable cases}}{\text{Exhaustive cases}} = \frac{6}{36} = \frac{1}{6}$$

(v) Total sum of numbers shown on both dice ranges from 2 to 12. So the required Probability is 1 always. It is called as *certain event.*

(vi) This is an *impossible event.* As the maximum value of the sum of numbers on both dice will always be less than or equal to 12. So the required Probability is 0.

Example 7.13: From among the five digits, 1, 2, 3, 4, 5; two digits are chosen one by one. Find the probability that (i) first digit will be always odd (ii) second digit will always be odd (iii) both times the digits will be odd?

Solution: There will be in total $5 \times 4 = 20$ exhaustive cases.

(i) Cases favourable to first time odd digit are: (1,2), (1,3), (1,4), (1,5), (3,1), (3,2), (3,4), (3,5), (5,1), (5,2), (5,3), (5,4); 12 cases. So the

$$\text{required probability} = \frac{\text{Favourable cases}}{\text{Exhaustive cases}} = \frac{12}{20} = \frac{3}{5}$$

(ii) Cases favourable to second time odd digit are: (2,1), (3,1), (4,1), (5,1), (1,3), (2,3), (4,3), (5,3), (1,5), (2,5), (3,5), (4,5); 12 cases. So

$$\text{the required probability} = \frac{\text{Favourable cases}}{\text{Exhaustive cases}} = \frac{12}{20} = \frac{3}{5}$$

(iii) Favourable cases to both times odd digit are: (1,3), (1,5), (3,1), (3,5), (5,1), (5,3); total 6 cases. So the required probability

$$= \frac{\text{Favourable cases}}{\text{Exhaustive cases}} = \frac{6}{20} = \frac{3}{10}$$

Example 7.14: Four cards are drawn from a deck of cards at random. Find the probability that they are (i) a king, a queen, a jack and an ace (ii) Two jacks and two queens (iii) Two are black and two are red (iv) Two spade cards and two club cards.

Solution: Four cards can be drawn from a deck of cards in $^{52}C_4$ Exhaustive cases

(i) A king can be drawn in $^4C_1$ ways. Similarly, a queen, a jack and an ace also can be drawn each in ways. So total favourable cases will be $^4C_1 \times {}^4C_1 \times {}^4C_1 \times {}^4C_1 = 256$ ways. Hence the required probability is: $({}^4C_1 \times {}^4C_1 \times {}^4C_1 \times {}^4C_1) \div {}^{52}C_4$.

(ii) Cases favourable to draw two jacks and two queens are: $^4C_2 \times {}^4C_2$. So the required probability will be: $({}^4C_2 \times {}^4C_2) \div {}^{52}C_4$.

(iii) There are in total 26 black (club + spade) cards and 26 red (heart + diamond) cards. Cases favourable to draw two are black and two are red are: ${}^{26}C_2 \times {}^{26}C_2$. So the required probability will be: $({}^{26}C_2 \times {}^{26}C_2) \div {}^{52}C_4$

(iv) There are in total 13 spades and 13 club cards. Cases favourable to draw two spades and two clubs are: ${}^{13}C_2 \times {}^{13}C_2$. So the required probability will be: $({}^{13}C_2 \times {}^{13}C_2) \div {}^{52}C_4$

Example 7.15: From 20 tickets marked from 1 to 20 numerals, one is drawn at random. Find the chance that (i) it is a multiple of 5 (ii) it is a multiple of either 4 or 5

Solution:

(i) Numbers within 1 to 20 with multiples of 5 are: 5, 10, 15, 20; there 4 favourable cases. So the required probability will be $P = \frac{4}{20} = \frac{1}{5}$

(ii) Numbers within 1 to 20 with multiples of 4 are: 4, 8, 12, 16, 20; total 5 favourable cases

(ii) Multiples of 5 are 5, 10, 15, 20; total 4 favourable cases. But 20 is common to both. Hence total favourable cases with multiple of 4 or 5 will be 5 + 4 – 1 = 8. So the required probability will be $P = \frac{8}{20} = \frac{2}{5}$

Example 7.16 In shuffling a deck of cards, two cards were accidentally dropped. Find the probability that the missing cards are one from each unit.

Solution: There will be ${}^{52}C_4$ of exhaustive cases that four cards will be dropped. Favourable cases of one coming from each suit will be $= {}^{13}C_1 \times {}^{13}C_1 \times {}^{13}C_1 \times {}^{13}C_1$. So the required probability is: $({}^{13}C_1 \times {}^{13}C_1 \times {}^{13}C_1 \times {}^{13}C_1) \div {}^{52}C_4$

Example 7.17: A box contains 8 white, 5 red and 7 black balls. If 3 balls are drawn at random, find the probability that (i) one is of each color (ii) two of the balls are white (iii) none is white (iv) at least one is white (v) at most two are white.

Solution: Exhaustive cases will be: 3 balls to be drawn out of total 8 + 5 + 7 = 20 balls in ${}^{20}C_3$ ways.

(i) Number of favourable cases of getting one ball of each color $= {}^{8}C_1$ ${}^{5}C_1 \times {}^{7}C_1$. So the required probability $= ({}^{8}C_1 \times {}^{5}C_1 \times {}^{7}C_1)\ ({}^{20}C_3)$

(ii) If two are white, these two balls will have to be drawn out of 8 white balls in ${}^{8}C_2$ ways. Another one ball will have to be drawn from the remaining 12 balls in ${}^{12}C_1$ ways. So the favourable cases will be: ${}^{8}C_2 \times {}^{12}C_1$. The required $({}^{8}C_2 \times {}^{12}C_1)\ ({}^{20}C_3)$

(iii) None will be white. So all the balls have to be drawn out of 12 red and black balls in $^{12}C_3$ ways. So the required probability will be $(^{12}C_3)$ $(^{20}C_3)$

(iv) At least one will be white. That means, one is white or two are white or three are white. So three cases can arise here. First- One white + two from red and black. Second- Two white + one from red and black. Third- Three are white and none from red and black.

Favourable cases for One white + two from red and black = $^8C_1 \times {}^{12}C_2$

Favourable cases for Two white + one from red and black = $^8C_2 \times {}^{12}C_1$

Favourable cases for three white + none from red and black = $^8C_3 \times {}^{12}C_0$

So the total favourable cases = $^8C_1 \times {}^{12}C_2 + {}^8C_2 \times {}^{12}C_1 + {}^8C_3 \times {}^{12}C_0$

Required probability = $(^8C_1 \times {}^{12}C_2 + {}^8C_2 \times {}^{12}C_1 + {}^8C_3 \times {}^{12}C_0) \div (^{20}C_3)$

This can be solved in the other way: P(at least one white) = 1 – P(none is white) = $1 - {}^{12}C_3$

Either way, the answer will be same.

(v) At most two will be white. That means, there will be either no white or one white or two whites. So the favourable cases of no white = $^{12}C_3$

Favourable cases of one white = $^8C_1 \times {}^{12}C_2$

Favourable cases of two whites = $^8C_2 \times {}^{12}C_1$

Total favourable cases = $^{12}C_3 + {}^8C_1 \times {}^{12}C_2 + {}^8C_2 \times {}^{12}C_1$

Required probability = $(^{12}C_3 + {}^8C_1 \times {}^{12}C_2 + {}^8C_2 \times {}^{12}C_1) \div {}^{20}C_3$

Example 7.18: A man got four diamond cards from a deck of cards. If another three cards are given to him, what is the probability that at least one additional card will also be diamond?

Solution: After the man is given four diamond cards, there remains a total of 52 – 4 = 48 cards.

Out of this 48 remaining cards 9 (=13 – 4) are diamonds and the rest 39 (=48 – 9) are heart, club and spade.

Three cards to be given out of this 48 remaining cards. So the exhaustive cases = $^{48}C_3$.

At least one will be diamond means, either 1 diamond + 2 other cards; 2 diamonds + 1 other cards; or 3 diamond cards. So the favourable cases will be respectively, $^9C_1 \times {}^{39}C_2$, $^9C_2 \times {}^{39}C_1$ and $^9C_3$. Hence the required probability will be $(^9C_1 \times {}^{39}C_2 + {}^9C_2 \times {}^{39}C_1 + {}^9C_3) \div (^{48}C_3) = 0.4718$.

This can be interpreted as, P(at least one is diamond) = 1 – P (none is diamond). That means all the three cards come from 39 non-diamond cards. Favourable cases of non-diamond cards will be ${}^{39}C_3$. So the required probability = 1 – (${}^{39}C_3$) (${}^{48}C_3$) = 0.4718

Example 7.19: What is the probability of getting 8 cards of same suit in one hand at a game of bridge?

Solution: One hand in a game of bridge receives 13 cards. So the exhaustive cases will be ${}^{52}C_{13}$. But out of this 13 cards, 8 should be from the same suit, i.e. either from spade, or club, or diamond or heart.

If 8 are from spade and 5 are from remaining, then favourable cases will be ${}^{13}C_8 \times {}^{39}C_5$.

If 8 are from club and 5 are from remaining, then favourable cases will be ${}^{13}C_8 \times {}^{39}C_5$.

If 8 are from diamond and 5 are from remaining, then favourable cases will be ${}^{13}C_8 \times {}^{39}C_5$.

If 8 are from heart and 5 are from remaining, then favourable cases will be ${}^{13}C_8 \times {}^{39}C_5$.

So total favourable cases will be ${}^{13}C_8 \times {}^{39}C_5 \times 4$.

Required probability = (${}^{13}C_8 \times {}^{39}C_5 \times 4$) ÷ (${}^{52}C_{13}$).

Example 7.20: A committee of four people is to be formed from 4 Directors, 6 Deputy Directors, 9 Assistant Directors and 1 Financial Advisor. Find the probability of forming the committee in the following manner.

(i) There must be one from each category

(ii) It should have at least one Director.

(iii) Financial Advisor must be in the committee.

Solution: There are in total 4 + 6 + 9 + 1 = 20 persons. Four persons will come from 20. So exhaustive cases is ${}^{20}C_4$ ways.

(i) Favourable cases of one coming from each category will be ${}^{4}C_1 \times {}^{6}C_1 \times {}^{9}C_1 \times {}^{1}C_1$. So the required probability is: (${}^{4}C_1 \times {}^{6}C_1 \times {}^{9}C_1 \times {}^{1}C_1$) ÷ (${}^{20}C_4$)

(ii) P(committee has at least one Director) = 1 – (committee has no Director).

Favourable cases of no Directors = ${}^{16}C_4$. So the required probability = 1 – (${}^{16}C_4$) ÷ (${}^{20}C_4$)

(iii) If Financial Advisor must be in the committee, the rest three persons will have to be chosen from among balance 19 persons in ${}^{19}C_3$ ways. So the required probability is ${}^{19}C_3$ ÷ (${}^{20}C_4$)

Example 7.21: From a pack of cards, two cards are drawn. What is the probability that,

(i) One is a king and the other is a queen, if cards are replaced after each drawal?

(ii) One is a king and the other is a queen, if cards are not replaced after each drawal?

(iii) One is a king and the other is a queen, in the same order, if cards are replaced after each drawal?

(iv) One is a king and the other is a queen, in the same order, if cards are not replaced after each drawal?

(v) Both are kings with replacement?

(vi) Both are kings without replacement?

(vii) Either a king or a heart?

Solution:

(i) If the first drawal is king and second is queen then $P = \frac{4}{52} \times \frac{4}{52}$

If the first drawal is queen and second is king then $P = \frac{4}{52} \times \frac{4}{52}$

So, the total probability $= \frac{4}{52} \times \frac{4}{52} + \frac{4}{52} \times \frac{4}{52}$

(ii) If the first drawal is king and second is queen then $P = \frac{4}{52} \times \frac{4}{51}$

If the first drawal is queen and second is king then $P = \frac{4}{52} \times \frac{4}{51}$

So, the total probability $= \frac{4}{52} \times \frac{4}{51} + \frac{4}{52} \times \frac{4}{51}$

This probability can also be found using Permutations and Combinations, i.e. $P = \frac{{}^{4}_{1}C \times {}^{4}_{1}C}{{}^{52}_{2}C}$. This is otherwise known as ***Hyper Geometric Distribution.***

(iii) Drawal of king and queen in the order, with replacement, then P $= \frac{4}{52} \times \frac{4}{52}$

(iv) Drawal of king and queen in the order, without replacement, then

$$P = \frac{4}{52} \times \frac{4}{51}$$

(v) Drawal of two kings with replacement, then $P = \frac{4}{52} \times \frac{4}{52}$

(vi) Drawal of two kings without replacement, then $P = \frac{4}{52} \times \frac{3}{31}$

(vii) Let probability of king is P(K) and probability of heart is P(H). So

$$P(KUH) = P(K) + P(H) - P(K \cap H) = \frac{4}{52} + \frac{13}{52} - \frac{1}{52}$$

Example 7.22: A problem in statistics is given to 3 students, A, B and C and their probability of solving are $\frac{1}{2}, \frac{1}{3}$ and $\frac{1}{4}$ respectively. If all of them try independently, what is the probability that

(i) The problem will be solved?
(ii) Exactly two of them will be solved?
(iii) At least two of them will be solved?
(iv) Either A or B can be solved?

Solution: Given That P(A) = ½; P(B) = 1/3 and P(C) = 1/4

(i) The problem can be solved, if any one solves it, or in other words it is solved by at least one of them, or in other words either one or two or all three solve it. That means either A or B or C. So we have to find $P(AUBUC) = P(A) + P(B) + P(C) - P(A \cap B) - P(B \cap C) - P(C \cap A) + P(A \cap B \cap C)$. As all the three are solving the problem independently, it is a case of statistical independence. So, $P(AUBUC) = P(A) + P(B) + P(C) - P(A) \times P(B) - P(B) \times P(C) - P(C) \times P(A) + P(A) \times P(B) \times P(C)$

$$= \frac{1}{2} + \frac{1}{3} + \frac{1}{4} - \frac{1}{2} \times \frac{1}{3} - \frac{1}{3} \times \frac{1}{4} - \frac{1}{4} \times \frac{1}{2} + \frac{1}{2} \times \frac{1}{3} \times \frac{1}{4} = \frac{3}{4}$$

Corollary 1: It can also be answered in another way. Probability that the problem is solved is equal to 1- Probability that nobody solves it.

Probability that A can't = $P(A^c) = 1 - P(A) = 1 - \frac{1}{2} = \frac{1}{2}$

Probability that B can't = P(B^c)= 1 – P(B) = $1-\frac{1}{3}=\frac{2}{3}$

Probability that C can't = P(C^c)= 1 – P(C) = $1-\frac{1}{4}=\frac{3}{4}$

Probability that nobody solves it is equal to A can't solve and B can't and C can't solve, i.e. $P(A^c\cap B^c\cap C^c)=P(A^c)\times P(B^c)\times P(C^c)=\frac{1}{2}\times\frac{2}{3}\times\frac{3}{4}=\frac{1}{4}$

So the required probability = $1-\frac{1}{4}=\frac{3}{4}$.

Corollary 2: It can also be answered by adding the probabilities of solving the problem by any one, any two and all three of them.

Solving by Any One: (Any one solves and other two can't solve)

Probability of solving by A:

$$P(A)\times P(B^c)\times P(C^c)=\left(\frac{1}{2}\right)\times\left(1-\frac{1}{3}\right)\times\left(1-\frac{1}{4}\right)=\frac{1}{4}$$

Probability of solving by B:

$$P(A^c)\times P(B)\times P(C^c)=\left(1-\frac{1}{2}\right)\times\left(\frac{1}{3}\right)\times\left(1-\frac{1}{4}\right)=\frac{1}{8}$$

Probability of solving by C:

$$P(A^c)\times P(B^c)\times P(C)=\left(1-\frac{1}{2}\right)\times\left(1-\frac{1}{3}\right)\times\left(\frac{1}{4}\right)=\frac{1}{12}$$

Solving by Any Two: - (Any two solve and other one doesn't solve)

Probability of solving by A&B:

$$P(A)\times P(B)\times P(C^c)=\left(\frac{1}{2}\right)\times\left(\frac{1}{3}\right)\times\left(1-\frac{1}{4}\right)=\frac{1}{8}$$

Probability of solving by B & C:

$$P(A^c)\times P(B)\times P(C)=\left(1-\frac{1}{2}\right)\times\left(\frac{1}{3}\right)\times\left(\frac{1}{4}\right)=\frac{1}{24}$$

Probability of solving by C & A:

$$P(A)\times P(B^c)\times P(C)=\left(\frac{1}{2}\right)\times\left(1-\frac{1}{3}\right)\times\left(\frac{1}{4}\right)=\frac{1}{12}$$

Solving by all three: Probability of solving by A&B&C:

$$P(A)\times P(B)\times P(C^c)=\left(\frac{1}{2}\right)\times\left(\frac{1}{3}\right)\times\left(\frac{1}{4}\right)=\frac{1}{24}$$

Adding all values, the required probability

$$=\frac{1}{4}+\frac{1}{8}+\frac{1}{12}+\frac{1}{8}+\frac{1}{24}+\frac{1}{12}+\frac{1}{24}=\frac{3}{4}$$

(ii) Exactly two will solve–means A and B will solve or B and C will solve or C and A will solve.

= A&B solve and C can't solve + B&C solve and A can't solve + C&A solve and B can't solve

$$=P(A\cap B\cap C^c)+P(B\cap C\cap A^c)+P(C\cap A\cap B^c)$$

$$=P(A)\times P(B)\times P(C^c)+P(B)\times P(C)\times P(A^c)+P(C)\times P(A)\times P(B^c)$$

$$=\frac{1}{2}\times\frac{1}{3}\times\frac{3}{4}+\frac{1}{3}\times\frac{1}{4}\times\frac{1}{2}+\frac{1}{4}\times\frac{1}{2}\times\frac{2}{3}$$

(iii) At least two of them will solve means- A&B will solve and C can't solve + B&C can solve and A can't solve + C&A will solve and B can't solve + ABC solve

$$=P(A\cap B\cap C^c)+P(B\cap C\cap A^c)+P(C\cap A\cap B^c)+P(ABC)$$

$= P(A)\times P(B)\times P(C^c) + P(B)\times P(C)\times P(A^c) + P(C)\times P(A)\times P(A)\times P(B^c) + P(A)\times P(B)\times P(C)$

$$=\frac{1}{2}\times\frac{1}{3}\times\frac{3}{4}+\frac{1}{3}\times\frac{1}{4}\times\frac{1}{2}+\frac{1}{4}\times\frac{1}{2}\times\frac{2}{3}+\frac{1}{2}\times\frac{1}{3}\times\frac{1}{4}=\frac{1}{2}$$

(iii) A or B can solve means the required probability is (PUB) = P(A) $+P(B)-A\cap B=P(A)+P(B)-P(A)\times P(B)=\frac{1}{2}+\frac{1}{3}-\frac{1}{2}\times\frac{1}{3}=\frac{2}{3}$

Example 7.23: Five numbers are randomly selected from the set of numbers from 1 to 20. Find the probability that the third smallest number is 8.

Solution: If the third smallest number is 8, then two numbers must be chosen from 1 to 7 and another two numbers must be chosen from 9 to 20. So two numbers to be chosen from 1-7 in ${}^{7}C_2$ ways and another two numbers to be chosen from 9-20 in ${}^{12}C_2$ ways. Exhaustive case is obviously ${}^{19}C_4$ ways.

Hence the required probability is $\frac{C_2^7\times C_2^{12}}{C_2^{19}}$.

Example 7.24: Two dice are tossed. Find the probability of getting a total score not exceeding 4.

Solution: Favorable cases of score not exceeding 4 are (1,3), (2,2) and (3,1). So the required probability is $\frac{3}{36}$.

Example 7.25: Three numbers are randomly selected from a set of numbers from 1 to 15. Find the probability that the sum of the numbers will be odd.

Solution: Sum of numbers can be odd, if either (i) all numbers are odd or (ii) two are even and one is odd. Three odd numbers can be chosen from 1-20 in ${}^{10}C_3$ ways. Two even numbers and one odd number from 1-20 can be chosen in ${}^{10}C_2 \times {}^{10}C$ ways. Exhaustive cases of choosing three numbers from 1-20 is ${}^{20}C_3$. Hence, the required probability is $\frac{C_3^{10} + C_2^{10} \times C_1^{10}}{C_3^{20}}$.

Example 7.26 Four balls are randomly chosen from a box containing 8 white, 6 black and 7 red balls. What is the probability that balls of all colours are included in the choice?

Solution: Balls of all colours can be included by either (i) 2 white, 1 black and 1 red or (ii) 1 white, 2 black and 1 red or (iii) 1 white, 1 black or 2 red. Hence the required probability is $\frac{C_2^8 \times C_1^6 \times C_1^7 + C_1^8 \times C_2^6 \times C_1^7 + C_1^8 \times C_1^6 \times C_2^7}{C_4^{21}}$

Example 7.27: Three balls are randomly selected from a bag containing 4 Red, 5 Black and 3 White balls. Find the probability that at least two of them will be of same colour.

Solution: At least two of them will be of same colour by choosing either (i) 2 red & 1 other color or (ii) 2 black & 1 other color or (iii) 2 white & 1 other color. So the required probability is $\frac{C_2^4 \times C_1^8 + C_2^5 \times C_1^7 + C_2^3 \times C_1^9}{C_3^{12}}$.

Example 7.28: The probabilities of getting medicine in three stores A, B and C are 0.8, 0.5 and 0.4 respectively. If a boy visits the stores in this order, find the probability that he will get the medicine.

Solution: He will get the medicine, if he gets in either (i) A or (ii) not in A and in B or (iii) Not in A and not in B and in C. That means $P(A) + P(A^c) \times P(B) + P(A^c) \times P(B^c) \times P(C) = 0.8 + (1\text{-}0.8) \times 0.5 + (1\text{-}0.8)\ (1\text{-}0.5) \times 0.4$

Example 7.29: A, B, C and D are mutually independent events with probabilities P_1, P_2, P_3 and P_4 respectively. Find the probability that (i) exactly two of them happen (ii) Three or more of them happen.

Solution:

(i) Exactly two of them can happen in the following ways:

A&B happen and C&D not happen = $P_1 \times P_2 + (1\text{-}P_3) \times (1\text{-}P_4)$

A&C happen and B&D not happen = $P_1 \times P_3 + (1\text{-}P_2) \times (1\text{-}P_4)$

A&D happen and B&C not happen = $P_1 \times P_4 + (1\text{-}P_2) \times (1\text{-}P_3)$

B&C happen and A&D not happen = $P_2 \times P_2 + (1\text{-}P_1) \times (1\text{-}P_4)$

B&D happen and A&C not happen = $P_2 \times P_4 + (1\text{-}P_1) \times (1\text{-}P_3)$

C&D happen and A&B not happen = $P_3 \times P_2 + (1\text{-}P_1) \times (1\text{-}P_2)$

So the required probability is $[P_1 \times P_2 + (1\text{-}P_3) \times (1\text{-}P_4)] + [P_1 \times P_3 + (1\text{-}P_2) \times (1\text{-}P_4)] + P_1 \times P_4 + (1\text{-}P_2) \times (1\text{-}P_3)] + [P_2 \times P_2 + (1\text{-}P_1) \times (1\text{-}P_4)] + [P_2 \times P_4 + (1\text{-}P_1) \times (1\text{-}P_3)] + [P_3 \times P_2 + (1\text{-}P_1) \times (1\text{-}P_2)]$

(ii) Three or more will happen means (i) any three will happen & fourth one will not happen and (ii) all four will happen. So the required probability will be $P_1P_2P_3(1\text{-}P_4) + P_1P_3P_4(1\text{-}P_2) + P_2P_3P_4(1\text{-}P_1) + P_1P_2P_4(1\text{-}P_3) + P_1P_2P_3P_4$

Example 7.30: In a hardware store, 2 inches size iron nails are received from 3 manufacturers say A, B and C in proportions 60:25:15. 5%, 6% and 8% of the nails from A, B and C respectively are defective. If a nail is randomly selected from the entire lot supplied by A, B and C and found to be of defective, find the probability that it was supplied by A.

Solution: From the given data, it can be seen that $P(A) = \frac{60}{100} = 0.6$; $P(B) = \frac{25}{100} = 0.25$ and $P(C) = \frac{15}{100} = 0.15$

Let D denotes defective and ND denotes non-defective items. So the various conditional probabilities are worked out as under:

Probabily that item is defective, subject to condition that it was supplied by A is $P(D|A) = 0.05$ and so probability of getting non-defective, subject to condition that it was supplied by A is obviously $P(ND|A) = 1 - 0.05 = 0.95$. With the same notations, we can have $P(D|B) = 0.06$, $P(ND|B) = 0.94$ and $P(D|C) = 0.08$, $P(ND|C) = 0.92$. The probability tree is shown here in.

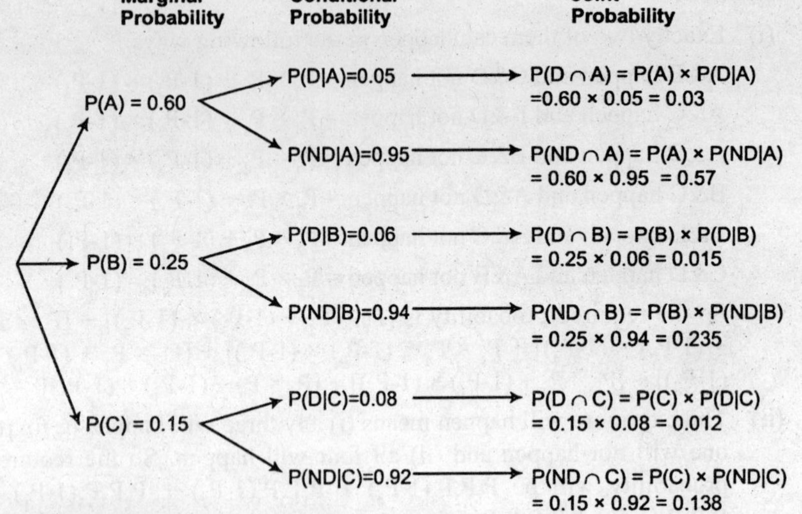

$$P(D) = P(D \cap A) + P(D \cap B) + P(D \cap C) = 0.03 + 0.015 + 0.012 = 0.057$$

So, the probability that the item was supplied by A subjected to condition that it is found to be defective is $P(A|D) = \dfrac{P(D \cap A)}{P(D)} = \dfrac{0.03}{0.057} = 0.526$

Example 7.31: Urn-1 has twice as many red balls as white balls and Urn-2 has 2 white and 1 red balls. One ball is randomly transferred from Urn-1 to Urn-2 and then a ball is randomly chosen from Urn-2. If it is found to be white ball, find the probability that the transferred ball was also white.

Solution: As per the given data, probability of white (w) in Urn-1 is $\frac{1}{3}$ = 0.33 and that of red (r) is $\frac{2}{3}$ = 0.67. So, when a ball is transferred from Urn-1, White will go with a probability of 0.33 and red with 0.67.

If White ball goes to Urn-2, there will be 3 white and 1 red ball in Urn-2 and if a red ball is transferred to Urn-2, there will be 2 white and 2 red balls in Urn-2.

Under the condition that a white (w) is transferred to Urn-2, probability of White (W|w) in Urn-2 will be $\frac{3}{4}$ = 0.75 and probability of red (R|w) will obviously be $\frac{1}{4}$ = 0.25.

Under the condition that a red (r) is transferred to Urn-2, probability of white (W|r) in Urn-2 will be $\frac{2}{4}$ = 0.50 and probability of red (R|r) will be $\frac{2}{4}$ = 0.5.

The probability tree is shown below, where the notations are self-explanatory. Small letters belong to Urn-1 and capital letters belong to Urn-2.

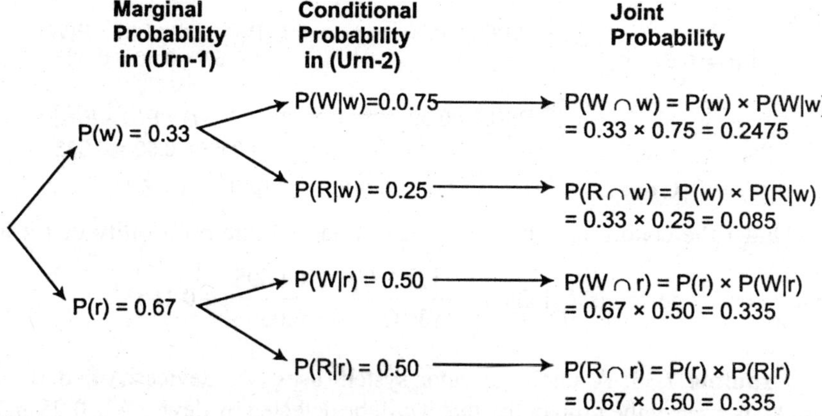

$$P(W) = P(W \cap w) + P(W \cap r) = 0.2475 + 0.335 = 0.5825$$

Under the condition that the ball chosen from Urn-2 is found to be white (W), the probability that a white (w) ball was transferred from Urn-1 is

$$P(w|W) = \frac{(w \cap W)}{P(W)} = \frac{0.2475}{0.5825} = 0.425$$

Example 7.32: A dam may collapse due to faulty design. There is 99% chance that the design is done correctly. The probability of dam collapse due to faulty design is 0.98, or else it is 0.05. If the dam has collapsed, what is the probability that it is due to faulty design?

Solution: Probability tree is shown below. Notations are self-explanatory.

FD- Faulty Design; CD- Correct Design; C- Collapse; NC- Not Collapse

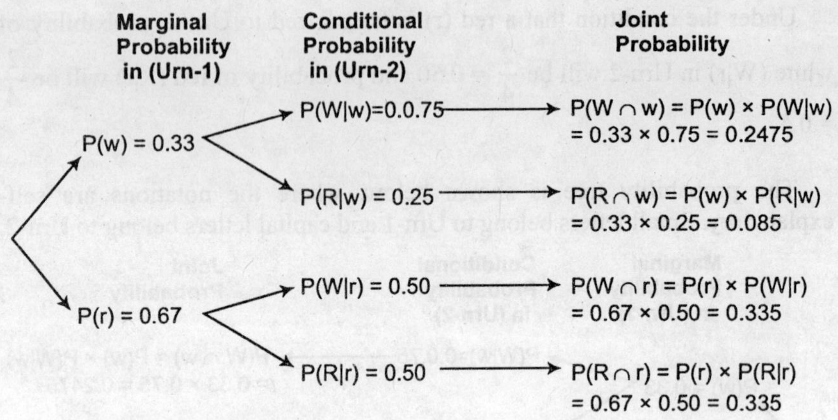

$P(C) = P(C \cap FD) + P(C \cap CD) = 0.0098 + 0.0495 = 0.0593$

Under the condition that the dam has collapsed, the probability that it is due to faulty design is $P(FD|C) = \dfrac{(FD \cap C)}{P(C)} = \dfrac{0.0098}{0.0593} = 0.165$

Example 7.33: A smoke detector system uses two devices A and B. If smoke is present the probability that it will be detected by device A is 0.95 and by device B is 0.98 and for both the devices it is 0.94, then find the probability that it will not be detected.

Solution: $P(A) = 0.95$; $P(B) = 0.98$; $P(A \cap B) = 0.94$

Now, $P(AUB) = P(A) + P(B) - P(A \cap B) = 0.95 + 0.98 - 0.94 = 0.99$

The probability that it will not be detected by both A and B is $P(A^c \cap B^c)$ $= 1 - P(AUB) = 1 - 0.99 = 0.01$

Example 7.34: Two people enter in a room and their birthdays will be recorded, then find the probability that the two people have (i) a specific pairs of birthdays (ii) birthdays on same day of the year.

Solution: A year has 365 days. So there can be 365 × 365 exhaustive number of possibilities. Out of these 365 × 365 pair of days, there are 365 days when the two birthdays may coincide. The remaining number of pairs is 365 × 365 – 365, which are the specific days when birthdays will not coincide.

(i) So the required probability that they have a specific pair of birthdays is obviously $\dfrac{365 \times 365 - 365}{365 \times 365} = 1 - \dfrac{1}{365} = \dfrac{364}{365}$.

(ii) The probability that they have same birthday in a year is $\dfrac{1}{365}$.

EXERCISES

1. Determine the probabilities of the following events in drawing a card from a pack of 52 cards.

   (a) A six (b) A red card (c) An ace or a queen (d) A black three or a black five (e) A red face card (king, queen or jack)

2. During a bridge game, once the lead card had been played and the dummy's hand revealed, the declarer took a moment to count up the number of cards in each suit with the results given below:

Suit	**A**	**B**
Spades	7	6
Hearts	5	8
Diamonds	9	4
Clubs	5	8
	26	26

   (a) What is the probability that a card randomly selected from the A's team is a club?

   (b) What is the probability that a card randomly selected from all the cards is either a heart or a diamond?

   (c) What is the probability that a card randomly selected from B's team is a heart?

   (d) If this type of analyses were repeated for every hand many times, what would be the long-run probability that a card drawn from the B's team is a club?

3. The inspector of Bharat Petroleum undertook a study of two filling stations regarding their susceptibility of being shut down either due to pump failure or leakage or both. The following probabilities are the result:

Station	**P(Pump Failure)**	**P(Leakage)**	**P(Both)**
1	0.06	0.11	0
2	0.08	0.12	0.07

   Which station has the higher probability of closure?

4. A bag contains 78 marbles: 36 are red and 26 of these red marbles are swirled. The rest of them are blue and 32 of the blue ones are swirled. The marbles that are not swirled are clear. What is the probability of the following:

(a) A red marble from the bag
(b) A clear marble from the bag
(c) A red, swirled marble
(d) A blue, clear marble
(e) A swirled marble

5. An employee at Infratech must enter product information into the computer. The employee may use a light pen that transmits the information to the PC along with the keyboard to issue commands, or fill out an OMR sheet and feed it directly into the old mainframe. Historically, the following are the probabilities:

   P (Light pen will fail) = 0.022

   P (PC keyboard will fail) = 0.15

   P (Light pen or PC keyboard will fail) = 0.006

   P (Mainframe will fail) = 0.25

   Data can be entered into the PC only if the light pen and keyboard function properly. What is the probability that (a) the employee can use the PC to enter the data? (b) either PC fails or the mainframe fails? Assume that they both cannot fail at the same time.

6. A bag contains 36 balls: 5 are red, 10 are black, 15 are blue and 6 are yellow. The balls are taken out one at a time with replacement. What is the probability that (a) The second ball is red given that the first one was blue (b) The second ball is red given that the first one was black (c) The third ball is blue given that both the first and second were black?

7. Four floodgates of a dam failed and were repaired independently of each other. From experience, it is known that each floodgate is out of order 5% of the time. (a) If floodgate 1 is out of order, what is the probability that floodgates 2 and 3 are out of order (b) During a tour of the dam site, it became known that the chances of all four floodgates being out of order are less than 1 in 6,000,000. Is this statement true?

8. A grocery store was examining its stock policies and found the number of 1kg Pepsi bottle sold each day for the past month. The data are given below:

Number sold	Morning	Afternoon	Evening
0-19	5	10	3
20-39	4	6	5
40-59	16	8	6
60-79	5	12	10
80-99	6	4	7
100 or more	4	0	9
	40	40	40

(a) What is the probability that on a randomly selected day the number of Pepsi bottles sold in the morning is between 60 and 79?

(b) What is the probability that 59 or fewer bottles were sold during a randomly selected afternoon?

(c) What is the probability that either 0-19 or 100 or more bottles were sold in a randomly selected evening?

9. In a survey conducted regarding auto accidents on highways, it was found that 65% of all accidents occurred at night, 50% were alcohol-related and 35% occurred at night and alcohol-related. What is the probability that (a) an accident was alcohol-related, given that it occurred at night? (b) an accident occurred at night, given that it was alcohol-related.

10. A University library was surveying as to the number of use of library books, reference and periodicals by its students (graduate, undergraduate) and faculty members and the following is the findings:

	References	Periodicals	Books
Undergraduate	42	25	72
Graduate	26	62	20
Faculty	16	69	18
Total:	84	156	110

Find the probability that a randomly chosen person (a) is a graduate (b) visited the periodical section, given that he is a graduate student (c) is an undergraduate who visited the book section.

11. A consignment of 11,000 boxes from Cairo and Damascus containing dates had the following information:

	Number of Boxes	Number of Boxes with Damaged Fruit	Number of Boxes with Overripe Fruit
Cairo	6000	150	850
Damascus	5000	360	290

(a) What is the probability that a box selected at random will contain damaged fruit?

(b) What is the probability that a randomly selected box is from Cairo or Damascus?

(c) Given that a randomly selected box contains overripe fruit, what is the probability that it came from Damascus?

(d) If damaged fruit and overripe fruit are mutually exclusive, what is the probability that a box contains damaged or overripe fruit? What if they are not mutually exclusive?

12. A physical therapist, interested in studying the relationship between foot injuries and the position of a player in the field, collected the following data of past few years.

	Offensive Line	Defensive Line	Offensive Backfield	Defensive Backfield
Number of players:	48	55	25	20
Number injured:	30	38	10	8

Given that a randomly selected player incurred a foot injury, what is the probability that he plays in the (a) offensive line (b) defensive line (c) offensive backfield (d) defensive backfield?

13. Given: The probabilities of three events, A, B and C, occurring are P(A)=0.32, P(B) = 0.40 and P(C) = 0.28. Assuming that A, B or C has occurred, the probabilities of another event, X, occurring are P(X|A) = 0.75, P(X|B) = 0.65 and P(X|C) = 0.35. Find P(A|X), P(B|X) and P(C|X).

14. Given: that P(A) = 3/14, P(B) = 1/6, P(C) = 1/3, P(AC) = 1/7 and P(B|C) = 5/21, find the following probabilities: P(A|C), P(C|A), P(BC) and P(C|B).

15. Two events X and Y are statistically dependent. If P(X) = 0.38, P(Y) = 0.20 and P(X or Y) = 0.45, find the probability that (a) Neither X nor Y will occur (b) Both X and Y will occur (c) Y will occur, given that X has already occurred (d) X will occur, given that Y has already occurred.

PROBABILITY DISTRIBUTION

INTRODUCTION

It has already been discussed about various ways of representing the data such as histograms, frequency distribution tables, etc. Number of variables based on actual observations called the data are arranged and classified and finally grouped into different groups or classes in order to form the frequency distribution table. In actual frequency distribution the number of observations of a variable is generally limited which usually forms a sample and all the items included in it are actually observed, classified and grouped into classes.

On the contrary, in case of a frequency distribution on theoretical basis, the actual observations are not taken. Here the frequency table is prepared purely on theoretical basis. The frequencies are obtained by mathematical computation. It is considered to be ideal and reasonable distribution, as it reflects a very close approximation to the actual distribution of the variable in the population.

EXPECTED VALUE OF A VARIABLE

In fact, we can think of a probability distribution as a theoretical frequency distribution. This describes how outcomes are expected to vary. Because these distributions deal with expectations, they are useful models in making inferences and decisions under situations of uncertainty. Let us take an example of tossing of two coins, where it is required to know a probability distribution of the number of heads that could possibly result. Here there can be four possibilities in total e.g. (H,H), (H,T), (T,H) and (T,T). The results are represented in tables and a graph below.

1st Toss	2nd Toss	Number of Heads in two tosses	Probability of outcome P(Head)
H	H	2	0.5 0.5 = 0.25
H	T	1	0.5 0.5 = 0.25
T	H	1	0.5 0.5 = 0.25
T	T	0	0.5 0.5 = 0.25
			1.00

Number of Heads	Tosses	Probability of outcome P(H)
2	(H,H)	0.25
1	(H,T) + (T,H)	0.50
0	(T,T)	0.25

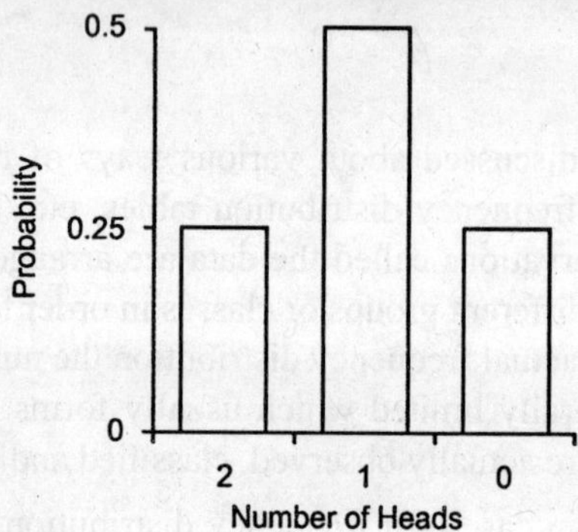

It is to be noted that it is not the actual outcome, rather it is a theoretical distribution. It represents the way in which we would expect our two-toss experiment to behave. So a frequency distribution is a listing of the observed frequencies of all the outcomes of an experiment that actually occurred when the experiment was done, whereas a probability distribution is a listing of the probabilities of all the possible outcomes that could result if the experiment were done.

Let us take another example. Suppose the coin is tossed 10 times and 7 heads occurred like:

Heads	Tails	Total
7	3	10

Now suppose the coin is tossed 20 times, where 15 heads occurred. Then again it is tossed 30 times and say 23 heads occurred. So, in practice the theoretical outcome may not always occur as it is expected that half the number of tosses will be heads as well as tails. So the outcome may not be up to expectation. So the question arises what is the expected value of a variable? To obtain the expected value of a random variable, we multiply each value that the random variable can assume with the probability of occurrence of that value and then add these products. Mathematically,

E(X) = x × P(X)

Where, E(X) = Expected value of the random variable X

x = Value of the variable

P(X) = Probability of occurrence of x

Considering an example of sale record of a particular item in a departmental store in the past 100 days.

No of items sold	102	105	108	111	114	117
No of days sold	8	14	30	15	19	14

The owner now wants to know as to the number of items those are expected to be sold in the next one week so that he will verify the stock and will be in readiness. Look at the following calculation table:

No of Items sold (x)	**No of days sold (frequency)**	**Probability (P(X))**		**Expected Value (1) (3) x P(X)**	
(1)	**(2)**	**(3)**		**(4)**	
102	8	8 100 =	0.08	102 0.08 =	8.16
105	14	14 100 =	0.14	105 0.14 =	14.7
108	30	30 100 =	0.30	108 0.30 =	32.4
111	15	15 100 =	0.15	111 0.15 =	16.65
114	19	19 100 =	0.19	114 0.19 =	21.66
117	14	14 100 =	0.14	117 0.14 =	16.38
Total	100		1.00	$\sum x \times P(X) =$	109.95

So the expected value of the variable is 109.95 = say 110. The owner will now think that in the coming days his daily sale will be on an average 110 number of items only and he can then go for keeping the required number of items in store for next week accordingly. But it is not necessary that immediately next day the sale will be exactly 110 number of items or in coming days it will be 110 every day.

In real world situations, decision makers combine probabilities that a random variable will take on certain values with monetary gain or loss that results when it does take on those values. This helps them making smart decisions under conditions of uncertainty. Let us take a very simple example to fully understand this:

Example 8.1: A newspaper boy has observed the daily sales of a newly published newspaper for the last 100 days and following is the result.

No of copies sold	10	11	12	13	14
No of days sold	10	15	20	25	30

Cost of a copy is 30 paisa and the sale price is 50 paisa. He cannot return unsold copies. How many copies should he order ?

Solution: In business two types of losses are incurred e.g. (1) *Obsolescence losses*, caused by stocking too much commodities on any one day and having to throw it away the next day (2) *Opportunity losses*, caused by being out of stock any time that costumers call for them. So the newspaper boy will always want either to minimize the losses or maximize the profit. The problem can be solved either way. Let us first solve it by constructing conditional loss table.

The probability of 10, 11, 12, 13 & 14 number of copies being sold are calculated as follows:

Daily Sales	Number of Days Sold	Probability of Each Number Being Sold	
10	10	10 100 =	0.10
11	15	15 100 =	0.15
12	20	20 100 =	0.20
13	25	25 100 =	0.25
14	30	30 100 =	0.30
	100		1.00

Loss due to a copy not being sold = 30 paisa.

Profit in one copy being sold = 50 – 30 = 20 paisa.

So the conditional loss table is constructed as follows:

Possible Demand of Newspaper	Possible Stock Options				
	10	11	12	13	14
10	0	30	60	90	120
11	20	0	30	60	90
12	40	20	0	30	60
13	60	40	20	0	30
14	80	60	40	20	0

Sample calculation in the above table is:

Suppose stock position is 13 and copies sold is 11. So the conditional loss = (13 – 11) × 30 = 60 paisa i.e. loss due to two unsold copies. (*Obsolescence loss*)

Suppose stock position is 11 and copies in demand is 14. Conditional loss = (14 – 11) × 20 = 60 paisa i.e. the boy suffers 60 paisa for not having enough stock for selling on the day. (*Opportunity loss*)

Now the expected losses for various stock positions and copies in demand are calculated in the following table:

Possible Demand of newspaper	**Probability of this Many Demands**	**Possible Stock Options**														
		10			**11**			**12**			**13**			**14**		
10	0.10	0	0 × 0.10 =	0	30	30 × 0.10 =	3	60	60 × 0.10 =	6	90	90 × 0.10 =	9.00	120	120× 0.10=	12
11	0.15	20	20 × 0.15 =	3	0	0 × 0.15 =	0	30	30 × 0.15 =	4.50	60	60 × 0.15 =	9.00	90	90× 0.15=	13.50
12	0.20	40	40 × 0.20 =	8	20	20 × 0.20 =	4	0	0 × 0.20 =	0	30	30 × 0.20 =	6.00	60	60× 0.20=	12
13	0.25	60	60 × 0.25 =	15	40	40 × 0.25 =	10	20	20 × 0.25 =	5	0	0 × 0.25 =	0.00	30	30× 0.25=	7.50
14	0.30	80	80 × 0.30 =	24	60	60 × 0.30 =	18	40	40 × 0.30 =	12	20	20 × 0.3 =	6.00	0	0 × 0.30=	0
				50			35			27.50			30			45

It is seen that for a stock position of 12 numbers of newspaper, the expected loss is minimum, i.e. 27.50 paisa. Hence the newspaper boy should purchase 12 newspapers per day for his business.

It may be noted that the problem can be solved by taking the alternative approach that is, maximizing expected gain instead of minimizing expected loss. The answer 12 would have been the same.

LAW OF LARGE NUMBERS

The law of large numbers is closely related to what is commonly called the law of averages. In coin tossing, the law of large numbers stipulates that the fraction of heads will eventually be close to 1/2. Hence, if the first 10 tosses produce only 3 heads, it seems that some mystical force must somehow increase the probability of a head, producing a return of the fraction of heads to its ultimate limit of 1/2. Yet the law of large numbers requires no such mystical force. Indeed, the fraction of heads can take a very long time to approach 1/2. For example, to obtain a 95 percent probability that the fraction of heads falls between 0.47 and 0.53, the number of tosses must exceed 1,000. In other words, after 1,000 tosses, an initial shortfall of only 3 heads out of 10 tosses is swamped by results of the remaining 990 tosses.

In probability theory, the **Law of Large Numbers (LLN)** is a theorem that describes the result of performing the same experiment a large number of times. According to the law, the average of the results obtained from a large number of trials should be close to the expected value, and will tend to become closer as more trials are performed.

The **LLN** is important because it "guarantees" stable long-term results for the averages of some random events. For example, while a casino may lose money in a single spin of the roulette wheel, its earnings will tend towards a predictable percentage over a large number of spins. Any winning streak by a player will eventually be overcome by the parameters of the game. It is important to remember that the LLN only applies (as the name indicates) when a large number of observations are considered. There is no principle that a small number of observations will coincide with the expected value or that a streak of one value will immediately be "balanced" by the others.

For example, a single roll of a fair, six-sided dice produces one of the numbers 1, 2, 3, 4, 5, or 6, each with equal probability. Therefore, the expected value of a single dice roll is $\frac{1+2+3+4+5+6}{6} = 3.5$

According to the law of large numbers, if a large number of six-sided dice are rolled, the average of their values (sometimes called the sample mean) is likely to be close to 3.5, with the precision increasing as more dice are rolled.

It follows from the law of large numbers that the empirical probability of success in a series of Bernoulli trials will converge to the theoretical probability. For a Bernoulli random variable, the expected value is the theoretical probability of success, and the average of n such variables (assuming they are independent and identically distributed (i.i.d.)) is precisely the relative frequency.

For example, a fair coin toss is a Bernoulli trial. When a fair coin is flipped once, the theoretical probability that the outcome will be heads is equal to 1/2. Therefore, according to the law of large numbers, the proportion of heads in a "large" number of coin flips "should be" roughly 1/2. In particular, the proportion of heads after n flips will almost surely converge to 1/2 as n approaches infinity.

Though the proportion of heads (and tails) approaches 1/2, almost surely the absolute (nominal) difference in the number of heads and tails will become large as the number of flips becomes large. That is, the probability that the absolute difference is a small number, approaches zero as the number of flips becomes large. Also, almost surely the ratio of the absolute difference to the number of flips will approach zero. Intuitively, expected absolute difference grows, but at a slower rate than the number of flips, as the number of flips grows.

Two different versions of the law of large numbers (LLN) are described below; they are called the strong law of large numbers, and the weak law of large numbers. Both versions of the law state that – with virtual certainty – the sample average $\overline{X}_n = \frac{1}{n}(X_1 + X_2 + ...X_n)$ converges to the expected value

$\overline{X}_n \to \mu$ for $n \to \infty$

where X_1, X_2, ...X_n is an infinite sequence of i.i.d. An assumption of finite variance $\mathrm{Var}(X_1) = \mathrm{Var}(X_2) = ... = \sigma^2 < \infty$ is not necessary. Large or infinite variance will make the convergence slower, but the LLN holds anyway. This assumption is often used because it makes the proofs easier and shorter.

The **Weak Law of Large Numbers** (also called **Khintchine's Law**) states that the sample average converges in probability towards the expected value $\overline{X}_n \xrightarrow{P} \mu$ when $n \to \infty$. That is to say that for any positive number ε, $\lim_{n \to \infty} \Pr\left(\left|\overline{X}_n - \mu\right| > \varepsilon\right) = 0$ Interpreting this result, the weak law essentially states that for any nonzero margin specified, no matter how small, with a sufficiently large sample there will be a very high probability that the average of the observations will be close to the expected value; that is, within the margin.

Convergence in probability is also called **weak convergence of random variables.** This version is called the weak law because random variables may converge weakly (in probability) as above without converging strongly (almost surely) as below.

The **Strong Law of Large Numbers** states that the sample average converges almost surely to the expected value $\overline{X}_n \xrightarrow{a.s.} \mu$ when $n \to \infty$. That is, $\Pr\left(\lim_{n\to\infty} \overline{X}_n = \mu\right) = 1$. The proof is more complex than that of the weak law. This law justifies the intuitive interpretation of the expected value of a random variable when sampled repeatedly as the "long-term average".

The difference between the strong and the weak version is concerned with the mode of convergence being asserted. Almost sure convergence is also called **strong convergence of random variables.** This version is called the strong law because random variables which converge strongly (almost surely) are guaranteed to converge weakly (in probability). The strong law implies the weak law but not vice versa, when the strong law conditions hold the variable converges both strongly (almost surely) and weakly (in probability). However the weak law may hold in conditions where the strong law does not hold and then the convergence is only weak (in probability).

Sometimes statisticians become interested in knowing the nature of distribution of the total items of a variable included in the population and in most cases it is neither possible nor desirable to have a complete enumeration of all the items included in the population. In reality, population is unlimited in number and is infinitely large. It is hypothetical in nature. An experiment of an event may be repeated for an unlimited number of times, which will produce an infinitely large number of possible results and of possible distribution. A coin may be tossed one time, two times, three times or any number of times. So tossing of a coin may be repeated for an infinitely large number of times, which have no end thus producing an infinitely large number of possible results and of possible distribution.

In case of a hypothetical population, where the number of items is infinitely large and where all the items in the population cannot be observed, the frequency distribution cannot be formed on actual observation basis. Here the question of theoretical distribution finds its place. In this hypothetical population the frequency distribution of the repeated or possible results is obtained on theoretical basis with the help of mathematical formulae.

A theoretical frequency though formed not on actual observation basis, represents a close approximation to the actual frequency distribution in most cases and provides very reliable results in such cases. Here also the statistical measures like mean, standard deviation, variance, etc. are computed and they give a very close approximation to such statistical measures as computed from the actual frequency distribution.

From among several theoretical distributions in statistics, three types of distribution are important and deserve special attention. They are as follows:

1. Binomial Distribution
2. Poisson Distribution
3. Normal Distribution

Binomial distribution and Poisson distribution are discrete distributions, which give us the theoretical probabilities of discrete variables. Normal distribution is a continuous distribution, which gives us the theoretical probabilities of a continuous variable.

BINOMIAL DISTRIBUTION

Binomial distribution is a type of distribution where there are only two possible outcomes. For example: (1) In tossing a coin, either a head will come up or a tail will come up. (2) In drawing a card from a pack of cards, either a black or a red card will result. (3) In throwing a dice, either an even or an odd number will come up. (4) In selling an article in the market, either there will be profit or loss. (5) In catching a fish from a fishpond, either it will be disease free or disease-prone. (6) In experiments on survivality of animals when they are exposed to a drug, either the specimen will survive or will die and so on.

It is seen that in each of the cases cited here, the result that is produced may be grouped into two classes: (i) happening of the event and (ii) not-happening of the event. The happening of the event is generally termed as success and not happening of the event as failure.

In the chapter on Probability we saw that the probability of happening of an event is known as success and the probability of not happening of an event as failure. So, binomial theorem can be conveniently applied in this case in obtaining the theoretical distribution of occurrence of such events.

Binomial distribution originated from a process known as Bernoulli trial, named in honor of the Swiss mathematician James Bernoulli (1654-1705), who made outstanding contributions in the fields of probability and fluid mechanics. A sequence of Bernoulli trials forms a Bernoulli process under the following conditions.

1. Each trial results in one of two possible, mutually exclusive, outcomes i.e. success or failure; yes or no; heads or tails. It indicates the outcome of the random experiment (trial) in the dichotomous classification of events.
2. The probability of success remains constant from trial to trial. With a fair coin, the probability of heads remains 0.5 for each toss regardless of the number of times the coin is tossed.
3. The trials are *statistically independent*; that is, the outcome of any particular trial does not affect or is not affected by the outcome of any other trial.

The probability of success is denoted by p and the probability of failure is denoted by q i.e. $q = 1 - p$

If X denotes the number of success in 'n' number of trials satisfying above conditions then X is the random variable which can take the values 0, 1, 2, ………., n. When X = 0, it indicates there is no success in 'n' trials, X = 1, it indicates one success in 'n' trials. Similarly we may get two successes, three successes etc. up to 'n' successes.

Hence the general expression for probability of x successes in 'n' trials is given by

$P(X) = P(X = x) = {}^nC_x\, p^x\, q^{n-x}$, x = 0, 1, 2, … n

= 0, otherwise

The probability function is represented by the table given below:

Number of Successes, x	Probability, P (x)
1	${}^nC_0 p^0 q^{n-0}$
1	${}^nC_1 p^1 q^{n-1}$
1	${}^nC_2 p^2 q^{n-2}$
1	${}^nC_3 p^3 q^{n-3}$
.	.
.	.
x	${}^nC_x p^x q^{n-x}$
.	
.	
n	${}^nC_n p^n q^{n-n}$
Total	1

As discussed earlier, the sum of all possible probability of a distribution is 1.

Now considering the Binomial distribution,

$$\sum_{x=0}^{n} P(x) = P(0) + P(1) + \ldots + P(n) = {}^{n}C_0 p^0 q^{n-0} + {}^{n}C_1 p^1 q^{n-1} + \ldots + {}^{n}C_n p^n q^0$$

$= (p + q)^n = 1$

Hence P(x) is a probability distribution. Binomial distribution has an expected value or mean (μ) and standard deviation (σ), which are calculated by the formulae:

Mean of a Binomial distribution $= \mu = np$

Standard Deviation of a Binomial distribution $= \sigma = \sqrt{(npq)}$

Where, n = number of trials

p = probability of success

q = probability of failure = 1 – p

It will be clear by some examples given below.

Example 8.2: In tossing of 5 coins, what is the probability of getting 3 heads?

Solution: In tossing a coin, either a head or a tail is obtained (Only two possible outcomes). So it is a case of binomial distribution. Let the probability of getting head be p (success) and tail be q (failure). Now tossing of 5 coins is same as one coin tossed 5 times. Hence n = 5. x = 3.

Probability of head, p = 0.5 Probability of tail, 1 – p = q = 0.5

So, $P(X = 3) = {}^{5}C_3 (0.5)^3 (0.5)^{5-3}$

$$= \frac{5!}{3! \times (5-3)!} \times (0.5)^3 \times (0.5)^2 = \frac{5 \times 4 \times 3 \times 2 \times 1}{3 \times 2 \times 1 \times 2 \times 1} \times (0.5)^5 = 0.3125$$

Example 8.3: An experiment was carried out to study the effect of one pesticide in relation to survival of a certain plant. It was found that the probability of survival was 20%. If 30 batches of 6 plants each are subjected to such experiment, in how many batches 4 or more plants would die.

Solution: The probability of survival is 20%. Hence the probability of death would be 80%.

Let p denotes probability of death and q denote probability of survival. The values of p and q would be 0.8 and 0.2 respectively.

The probability of death in 4 or more cases out of 6 means probabilities of 4, 5 and 6 deaths.

Probability of 4, 5 and 6 deaths would be = P(X = 4) + P(X = 5) + P(X = 6)

$= {}^6C_4\, p^4\, q^2 + {}^6C_5\, p^5\, q^1 + {}^6C_6\, p^6$

$= 15 \text{ X } (0.8)^4 \times (0.2)^2 + 6 \times (0.8)^5 \times (0.2)^1 + (0.8)^6$

= 0.901.

So the probability of 4, 5 and 6 deaths is 0.901 in one batch. Since there are 30 batches, 6 plants for the test to be carried out, thus probability of death would happen in 30 × 0.901 = 27.3 batches.

Thus in 27 batches out of 30 batches 4 or more number of plants would die on application of the pesticide.

Example 8.4: A National Center for Health Statistics report based on 1985 data states that 30% of American adults smoke. Consider a simple random sample of 15 adults selected at that time. Find the probability that the number of smokers in the sample would be (a) Three, (b) Less than three, (c) Between three and five, inclusive (d) More than four, but less than eight (e) Four or more.

Solution: We have probability of finding a smoker (success) = p = 30% = 0.30.

So, q = 1 – 0.30 = 0.70.

n = 15

(a) $P(X = 3) = {}^nC_x\, p^x\, q^{n-x} = {}^{15}C_3\, (0.30)^3\, (0.70)^{15-3}$

$$= \frac{15!}{3! \times 12!} \times 0.027 \times 0.014 = 0.170$$

(b) $P(X < 3) = P(X = 0) + P(X = 1) + P(X = 2) = {}^{15}C_0\, (0.30)^0\, (0.70)^{15-0} + {}^{15}C_1\, (0.30)^1\, (0.70)^{15-1} + {}^{15}C_2\, (0.30)^2\, (0.70)^{15-2} = 0.005 + 0.030 + 0.092 = 0.127$

(c) $P(3 \le X \le 5) = P(X = 3) + P(X = 4) + P(X = 5) = {}^{15}C_3\, (0.30)^3\, (0.70)^{15-3} + {}^{15}C_4\, (0.30)^4\, (0.70)^{15-4} + {}^{15}C_5\, (0.30)^5\, (0.70)^{15-5} = 0.170 + 0.219 + 0.206 = 0.595$

(d) $P(4 < X < 8) = P(X = 5) + P(X = 6) + P(X = 7) = {}^{15}C_5\, (0.30)^5\, (0.70)^{15-5} + {}^{15}C_6\, (0.30)^6\, (0.70)^{15-6} + {}^{15}C_7\, (0.30)^7\, (0.70)^{15-7} = 0.206 + 0.147 + 0.081 = 0.434$

(e) P(X 4) = P(X = 4) + P(X = 5) + P(X = 6) ....P(X = 15)

= 1 – [P(X = 0) + P(X = 1) + P(X = 2) + P(X = 3)]

= 1 – [0.005 + 0.030 + 0.092 + 0.170] = 0.703

Example 8.5: The probability that a person suffering from migraine headache will obtain relief with a particular drug is 0.9. Three randomly selected sufferers from migraine headache are given the drug. Find the probability that the number obtaining relief will be: (a) Exactly zero, (b) Exactly one, (c) More than one, (d) Two or fewer, (e) Two or three, (f) Exactly three.

Solution: In the said question, we have p = 0.90 and q = 1 – 0.90 = 0.10 n = 3

(a) $P(X = 0) = {}^3C_0\,(0.90)^0\,(0.10)^{3-0} = 0.001$

(b) $P(X = 1) = {}^3C_1\,(0.90)^1\,(0.10)^{3-1} = 0.027$

(c) $P(X > 1) = P(X = 2) + P(X = 3) = {}^3C_2\,(0.90)^2\,(0.10)^{3-2} + {}^3C_3\,(0.90)^3\,(0.10)^{3-3} = 0.243 + 0.729 = 0.972$

(d) $P(X \le 2) = P(X = 0) + P(X = 1) + P(X = 2) = 0.001 + 0.027 + 0.243 = 0.271$

(e) $P(X = 2 \text{ or } 3) = P(X = 2) + P(X = 3) = 0.243 + 0.729 = 0.972$

(f) $P(X = 3) = 0.729$

Important Note

Extreme care is to be taken while applying Binomial Distribution to be certain that the three conditions stated earlier are met for a Bernoulli process. The condition that the outcome of any trial to remain fixed over time, may not be guaranteed in many industrial processes. Each time an industrial machine produces a part, for instance, there is some infinitesimal wear on the machine, which when accumulated beyond a reasonable time, the proportion of acceptable parts produced by the machine may be altered and hence the condition may be violated. Although this problem is not encountered in a coin-toss experiment, but it is an integral consideration in all real applications of the binomial probability. Going from textbook to the real world is often difficult, and the managers use their intuitive power through experience only to know when a Bernoulli process is applicable.

POISSON DISTRIBUTION

Binomial distribution is a form of distribution where the outcomes (either successes or failures) are known beforehand. But there are some occasions where the number of times an event does occur (success) may be known, but the number of times an event does not occur (failure) cannot be known. This happens in those where the event is rare and few, and is casual in nature. This distribution is credited after the French mathematician Simeon Denis Poisson (1781-1840), who first published its derivation in 1837.

To take a few real world examples-Distribution of telephone calls going through a switchboard system, the demand of patients for service at a health care institution, arrival of freight movers at a toll booth, the number of accidents at a road junction, the number of babies born blind per year in a large city, the number of printing mistakes per page in a large volume of book and the number of air-pockets in glass sheets etc. A very common element that describes the variable in such examples is that it takes on only integer values (0, 1, 2, 3, ... and so on). Such events are known as occurrence of casual events in a continuum of time. These phenomena are rare in the sense that the probabilities of their happening are very small. In each of these cases we may know the number of times an event occurs, but we cannot know the number of times an event does not occur. For example, we do not know how many people did not die by accidentally falling from a roof. But we know the number of people who died in this manner. In such cases it is not possible to obtain the theoretical probabilities and the theoretical frequencies with the help of the binomial expansion formula, as the value of n is not precisely known and the binomial formula $(p+q)^n$ cannot be expanded. Here lies the importance of the Poisson distribution. In such cases where the binomial expansion formula cannot be used in obtaining the theoretical probabilities and the theoretical frequencies of certain events, the Poisson distribution may conveniently be used in getting those probabilities and frequencies.

Poisson distribution is a limiting form of binomial distribution as n moves towards infinity (i.e. very large number of observations) and p moves towards zero (i.e. very rare successes) and np remains constant (= m).

It has been seen in case of the binomial distribution, that it results from a set of assumptions about an underlying process yielding a set of numerical observations. Same also is the case with the Poisson distribution and the following describe what is known as *Poisson process*:

1. Poisson distribution is a discrete probability distribution with a single parameter λ.
2. Both mean and variance of Poisson distribution are equal to λ.
3. It is a limiting case of Binomial distribution, where p is small, n is large and np is finite.
4. Normal distribution is a limiting case of Poisson distribution as $\lambda \rightarrow \infty$
5. The occurrences of the events are independent. The occurrence of an event in an interval of space and time has no effect on the probability of a second occurrence of the event in the same, or any other, interval.
6. Theoretically, an infinite number of occurrences of the event must be possible in the interval.

The formula of the Poisson distribution for computing the theoretical probabilities and theoretical frequencies of random variable x is:

$$P(X) = \frac{e^{-m} \times m^{x}}{x!} \text{ for } x = 0, 1, 2, \text{ ...... } \infty$$

= 0; otherwise

where, m = mean number of successes,

x = number of successes,

e = 2.71828

Example 8.6: In a printing machine, it was found that the average number of printing mistakes per page in a large volume is three. Find the probability that a (i) page contains more than four printing mistakes (ii) at best two mistakes.

Solution: Data given: Mean m = 3

(i) $P(X > 4) = P(X = 5) + P(X = 6) + P(X = 7) + \ldots\ldots$

$= 1 - [P(X = 0) + P(X = 1) + P(X = 2) + P(X = 3) + P(X = 4)]$

$$= 1 - \left[\frac{e^{-3}3^0}{0!} + \frac{e^{-3}3^1}{1!} + \frac{e^{-3}3^2}{2!} + \frac{e^{-3}3^3}{3!} + \frac{e^{-3}3^4}{4!}\right] = 0.1847$$

(ii) $P(X \leq 2) = P(X=0) + P(X=1) + P(X=2) = \frac{e^{-3}3^0}{0!} + \frac{e^{-3}3^1}{1!} + \frac{e^{-3}3^2}{2!} = 0.4232$

Example 8.7: In a certain population an average of 13 new cases of esophageal cancer are diagnosed each year. If the annual incidence of esophageal cancer follows a Poisson distribution, find the probability that in a given year the number of newly diagnosed cases of esophageal cancer will be:

(a) Exactly 10

(b) At least 8

(c) No more than 4

(d) Between 9 and 11, inclusive

(e) Fewer than 7

Solution: We have

m = 13; e = 2.71828

(a) $P(X = 10) = e^{-m} . m^x / x! = (2.718)^{-13} . 13^{10} / 10! = 0.086$

(b) $P(X \geq 8) = 1 - [P(X = 0) + P(X = 1) + P(X = 2) + P(X = 3) + P(X = 4) + P(X = 5) + P(X = 6) + P(X = 7)]$

$P(X = 0) = (2.718)^{-13} . 13^0 / 0! = 0.0000022$

$P(X = 1) = (2.718)^{-13} . 13^1 / 1! = 0.000029$

$P(X = 2) = (2.718)^{-13} . 13^2 / 2! = 0.00019$

$P(X = 3) = (2.718)^{-13} . 13^3 / 3! = 0.00083$

$P(X = 4) = (2.718)^{-13} . 13^4 / 4! = 0.0027$

$P(X = 5) = (2.718)^{-13} . 13^5 / 5! = 0.007$

$P(X = 6) = (2.718)^{-13} . 13^6 / 6! = 0.015$

$P(X = 7) = (2.718)^{-13} . 13^7 / 7! = 0.028$

0.054

Hence Required Probability = 1 – 0.054 = 0.946

(c) $P(X \leq 4) = P(X = 0) + P(X = 1) + P(X = 2) + P(X = 3) + P(X = 4)$

$= 0.0000022 + 0.000029 + 0.00019 + 0.00083 + 0.0027 = 0.00375$

(d) $P(X = 9) + P(X = 10) + P(X = 11)$

$= (2.718)^{-13} . 13^9 / 9! + (2.718)^{-13} . 13^{10} / 10! + (2.718)^{-13} . 13^{11} / 11!$

$= 0.066 + 0.086 + 0.102 = 0.254$

(e) $P(X < 7) = P(X = 0) + P(X=1) + P(X=2) + P(X=3) + P(X=4) + P(X=5) + P(X=6)$

$= 0.0000022 + 0.000029 + 0.00019 + 0.00083 + 0.0027 + 0.007 + 0.015 = 0.026$

Important Note

In the foregoing discussions, it can be seen that the calculations in Binomial distribution is very tedious, which can be avoided if Poisson distribution is used instead. Poisson distribution can be a reasonable probability approximation of the Binomial distribution, but under certain conditions. When n is large, p is small, i.e. when number of trials is large and binomial probability of success is small, Poisson distribution can be applied in place of Binomial distribution. Statisticians most often use Poisson distribution when n is greater than or equal to 20 and p is less than or equal to 0.05. In cases that meet these conditions, we can substitute the mean of the Binomial distribution (np) in place of the mean of the Poisson distribution (m) so that the formula becomes:

$$P(x) = \frac{e^{-np}(np)^x}{x!}$$

Take for example in a production factory there are 20 lathe machines and that the chance of any one of them giving trouble during operation in any day is 0.02. What is the probability that exactly three machines will be out of service on the same day?

Binomial Approach	Poisson Approach
$P(r)=\frac{n!}{r!\times(n-r)!}p^{r}q^{n-r}$	$P(x)=\frac{e^{-np}\times(np)^{x}}{x!}$
$P(3)=\frac{20!}{3!\times(20-3)!}0.02^{3}0.98^{17}=0.0065$	$P(3)=\frac{e^{-(20\times0.02)}\times(20\times0.02)^{3}}{3!}=0.00715$

It can be seen that the difference between the two probability distributions is slight (only about 10% error). Even though a slight bit of accuracy is often lost in doing this, the trade-off is affordable.

NORMAL DISTRIBUTION

The Normal distribution is a theoretical distribution. It is probably the most important of all the theoretical distributions for the reason that many natural phenomenon which exhibit frequency distributions closely resemble the normal distribution. Moreover, the normal distribution unlike the Binomial and Poisson distributions is a continuous frequency distribution. This distribution is also called the *normal probability distribution.* The theory of sampling and its application in statistics are mainly based on the normal distribution. As the normal distribution plays a very important role in the theory and practice of sampling, it has occupied an indispensable and dominant position in modern statistics. It is often said to be the cornerstone of statistical analysis and drawing of inferences.

The normal distribution of a variable, when represented by a graph, takes the form of a symmetrical bell shaped curve, which is known as the *Normal Curve.* Observe the figure below and the following are some important properties of a normal curve.

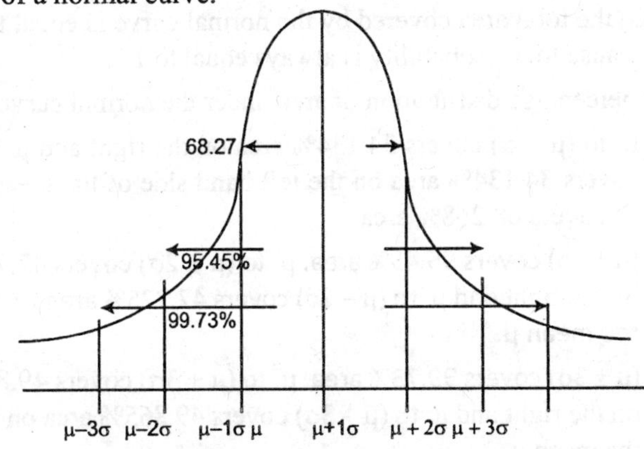

Percentage of Area Under Normal Curve

Properties and Features of Normal Distribution

In most scientific analyses, values are often distributed in accordance with the normal distribution. The normal distribution curve has the following properties:

1. The normal distribution is a continuous probability distribution.
2. The normal distribution of a continuous variable, represented by a graph, takes the form of a smooth curve. The smooth curve is known as the normal curve, which is perfectly symmetrical and continuous.
3. The normal curve is bell shaped and it extends on both sides indefinitely. It tends to approach the base line of the curve as the distance from the mean increases, but never meets the base line.
4. The normal distribution is perfectly symmetrical about the mean μ. So mean, median and mode coincide. Hence Mean = Median = Mode.
5. The maximum ordinate of the normal curve is at mean μ and the value of its maximum ordinate is $1/[\sigma\sqrt{(2\pi)}]$.
6. The maximum ordinate divides the normal curve into two equal and exactly similar parts.
7. The normal distribution can be fully determined and the normal curve can be completely specified if its mean μ and standard deviation σ are known.
8. Though theoretically the normal curve can take any value between $-\infty$ and $+\infty$, but practically most of the values lie between two limits $\mu - 3\sigma$ and $\mu + 3\sigma$. Hence the interval between the two limits $\mu - 3\sigma$ and $\mu + 3\sigma$ is called the effective range of normal curve.
9. The normal distribution is a continuous probability distribution and hence the total area covered by the normal curve is equal to unity or 1, because total probability is always equal to 1.
10. The percentage distribution of area under the normal curve is
    (a) μ to $(\mu + \sigma)$ covers 34.134% area on the right and μ to $(\mu - \sigma)$ covers 34.134% area on the left hand side of the mean. So $(\mu \pm \sigma)$ covers 68.268% area.

    (b) $(\mu \pm 2\sigma)$ covers 95.45% area. μ to $(\mu + 2\sigma)$ covers 47.725% area on the right and μ to $(\mu - 2\sigma)$ covers 47.725% area on the left of the mean μ.

    (c) $(\mu \pm 3\sigma)$ covers 99.73% area. μ to $(\mu + 3\sigma)$ covers 49.865% area on the right and μ to $(\mu - 3\sigma)$ covers 49.865% area on the left of the mean μ.

So it shows that $(\mu \pm 3\sigma)$ covers almost the total area of the normal curve, excepting only 0.27% area which remains outside the limits .

Besides the above, the following facts are also to be considered in the context of curve area.

$\mu \pm 1.96\ \sigma$ covers 95% area

$\mu \pm 2.58\ \sigma$ covers 99% area

$\mu \pm 0.67\ \sigma$ covers 50% area

These figures are extremely useful in statistical analyses. In most experiments 95% significance is accepted level of accuracy. Here in the theoretical normal probability distribution curve 95% area are covered between $(\mu - 1.96\sigma)$ and $(\mu + 1.96\sigma)$. The rest 5% area falls outside this range. So one value of a variable in any population study can deviate from the theoretical mean μ. But what is the chance that it will deviate from this theoretical mean μ by an amount of 2σ on either side? That means what is the probability that a value x will deviate from the mean either positively or negatively by an amount 2σ? The area lying away from mean μ by an amount of 2σ is only 5% i.e. 0.05. But $0.05 = 1 \div 20$. That means the value of x deviating from the mean either positively or negatively by an amount greater than 2 standard deviation will occur 1 in 20 times.

Similarly, area outside 3σ distance from μ is only 0.27 i.e. 0.0027. But $0.0027 = 1 \div 370$. That means the value of x deviating from the mean either positively or negatively by an amount greater than three standard deviation will occur 1 in 370 times.

The value of x deviating from the mean either positively or negatively by an amount greater than 1 standard deviation will occur 1 in 3 times. (About 32% area lies outside 1σ distance from μ.)

It will be easier to understand about Normal Distribution by taking an example below. Consider a cement manufacturing company producing cement bags each supposed to be weighing 50 kg. On a randomly selected 100 number of bags, the following are the weight of different bags represented in a frequency table, using which a histogram and then the frequency curve is prepared.

Weight of cement bag (x)	Number of bags (f)	f × x	f × x²
41	1	41	1681
42	1	42	1764
43	2	86	3698
44	2	88	3872
45	4	180	8100
46	5	230	10580
47	7	329	15463
48	8	384	18432
49	9	441	21609
50	10	500	25000
51	10	510	26010
52	9	468	24336
53	8	424	22472
54	7	378	20412
55	5	275	15125
56	4	224	12544
57	3	171	9747
58	2	116	6728
59	2	118	6962
60	1	60	3600
Total:	100	5065	258135

$$\text{Mean} = \bar{x} = \frac{\sum(f \times x)}{\sum f} = \frac{5065}{100} = 50.65$$

$$\text{Standard Deviation} = s = \sqrt{\frac{\sum fx^2}{\sum f} - \left(\frac{\sum fx}{\sum f}\right)^2} = \sqrt{\frac{258135}{100} - \left(\frac{5065}{100}\right)^2} = 3.99$$

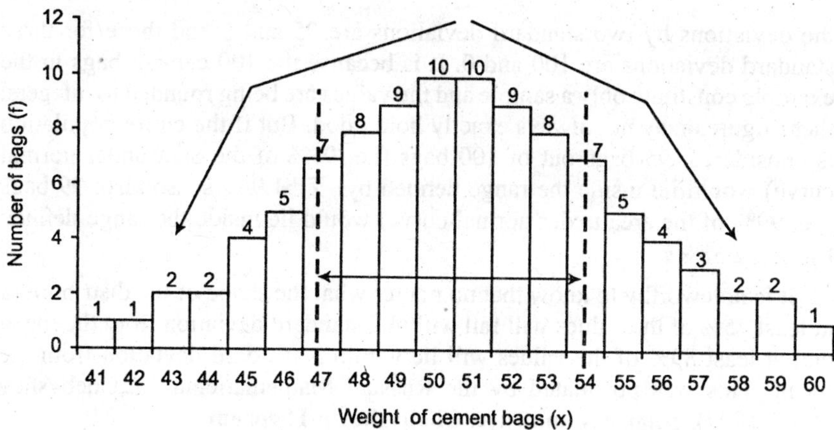

Let us now calculate the values of the variable when deviated from the mean value by amounts of ±1s; ±2s; ±3s; ±1.96s & ±2.58s respectively as shown in the following table:

		Range	Outside	Inside
$\overline{x} \pm 1 \times s$	50.65 – 1 × 3.99 & 50.65 + 1 × 3.99 i.e.46.66 Rounded to integer 47 to 54	54.64	32	68
$\overline{x} \pm 2 \times s$	50.65 – 2×3.99 & 50.65 + 2 × 3.99 i.e. 42.67 Rounded to integer 43 to 58	58.63	5	95
$\overline{x} \pm 3 \times s$	50.65 – 3 × 3.99 & 50.65 + 3 × 3.99 i.e. 38.68 Rounded to integer 39 to 62	62.62	0	100
$\overline{x} \pm 1.96 \times s$	50.65 – 1.96 × 3.99 & 50.65 + 1.96 × 3.99 i.e. 42.82 Rounded to integer 43 to 58	58.47	5	95
$\overline{x} \pm 2.58 \times s$	50.65 – 2.58 × 3.99 & 50.65 + 2.58 × 3.99 i.e. 40.3 Rounded to integer 40 to 61	60.94	0	100

In this example of a real world situation, on counting there lies 68 number of cement bags inside the range defined by $\overline{x} \pm 1 \times s$ i.e. 47 to 54 and 32 bags lie outside this range (See the figure above). Similarly, the figures for the deviations by two standard deviations are 95 and 5 and those for three standard deviations are 100 and 0. It is because the 100 cement bags in the example constitute only a sample and the values are being rounded to integers, these figures may not always exactly hold good. But if the entire population is considered, 95 bags out of 100 bags (i.e. 95% of the area under normal curve) would lie inside the range defined by $\overline{x} \pm 1.96 \times s$. So also, 99 bags (i.e. 99% of the area under normal curve) would lie inside the range defined by $\overline{x} \pm 2.58 \times s$.

It is noteworthy to know that no matter what the shape of the distribution, at least 75% of the values will fall within 2 standard deviation from the mean and at least 89% of the values will lie within 3 standard deviation from the mean. This was postulated by the Russian Mathematician P.L.Chebyshev (1812-1894), popularly known as **Chebyshev's Theorem.**

USES OF NORMAL DISTRIBUTION

1. The normal distribution can be used to approximate the Binomial and Poisson distribution.
2. It has wide application in sampling theory. Any statistic based on a large sample generally follows normal distribution. For example- the means of a large numbers of samples will follow a normal distribution. Hence it helps us to estimate population parameters from sample statistic and to find the confidence limits of population parameters.
3. The normal distribution has wide use in testing statistical hypothesis and test of significance in which the assumption is that the population from which the sample has been drawn is in normal distribution. We will study this in detail in later chapters.
4. The normal distribution may be used for smoothing a distribution which is not normal, by constructing a normal curve.

STANDARD NORMAL DISTRIBUTION

As it is apparent from the above discussion that the normal distribution is a continuous probability distribution, which wholly depends on the values of μ and σ. Such type of curves can thus be drawn with various μ and σ. Different normal distributions may have different means and standard deviations. In such cases the normal curves drawn will be symmetrical and bell shaped, but with different shapes and different sizes. In order to avoid this difficulty, these different normal curves may be standardized and converted to one *standard normal distribution curve.*

The standard normal distribution has mean zero ($\mu = 0$) and standard deviation ($\sigma = 1$). The standard normal curve drawn from such distribution covers the total area of unity. That means the total area under the standard normal curve is 1. *Such a normal curve with mean 0 and unit standard deviation is known as the standard normal curve.* So the areas under the standard normal curve represent probabilities.

A normal curve with mean μ and standard deviation σ can be converted into a standard normal distribution curve by performing the change of scale and origin. In the original scale (x-scale) the mean and standard deviation are μ and σ. In the new scale (z-scale) mean is 0 and standard deviation is1. The formula that enables us to change from x-scale to z-scale and vice versa is given by,

$$z = \frac{x - \mu}{\acute{o}}$$

where x is a normal variable with mean μ and standard deviation σ.

Consider a normal distribution having mean = 50 and standard deviation = 25. In the figure below, the x-values are converted to z-values by the above formula to obtain the standard normal variates. So any set of data following normal distribution can be converted to z-scale to obtain the standard normal variates.

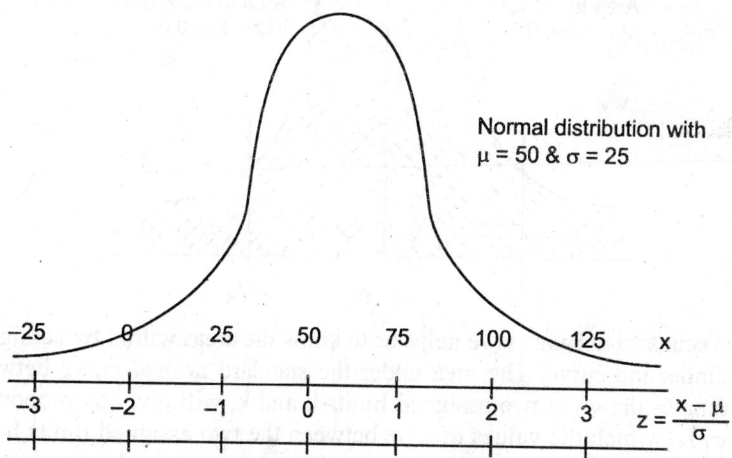

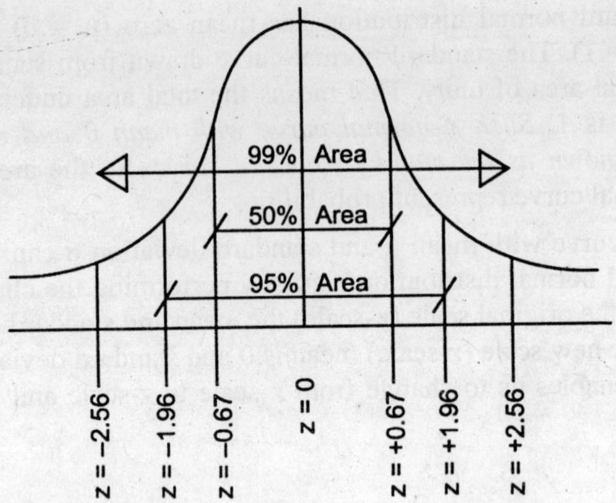

Now, "z" itself is a normal variable and is known as a standard normal variable or a normal variate with mean (μ = 0) and unit standard deviation (σ = 1). This standard normal distribution is also known as the *unit normal distribution* or *z-distribution*.

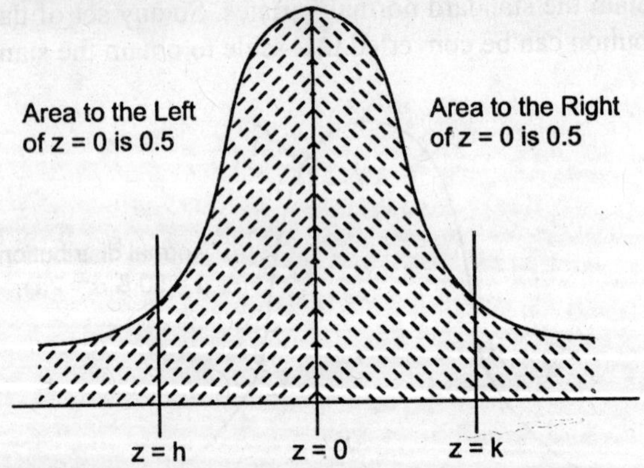

The standard normal curve helps us to know the areas within two assigned limits under the curve. The area under the standard normal curve between the ordinates drawn at two assigned limits h and k, will give the proportion of cases for which the values of z lie between the two assigned limits h and k. Hence the area between the two assigned limits h and k under the standard normal curve will represent the probability that z will be between the two assigned limits h and k.

Hence the probability that a standard normal variable z will be between two assigned values h and k is area under the standard normal curve between the two assigned values z =h and z = k.

Example 8.8: Find the probability that a standard normal variable is (i) greater than 1.68; (ii) less than –1.32 (iii) between 1.22 and 2.75.

Solution:

(i) Let the standard normal variable = z =1.68

Area between z = 0 and z = 1.68 is 0.4535 (Read from the table)

Area to the right of z = 0 is 0.5

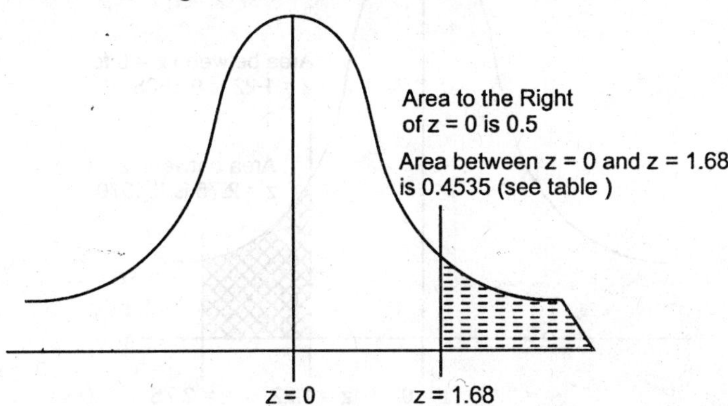

Hence area beyond z = 1.68 to the right of curve = 0.5 – 0.4535 = 0.0465

Probability that the normal variable is greater than 1.68 is 0.0465

(ii) Here z = –1.32

The area to the left of z = 0 is 0.5

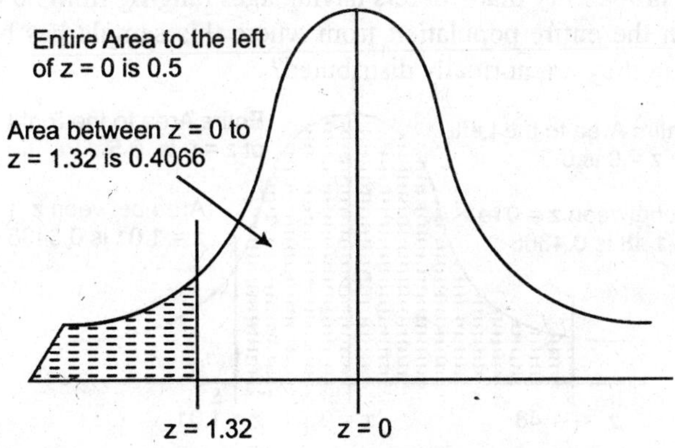

Area from z = 0 to z = –1.32 is 0.4066 ← Read from the table

(It is same as area from z = 0 to z = +1.32)

Area beyond this is = 0.5 – 0.4066 = 0.0934

Hence Probability that the normal variable is less than –1.32 is 0.0934

(iii) Area between z = 0 and z = 1.22 is 0.3888

Area between z = 0 and z = 2.75 is 0.4970

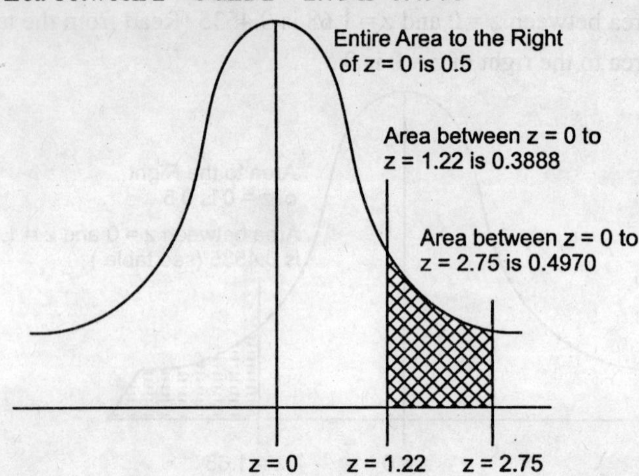

Hence the area between z = 1.22 and z = 2.75 is 0.4970 – 0.3888 = 0.1082

Hence the required probability is 0.1082.

Example 8.9: In one observation of 100 factories the average age of workers was found to be 28.9 years with a standard deviation of 6.03 years. What is the probability that workers having ages ranging from 20 to 35 years are found in the entire population from where this sample has been drawn assuming that they are normally distributed?

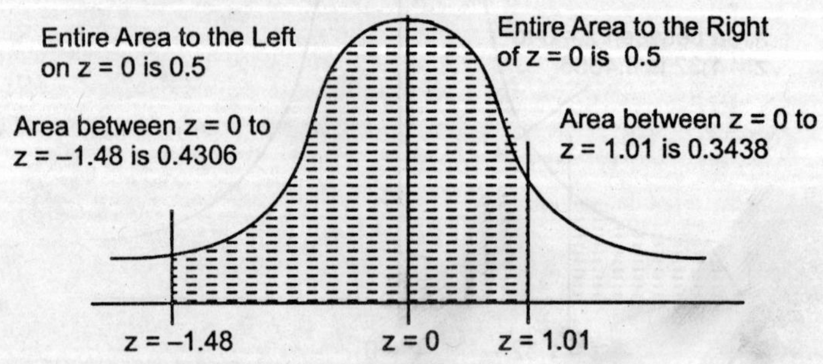

Solution: Here h = 20, $\mu = 28.9$, $\sigma = 6.03$

$$\text{Hence } z = \frac{20 - 28.9}{6.03} = -1.48$$

The area between z = 0 and z = -1.48 is 0.4306 (To the left of z = 0)

(It is same as area between z = 0 to z = +1.48)

And k = 35, $\mu = 28.9$, $\sigma = 6.03$

$$\text{Hence, } z = \frac{35-28.9}{6.03} = 1.01$$

The area between z = 0 and z = 1.01 is 0.3438 (To the right of z = 0)

So the total area between z = -1.48 and z = 1.01 is 0.4306 + 0.3438 = 0.7744

Hence there is 77.44% probability that workers having ages from 20 to 35 years will be found in the entire population.

MORE ILLUSTRATIVE EXAMPLES

Example 8.10: Suppose the ages at time of onset of a certain disease are approximately normally distributed with a mean of 11.5 years and a standard deviation of 3 years. A child has just come down with the disease. Find the probability that the child is:

(a) Between the ages of 8.5 and 14.5 years

(b) Over 10 years of age

(c) Under 12 years

Solution:

(a) Here, h = 14.5, $\mu = 11.5$, $\sigma = 3$

$$\text{Hence, } z = \frac{14.5-11.5}{3} = 1$$

The area between z = 0 and z = 1 is 0.3413 (To the right of z = 0)

And k = 8.5, $\mu = 11.5$, $\sigma = 3$

$$\text{Hence, } z = \frac{8.5-11.5}{3} = -1$$

The area between z = 0 and z = –1 is 0.3413 (To the left of z = 0)

(It is the same area as that between z = 0 and z = +1)

So the total area between z = –1and z = 1 is 0.3413+ + 0.3413 = 0.6826

(b) Here, h = 10, $\mu = 11.5$, $\sigma = 3$

$$\text{Hence, } z = \frac{10-11.5}{3} = -0.5$$

The area between z = 0 and z = –0.5 is 0.1915 (To the left of z = 0)

Now the total area will be 0.5 + 0.1915 = 0.6915

(c) Here, h = 12, μ = 11.5, σ = 3

$$\text{Hence, } z = \frac{12-11.5}{3} = 0.17$$

The area between z = 0 and z = 0.17 is 0.0675 (To the right of z = 0)

Now the total area will be 0.5 + 0.0675 = 0.5675

Example 8.11: In the study of fingerprints an important quantitative characteristic is the total ridge count for the 10 fingers of an individual. Suppose that the total ridge counts of individuals in a certain population are approximately normally distributed with mean of 140 and a standard deviation of 50. Find the probability that an individual picked at random from this population will have a ridge count:

(a) Of 200 or more
(b) Less than 100
(c) Between 100 and 200
(d) Between 200 and 250
(e) In a population of 10,000 people how many would you expect to have a ridge count of 200 or more?

Solution:

(a) Here, h = 200, μ = 140, σ = 50

$$\text{Hence, } z = \frac{200-140}{50} = 1.2$$

The area between z = 0 and z = 1.2 is 0.3849 (To the right of z = 0)

Now the total area will be 0.5 – 0.3849 = 0.1151

(b) Here, h = 100, μ = 140, σ = 50

$$\text{Hence, } z = \frac{100-140}{50} = -0.8$$

The area between z = 0 and z = –0.8 is 0.2881 (To the left of z = 0)

Now the total area will be 0.5 – 0.2881 = 0.2119

(c) Here, h = 200, μ = 140, σ = 50

$$\text{Hence, } z = \frac{200-140}{50} = 1.2$$

Here, k = 100, $\mu = 140$, $\sigma = 50$

Hence, $z = \dfrac{100-140}{50} = -0.8$

The area between z = 0 and z = 1.2 is 0.3849 (To the right of z = 0)

The area between z = 0 and z = –0.8 is 0.2881 (To the left of z = 0)

Now the total area will be 0.3849 + 0.2881 = 0.6730

(d) Here, h = 200, $\mu = 140$, $\sigma = 50$

Hence, $z = \dfrac{200-140}{50} = 1.2$

The area between z = 0 and z = 1.2 is 0.3849 (To the right of z = 0)

Here, k = 250, $\mu = 140$, $\sigma = 50$

Hence, $z = \dfrac{250-140}{50} = 2.2$

The area between z = 0 and z = 2.2 is 0.4861 (To the right of z = 0)

Now the total area will be 0.4861 – 0.3849 = 0.1012

(e) We have already calculated in part (a) above that the probability of getting a ridge count of 200 or more is 0.1151. So in a population of 10,000 people, this number will be

10000 × 0.1151 = 1151 number of persons.

Example 8.12: If the capacities of the cranial cavities of a certain population are approximately normally distributed with a mean of 1400 cc and a standard deviation of 125, find the probability that a person randomly picked from this population will have a cranial cavity capacity:

(a) Greater than 1450 cc

(b) Less than 1350 cc

(c) Between 1300 cc and 1500 cc

Solution:

(a) Here, h = 1450, $\mu = 1400$, $\sigma = 125$

Hence, $z = \dfrac{1450-1400}{125} = 0.4$

The area between z = 0 and z = 0.4 is 0.1554 (To the right of z = 0)

Now the total area will be 0.5 – 0.1554 = 0.3446

(b) Here, h = 1350, μ = 1400, σ = 125

$$\text{Hence, } z = \frac{1350 - 1400}{125} = -0.4$$

The area between z = 0 and z = –0.4 is 0.1554 (To the left of z = 0)

Now the total area will be 0.5 – 0.1554 = 0.3446

(c) Here, h = 1300, μ = 1400, σ = 125

$$\text{Hence, } z = \frac{1300 - 1400}{125} = -0.8$$

Here, k = 1500, μ = 1400, σ = 125

$$\text{Hence, } z = \frac{1500 - 1400}{125} = 0.8$$

The area between z = 0 and z = –0.8 is 0.2881 (To the left of z = 0)

The area between z = 0 and z = 0.8 is 0.2881 (To the right of z = 0)

Now the total area will be 0.2881 + 0.2881 = 0.5762

Example 8.13: If the total cholesterol values for a certain population are approximately normally distributed with a mean of 200 mg/ 100 ml and a standard deviation of 20 mg/ 100 ml, find the probability that an individual picked at random from this population will have a cholesterol value:

(a) Between 180 and 200 mg/ 100 ml

(b) Greater than 225 mg/ 100 ml

(c) Less than 150 mg/ 100 ml

(d) Between 190 and 210 mg/ 100 ml

Solution:

(a) Here, h = 180, μ = 200, σ = 20

$$\text{Hence, } z = \frac{180 - 200}{20} = -1.0$$

Here, k = 200, μ = 200, σ = 20

$$\text{Hence, } z = \frac{200 - 200}{20} = 0.0$$

The area between z = 0 and z = –1.0 is 0.3413 (To the left of z = 0)

The area between z = 0 and z = 0 is = 0

Now the total area will be 0 + 0.3413 = 0.3413

(b) Here, $h = 225, \mu = 200, \sigma = 20$

$$\text{Hence, } z = \frac{225 - 200}{20} = 1.25$$

The area between $z = 0$ and $z = 1.25$ is 0.3944 (To the right of $z = 0$)

Now the total area will be 0.5 - 0.3944 = 0.1056

(c) Here, $h = 150, \mu = 200, \sigma = 20$

$$\text{Hence, } z = \frac{150 - 200}{20} = -2.5$$

The area between $z = 0$ and $z = -2.5$ is 0.4938 (To the left of $z = 0$)

Now the total area will be 0.5 – 0.4938 = 0.0062

(d) Here, $h = 190, \mu = 200, \sigma = 20$

$$\text{Hence, } z = \frac{190 - 200}{20} = -0.5$$

Here, $k = 210, \mu = 200, \sigma = 20$

$$\text{Hence, } z = \frac{210 - 200}{20} = 0.5$$

The area between $z = 0$ and $z = -0.5$ is 0.1915 (To the left of $z = 0$)

The area between $z = 0$ and $z = 0.5$ is 0.1915 (To the right of $z = 0$)

Now the total area will be 0.1915 + 0.1915 = 0.3830

Example 8.14: Suppose a variable X is normally distributed with a standard deviation of 10. Given that 0.0985 of the values of X are greater than 70, what is the mean value of x?

Solution: We have $h = 70, \sigma = 10; x = ?$

Let the value $h = 70$ fall to the right of the mean value. So area between the mean value and the given value 70 will be equal to 0.5 – 0.0985 = 0.4015 which is the area between $z = 0$ and $z = 1.29$ as read from the table.

$$\text{But } z = \frac{70 - x}{10} = 1.29$$

So x = 70 – 1.29 10 = 57.10

Important Note

It is interesting to note that normal distribution can be regarded as an approximation to the binomial distribution. Let us consider the probability of getting 5, 6, 7 or 8 heads in 10 tosses of a fair coin. Using binomial distribution, the answer would be:

$$= \sum_{r=}^{r=8} C_r^n p^r q^{n-r} \quad P(r = 5, 6, 7 \text{ or } 8) = P(r = 5) + P(r = 6) + P(r = 7) + P(r = 8)$$

$$= {}^{10}C_5 \times 0.5^5 \times 0.5^{10-5} + {}^{10}C_6 \times 0.5^6 \times 0.5^{10-6} + {}^{10}C_7 \times 0.5^7 \times 0.5^{10-7} + {}^{10}C_8 \times 0.5^8 \times 0.5^{10-8} = 0.2461 + 0.2051 + 0.1172 + 0.0439 = 0.6123$$

Here in the binomial distribution, for n=10 and p=0.5 with a normal distribution superimposed on it with the same mean $\left(\mu = np = 10 \times \frac{1}{2} = 5\right)$ and the same standard deviation $\sigma = \sqrt{npq} = \sqrt{10 \times 0.5 \times 0.5} = \sqrt{2.5} = 1.581$

In the figure shown below, let us find out the area bounded by the two limits x = 4.5 and x = 8.5 of the normal curve (Class boundaries of 5 and 8 respectively).

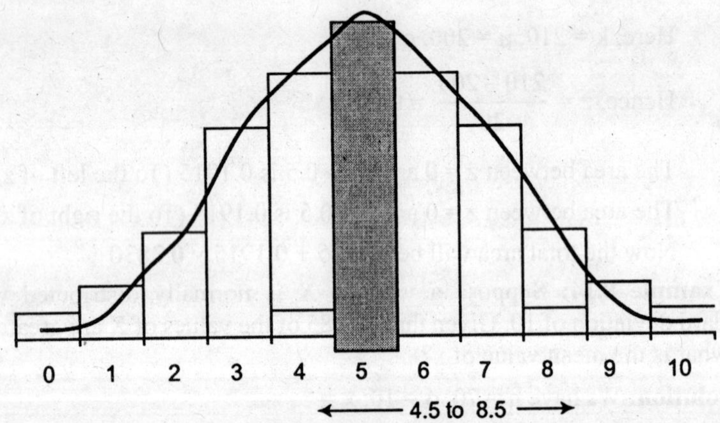

At x = 4.5; $z = \dfrac{x - \mu}{\sigma} = \dfrac{4.5 - 5}{1.581} = -0.32$

At x = 8.5; $z = \dfrac{x - \mu}{\sigma} = \dfrac{8.5 - 5}{1.581} = 2.21$

From the standard normal distribution table, the probability that z will lie between -0.32 and z = 0 (or correspondingly x lies between 4.5 and 5) is 0.1255. Probability that z will lie between 0 and 2.21 (or correspondingly x lies between 5 and 8.5) is 0.4864. Hence total probability that, x lies between 4.5 and 8.5 is 0.1255 + 0.4864 = 0.6119.

Comparing the binomial probability of 0.6123 with this normal approximation of 0.6119, we see that the error in the approximation is less than 0.1% only. So the normal approximation to the binomial distribution is very convenient because it enables us to solve the problem without extensive calculations.

It is noteworthy to mention that the normal probability distribution is so far known to be the only continuous distribution. But we should soon realize that there are other useful continuous distributions like Student's t distribution, Chi-Square distribution and F-distribution, which are of interest to decision makers who solve problems using statistics.

EXERCISES

1. Describe Normal distribution and its importance in statistics.
2. State the properties of Normal distribution.
3. Give a comparison among Binomial, Poisson and Normal distributions.
4. What are the advantages of using standard normal distribution tables over normal distribution tables?
5. Describe the properties of normal distribution.
6. Define Binomial distribution with parameter 'n' and 'p'. Write its mean and variance.
7. Define Poisson distribution with parameter 'λ'. Write its mean and variance.
8. Construct a probability distribution based on the following frequency distribution:

Outcome:	2	4	6	8	10	12	14
Frequency:	24	20	16	10	8	2	1

   Draw a graph of the hypothetical probability distribution and compute the expected value of the outcome.

9. District Fire Officer compiled the following data regarding the number of fire outbreaks in the apartments in different months of year.

Year	Jan	Feb	Mar	Apr	May	Jun	Jul	Aug	Sep	Oct	Nov	Dec
2001	25	30	15	10	10	4	2	2	1	5	8	10
2002	20	25	10	8	5	2	4	0	4	7	10	12

   (a) What is the expected number of fires per month?

   (b) What is the expected number of fires per summer month (Apr, May & Jun)?

10. Past record reveals the following probability of the sale of cake. Cost of preparation of each cake is Rs.8 and the selling price is Rs. 10 each. Due to inadequate facility for storage, any unsold cake is disposed on the next day with a reduced cost of Rs. 6 each. Calculate the number of cakes that the owner will stock for minimising the loss.

Number of cake:	1	2	3	4	5	6	7	8
Probability:	0.01	0.08	0.12	0.24	0.35	0.06	0.10	0.04

11. A publishing company wants to decide how many programs to print for a team's upcoming series. Each program costs Rs.56 for printing and sells for Rs. 82. Any programs unsold at the end of the series must be discarded. The owner of the publishing company has estimated the following probability distribution for programme sales using past sale data:

Programme sold:	30,000	40,000	50,000	60,000
Probability:	0.10	0.30	0.45	0.15

Calculate the number of programmes the owner has to print in order to minimise the expected losses.

12. In a sample survey it was revealed that 55% favoured local political party, 30% favoured national political party and 15% favoured independent candidates. From a randomly selected 10 persons, what is the probability that (a) four favoured local party (b) none favoured national party (c) at least eight favoured independent candidates? [Use binomial distribution]

13. In a sample survey conducted by an NGO, it was revealed that 40% of unemployed youths getting unemployment bonus from the Government are secretly employed somewhere. But the city mayor was not ready to accept this fact. He investigated 10 randomly selected unemployed youths. If the NGO's findings are correct, find the probability that (a) more than 7 have jobs (b) only four investigated youths have jobs?

14. For a binomial distribution with $n = 8$ and $p = 0.3$, find (a) $P(r=6)$ (b) $P(r > 3)$ (c) $P(r < 6)$ (d) $P(r \geq 5)$

15. Given $\lambda = 4.5$, for a Poisson distribution, find (a) $P(x \leq 3)$ (b) $P(x \geq 5)$ (c) $P(x = 6)$

16. A printing company has a reputation of printing of materials almost error free. But due to shortage of manpower recently, it has started exhibiting errors in 0.5% of cases. What is the probability that out

of a batch of 1000 bills (a) None have errors (b) Ten have errors (c) Fifteen have errors.

17. On an average five street dogs, who create traffic obstacles, are caught daily by the municipal authorities. But the commissioner ordered that no more funds can be released for such activities unless the probability of more than three dogs being caught on any day exceeds 0.8. Will the funds be released?

18. Medical emergency centre receives injured patients, but almost all are accident victims. In fact, 96% of the cases handled by the medical emergency staff are hit by accidents. What is the probability that (a) out of 80 patients handled by the medical staff, exactly 8 are not related to accidents (b) none are not due to accidents?

19. Given that a random variable, X has a normal distribution with mean 6.5 and standard deviation 2.8, find (a) $P(4.0 < x < 5.0)$ (b) $P(x > 2.0)$ (c) $P(x < 7.3)$ (d) $P[(x < 3.0) \text{ or } (x > 8.0)]$

20. Johnson and Johnson is developing a kidney dialysis machine, but its managing director faces trouble controlling the variability of the rate at which fluid moves through the device. Medical standards stipulate that the hourly flow be 4 liters, plus or minus 0.1 liter, at 80% of the time. The director observed a prototype machine and found that 68% of the time, the hourly flow is within 0.08 liters of 4.02 liters. Does the prototype satisfy the medical standards?

21. Owner of an electric farm assumed that the weekly demand of CFL bulbs is normally distributed. He knows that on average 100 bulbs are sold weekly and 90% of the time, weekly demand is below 115. (a) What is the standard deviation of this distribution? (b) The owner wants to stock enough bulbs each week so that the probability of running out of bulbs is no higher than 0.05. What is the lowest such stock level?

22. On the basis of past experience, the motor vehicle inspector in a district noticed that 5% of all four wheeled transport vehicles coming in to appear the test for extension of their annual permit fail to pass. Using the normal approximation to the binomial, find the probability that between 7 and 20 of the next 200 vehicles coming for the test will fail in the inspection.

DESIGN OF EXPERIMENTS

We have studied about one or more of the possible outcomes of doing something called trial or experiment. An experiment is an activity that would produce such events. So an experiment is conducted to produce meaningful results those can be useful to the decision makers. The way in which these experiments are designed, called research design, is of utmost importance.

Decisions regarding what, where, when, how much, by what means concerning an enquiry or a research study constitute a research design. A research design is the arrangement of conditions for collection and analysis of data in a manner that aims to combine relevance to the research purpose with economy in procedure. More explicitly, the decisions happen to be in respect of the following enquiries:

- Why is the study being made?
- What is the study about?
- Where will the study be carried out?
- What type of data is required?
- Where can the required data be found?
- What periods of time will the study include?
- What will be the sample design?
- What techniques of data collection will be used?
- How will the data be analysed?
- In what style will the report be prepared?

In brief, research design must, at least, contain- (a) a clear statement of the research problem; (b) procedures and techniques to be used for gathering information; (c) the population to be studied; and (d) methods to be used in processing and analyzing data.

FEATURES OF A GOOD DESIGN

A good design must be:

- Flexible, appropriate, efficient, economical etc.
- Should minimize bias and maximize reliability of data collected and analysed.
- Should give smallest experimental error.
- Should yield maximal information.

Besides, the availability of time, money, skills of the research staff and the means of obtaining the information must be given due weightage while working out the relevant details of the research design such as experimental design, survey design, sample design and alike.

IMPORTANT CONCEPTS RELATING TO RESEARCH DESIGN

A brief explanation of different terms concerned with research design are enumerated below:

1. **Dependent and Independent Variables:** The concepts which take on different quantitative values is called a variable e.g. number of Basophiles, WBC count, height, weight etc. Qualitative phenomena (or attributes) are also quantified on the basis of their presence or absence e.g. intelligence, courage, breeding capacity etc. Continuous variables can take up any value in decimal points e.g. age, height, etc, whereas discrete variables take up only integer values e.g. number of plants, number of seeds etc. Dependent variable is that which depends upon or is a consequence of the other variable and the variable that is antecedent to the dependent variable is termed as an independent variable.
2. **Extraneous Variable:** Independent variable affecting the dependent one, but is not related to the purpose of study is termed as extraneous variable. Children's self concept in understanding statistics will affect their result in examination, but intelligence surely plays an important role as an extraneous variable. Hence, study must always be so designed that the effect upon the dependent variable is attributed entirely to the independent variable and not to some extraneous variable or variables.
3. **Control:** The technical term 'control' is used to refer to restrain experimental conditions. It helps in minimizing the effects of extraneous independent variables.
4. **Confounded Relationship:** When the dependent variable is not free from the influence of extraneous variable, the relationship between the dependent and independent variables is said to be confounded by an extraneous variable.

5. **Research Hypothesis:** It is a predictive statement that relates an independent variable to a dependent variable and is to be tested by scientific methods.
6. **Experimental and Control Groups:** In an experimental hypothesis-testing research when a group is exposed to usual conditions, it is termed as 'control group', but when the group is exposed to some special conditions, it is termed as 'experimental group'.
7. **Treatment:** Different conditions, under which experimental and control groups are put, are usually referred to as 'treatments'.
8. **Experiment:** Process of examining the truth of a statistical hypothesis, relating to some research problem, is known as an experiment.
9. **Experimental Unit:** Pre-defined plots or the blocks, where different treatments are used are known as experimental units.

BASIC PRINCIPLES OF EXPERIMENTAL DESIGN

Three-fold principles of experimental designs are: Principle of Replication; Principle of Randomization and Principle of Local Control, which are enumerated below.

Principle of Replication

- The experiment should be repeated more than once.
- Each treatment is applied in many experimental units instead of one.
- Replication is introduced in order to increase the precision of a study.
- Entire experiment can be repeated several times for better results.
- For example, it is required to examine the effect of two varieties of rice. So we may divide the field into two parts and grow one variety in one part and the other variety in the other part. We can then compare the yields of the two parts and draw conclusion on that basis. But if the principle of replication is applied, then the entire field is divided into several parts, growing one variety in half of these parts, and the other variety in the remaining parts. We can then collect data of yield of the two varieties and draw conclusion by comparing the same. The result so obtained will be more reliable than what would have been obtained by without applying the principle of replication.

Principle of Randomization

- This provides protection against the effect of extraneous factors by randomization.

- Here the experiment or plan is so designed that the variations caused by extraneous factors can all be combined under the general heading of 'chance'.
- For example: if one variety of rice is grown in first half of the parts of field and the other variety is grown in the other half, then it is just possible that the soil fertility may be different in the first half in comparison to the other half. If so, then the results would be unrealistic. In such situation, the variety of rice may be assigned to grow in different parts of the field on the basis of some random sampling technique. e.g. randomization principle may be applied and effects of extraneous variable (soil fertility differences in the given case) may be guarded off.
- Here a better estimate of the experimental error can be had.

Principle of Local Control

- Here the extraneous factor, the known source of variability, is made to vary deliberately over as wide a range as necessary and this needs to be done in such a way that the variability it causes can be measured and hence eliminated from the experimental error.
- The field is first divided into several homogenous parts, known as blocks and then each such block is divided into parts equal to number of treatments. Then the treatments are randomly assigned to these parts of a block. Dividing the field into several homogenous parts is called 'blocking'. In general, blocks are the levels at which extraneous factor is fixed, so that its contribution to the total variability of the data can be measured by two-way analysis of variance. Three components of total variability are components attributed to treatments (varieties of rice in the instant case), the extraneous factor (soil fertility) and experimental error.
- Thus the variability due to extraneous can be eliminated from the experimental error through the principle of local control.

EXPERIMENTAL DESIGNS

Experimental design refers to the framework or structure of an experiment and as such there are several experimental designs.

(A) **Informal Experimental Designs:** Less sophisticated form of analysis is used based on differences in magnitudes. This is categorized as:

1. Before-and-After Without Control Design

2. After-only With Control Design
3. Before-and-After With Control Design

1. Before-and-After Without Control Design

A single test group or area is selected and the dependent variable is measured before introduction of the treatment. Dependent variable is again measured after introduction of the treatment. Effect of the treatment is equal to level of phenomenon after treatment minus that before treatment. Here the difficulty is that with the passage of time considerable extraneous variations may be there in its treatment effect.

Test Area:

Level of phenomenon before treatment (X) —Treatment introduced→ Level of phenomenon after treatment (Y)

Treatment effect = (Y) – (X)

2. After-only With Control Design

Two groups or areas (test area and control area) are selected and the treatment is introduced into the test area only. Dependent variable is measured in both the areas at the same time. Treatment effect is assessed by subtracting the value of the dependent variable in the control area from its value in the test area. Here it is assumed that the two areas are identical with respect to their behavior towards the phenomenon considered, or else there is every possibility of extraneous variation entering into the treatment effect. But data can be collected in such a design without introduction of problems with the passage of time. So the design is superior to before-and-after without control design.

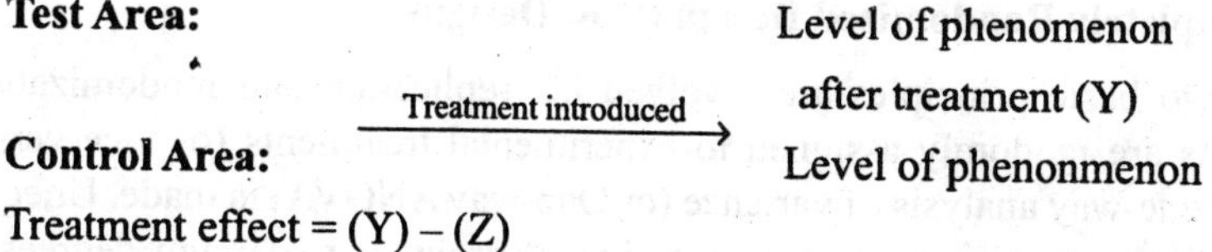

Treatment effect = (Y) – (Z)

3. Before-and-After With Control Design

Two areas are selected and dependent variable is measured in both areas for an identical time period before the treatment. Treatment is applied. Then dependent variable is measured in both for an identical time-period after the introduction of the treatment. Treatment effect is determined by subtracting the change in the dependent variable in control area from the change in that in the test area. This design is superior to the above two, because it avoids extraneous variations resulting both from the passage of time and from non-comparability of the test and control areas. Sometimes, for want of historical data, time or a comparable control area, any one of the above two informal designs is selected.

Test Area:

Level of phenomenon before treatment (X) —Treatment introduced→ Level of phenomenon after treatment (Y)

Control Area:

Level of phenomenon without treatment (A) — Level of phenomenon without treatment (Z)

$$\text{Treatment effect} = (Y - X) - (Z - A)$$

(B) Formal Experimental Designs: These offer relatively more control and use precise statistical procedures for analysis. This is categorized as:

1. Completely Randomized Design (C.R. Design)
   (i) Two Group Simple Randomized Design
   (ii) Random Replications Design
2. Randomized Block Design (R.B. Design)
3. Latin Square Design (L.S. Design)
4. Factorial Design
   (i) Simple Factorial Design
   (ii) Complex Factorial Design

1. 5. Split Plot Design (S.P. Design)

1. Completely Randomized Design (C.R. Design)

Here two basic principles are involved i.e. replication and randomization. Subjects are randomly assigned to experimental treatments (or vice-versa). Use of one-way analysis of variance (or One-way ANOVA) is made. Unequal replications can also work in this design. It provides maximum degrees of freedom to the error. Design is generally used in homogeneous experimental areas. Basically, when all the variations due to uncontrolled extraneous factors are included under the heading of chance variation, this C.R. design is referred to. Two forms of such a design are described below:

(i) Two-group Simple Randomized Design

As principle of randomization applies to this design, items, after being selected randomly from the population, are randomly assigned to the experimental and control groups. This design yields two groups as representatives of the population. Different treatments of the independent variable are applied to the two groups (experimental and control). The advantage of this design is that it is simple and randomizes the differences among those simple items and is quite common in behavioral sciences. But the drawback is the individual differences among those conducting the treatments are not eliminated i.e. it does not control the extraneous variable and as such the result of the experiment may not reveal the correct picture. Take for example- Two groups of animals are randomly selected and randomly administered two different treatments viz- Vitamin A and Vitamin B. It is hypothesized that Vitamin A yields better growth. To test this, each group is tested before and after administration of the vitamins and then amount of gain in growth for two groups are compared to accept or reject the hypothesis. This is an illustration of two-group randomized design, where individual differences among the animals are being randomized. But the differential effects of the extraneous independent variables (in this case individual differences between environmental conditions in which animals grow) are not controlled. The following diagram illustrates.

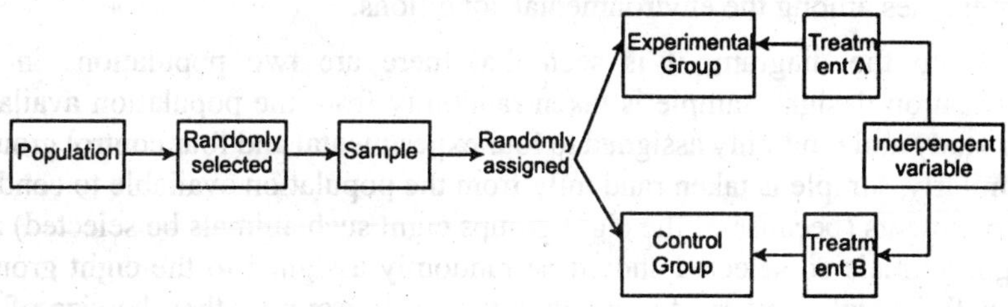

(ii) Random Replication Design

This random replication design eliminates the limitation of the two-group randomized design. In the above example, the environmental conditions under which animals grow are completely ignored, i.e. extraneous variable was not controlled. But in a random replications design, the effect of such differences are minimized (or reduced) by providing a number of repetitions for each treatment. Each repetition is technically called a 'replication'.

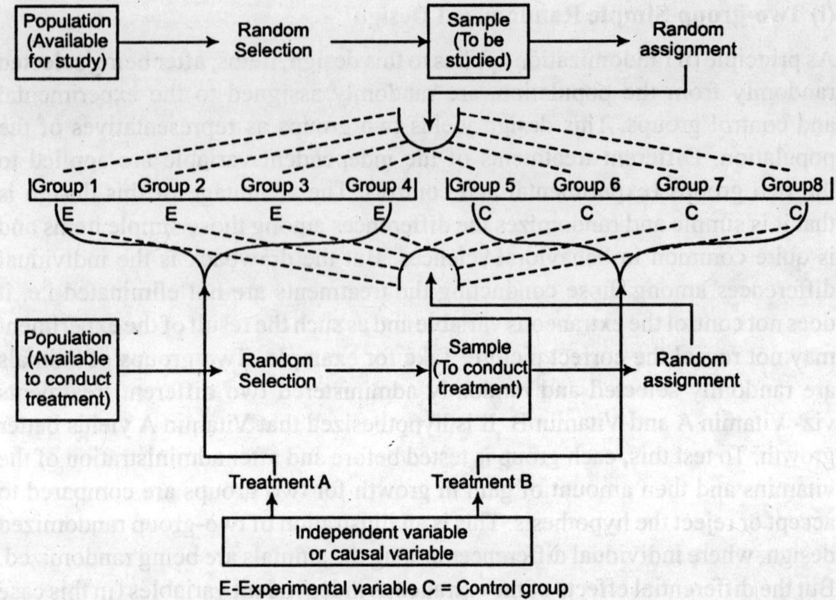

It serves two purposes: it provides controls for the differential effects of the extraneous independent variables and secondly, it randomizes any individual differences among the environmental conditions.

From the diagram, it is seen that there are two populations in the replication design. Sample is taken randomly from the population available for study and randomly assigned to four experimental and four control groups. Similarly, sample is taken randomly from the population available to conduct experiments (because of the eight groups eight such animals be selected) and eight animals so selected should be randomly assigned to the eight groups. Usually, equal number of items is put in each group so that the size of the group is not likely to affect the result of the study. Variables relating to both population characteristics are assumed to be randomly distributed among the two groups. Hence, this random replication design is, in fact, an extension of the two-group simple randomized design.

2. Randomized Block Design (R.B. Design)

In this design, principle of local control can be applied along with the other two principles of experimental designs. It is an improvement over the C.R. Design. Here, the subjects are first divided into groups, known as blocks, such that within each group the subjects are relatively homogenous in respect to some selected variable. The number of subjects in a given block would be equal to the number of treatments and one subject in each block would be randomly assigned to each treatment. On the whole, blocks are the levels at which extraneous factor is kept fixed, so that measurement of the total variability of its contribution to data can be made. The main feature of the R.B. design is that in this each treatment appears the same number of times in each block. R.B. design is analyzed by the two-way analysis of variance (Two-way ANOVA) technique. (ANOVA is explained in Chapter 12)

The following example illustrates. Let us suppose five students are selected each one from five I.Q. blocks and are subjected to four different forms of a standardized test in Biotechnology and following are the scores.

If each student separately randomized the order in which he or she took the four tests, then the design of experiment is referred to as R.B. design.

	Very low I.Q.	**Low I.Q.**	**Average I.O.**	**High I.W.**	**Very High I.Q.**
	Student X	**Student Y**	**Student Z**	**Student P**	**Student Q**
Form A	83	68	58	72	74
Form B	91	69	55	71	82
Form C	87	74	52	70	85
Form D	94	78	61	66	72

The main purpose of this randomization is to take care of such possible extraneous factors (say tiredness) or perhaps the experience gained from repeatedly taking the test.

3. Latin Square Design (L.S. Design)

This type of experimental design is mostly used in agricultural investigations, where nature plays a vital role. For example, if effect of five different varieties of fertilizers on the yield of a crop is to be studied, then the blocks, where soil with varying fertility levels is also to be considered, otherwise the results may not be dependable, because the output will be not only the effect of only fertilizers, but also initial fertility level may contribute to such outputs. Similarly, there may be impact of varying seeds on the yield too. For this, L.S. Design has an answer, where there are two major extraneous factors such as varying soil fertility and varying seeds. Here each fertilizer appears five times but is used only once in each row and in each column of the design. That means, treatments are so allocated among the plots that no treatment occurs more than once in any one row or any one column. The two blocking factors may be represented through rows and columns (one through rows and the other through columns). The following diagrammatic representation explains the design for five types of fertilizers A, B, C, D & E and the two blocking factors are varying soil fertility (I to V) & varying seeds(S_1 to S_5).

		FERTILITY LEVEL				
		I	II	III	IV	V
Seed differences	S1	A	B	C	D	E
	S2	B	C	D	E	A
	S3	C	D	E	A	B
	S4	D	E	A	B	C
	S5	E	A	B	C	D

The field is divided into as many blocks as there are varieties of fertilizers and then each block is again divided into as many parts as there are varieties of fertilizers in such a way that each of the fertilizer variety is used in each of the block (whether column-wise or row-wise) only once. This design is very similar to the two-way ANOVA technique.

The advantage of L.S. Design is that it enables differences in fertility gradients in the field to be eliminated in comparison to the effects of different varieties of fertilizers on the yield of the crop. The limitation of this design is that although each row and each column represents equally all fertilizer varieties, there may be considerable difference in the row and column means both up and across the field. That means, in L.S. Design it is assumed that there is no interaction between treatments and blocking factors. This demerit can be overcome by taking the means of rows and columns equal to the field mean by adjusting the results. Another limitation is that it requires number of rows, columns and treatments to be equal, thereby reducing the utility of this design. No degrees of freedom are available for the mean square error in case of 2 × 2 L.S. Design and hence the design cannot be used. Again, if treatments are more than 10, then each row and each column will be larger in size so that the rows and columns may not be homogenous, thereby making application of principle of local control ineffective. Therefore, L.S. Design of orders 5 × 5 to 9 × 9 are commonly used.

4. Factorial Designs

This type of design is used in experiments where effects of varying more than one factor are required to be determined. Several economic and social phenomena where a large number of factors affect a particular problem have drawn the attention of importance of this type of design. This is of two types (i) Simple Factorial Design (ii) Complex Factorial Design.

(i) Simple Factorial Design: Here, effects of varying two factors on the dependent variable is considered, as against effects of more than two factors in case of complex factorial design. Simple factorial design is often termed as two-factor-factorial design and complex factorial design as multi-factor-factorial design. Simple factorial design may either be 2 × 2, 3 × 4 or 5 × 3 or the like type of simple factorial design. The following illustrative examples make the things more clear.

Illustration 1: One 2 X 2 simple factorial design is depicted below.

		Experimental Variable	
		Treatment A	Treatment B
Control Variables	Level I	Cell 1	Cell 3
	Level II	Cell 2	Cell 4

In this design, experimental variable is the independent variable, which is manipulated and control variable is the extraneous variable to be controlled by homogeneity. There are two levels of control variable and two treatments of the experimental variable. So there are four cells into which the sample is divided, each one being accountable to one treatment or experimental condition. Then subjects are assigned at random to each treatment in the same manner as in a randomized group design. Means of four cells as well as means of rows and columns are found out. Means of different cells represent the mean scores for the dependent variable and the column means in the given design are termed the main effect for treatments without taking into account any differential effect that is due to the level of the control variable. So also, the row means in the design are the main effects for levels without regard to treatment. Hence, this design helps in studying the main effects of treatments and levels both. One can also examine the interaction between treatments and levels, by which it may be possible to say whether treatment and levels are independent of each other or not. The following 2 × 2 simple factorial design study explains this truth.

STUDY I				
		Experimental Variable		
		Treatment A	Treatment B	Row Mean
Control Variables (Intelligence)	Level I (Low)	10	22	16
	Level II (High)	40	36	38
	Level II (High)	25	29	

STUDY II				
		Experimental Variable (Training)		
		Treatment A	Treatment B	Row Mean
Control Variables (Intelligence)	Level I (Low)	12	15	13.5
	Level II (High)	36	45	40.4
	Column Mean	24	30	

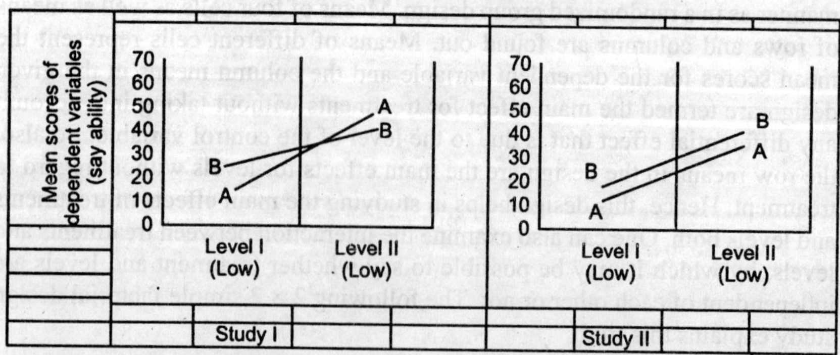

The above graphs indicate that in case of Study I, interaction between treatment and level exists; in other words, treatment and level are not independent of each other. But in case of Study II, there is no interaction effect, which means that treatment and level are relatively independent of each other.

The 2 × 2 design is not only confined in the manner as explained above i.e. comprising one experimental variable and one control variable, but it may also be of the type having two experimental variables or two control variables. For example, a professor wants to compare the effect of class size and introduction of new technique on learning of research methodology. As he conducts one study using 2 × 2 simple factorial design, his graphic form would take the following form:

		Experimental Variable I (Class Size)	
		Small	Usual
Experimental Variable II	New		
(Instruction technique)	Usual		

On the contrary, if the professor wants to compare males & females and senior & junior students in the college as they relate to the knowledge of research methodology, in that case a 2 × 2 simple factorial design would occur, wherein both the variables are control variables as no manipulation is involved in respect of both the variables.

Illustration 2: (4 × 3 Simple Factorial Design)

Graphically, the 4 × 3 simple factorial design can be represented as shown below, where four treatments of the experimental variable and three levels of control variable are included.

4 × 3 SIMPLE FACTORIAL DESIGN				
Control Variable	Experimental Variable			
	Treatment A	Treatment B	Treatment C	Treatment D
Level I	Cell 1	Cell 4	Cell 7	Cell 10
Level II	Cell 2	Cell 5	Cell 8	Cell 11
Level III	Cell 3	Cell 6	Cell 9	Cell 12

This simple factorial design model comprises four treatments of the experimental variable viz., A, B, C & D and three levels of the control viz., I, II & III and has thus 12 different cells as shown above. Hence, a 2 × 2 simple factorial design can be generalized to any number of treatments and levels and accordingly such (__ × __) design can be formulated. In such design, the means of columns provide an estimate of the main effects for treatments and means of rows provide an estimate for levels. Interaction between treatments and levels can also be determined by such design.

(ii) Complex Factorial Design: Complex factorial design is meant for experiments involving more than two factors at a time, where three or more independent variables are used. For example, three factors with one experimental variable having two treatments and two control variables, each one of which has two levels, the design is termed as 2 × 2 × 2 complex factorial design and this contains a total of eight cells as shown in the figure below.

2×2×2 COMPLEX FACTORIAL DESIGN					
		Experimental Variable			
		Treatment A		Treatment B	
		Control Variable 2 Level I	Control Variable 2 Level II	Control Variable 2 Level I	Control Variable 2 Level II
Control Variable 1	Level I	Cell 1	Cell 3	Cell 5	Cell 7
	Level II	Cell 2	Cell 4	Cell 6	Cell 8

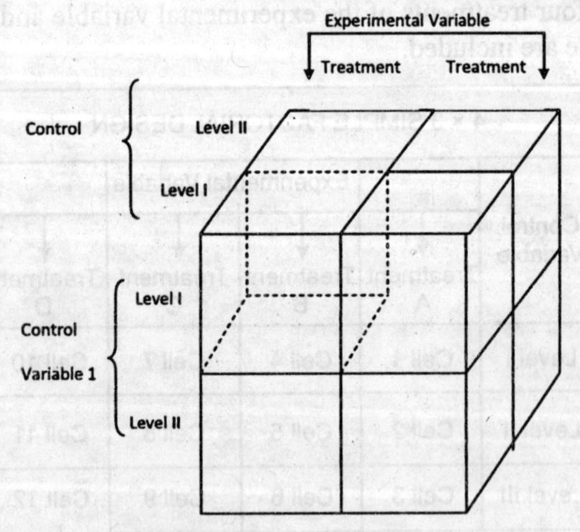

As it can be seen in the diagram above, the dotted line box corresponds to Cell 1 of the 2 × 2 × 2 design and is for Treatment A; level 1 of the control variable 1 and level 1 of the control variable 2. Hence, it is possible to determine the main effects for three variables i.e., one experimental and two control variables. It is also possible to determine the interactions between each possible pair of variables (such interactions are called 'first order interactions') and interaction between variable taken in triplets (such interactions are called Second Order interactions'). In case of a 2 × 2 × 2 design, further first order interactions are possible:

EV × CV 1 – Experimental Variable with Control Variable 1

EV × CV 2 – Experimental Variable with Control Variable 2

CV 1 × CV 2 – Control Variable 2 with Control Variable 2

There will be one second order interaction as well in the given design i.e. between all three variables – EV × CV 1 × CV 2.

It is necessary to compare the combined mean of data in cells 1, 2, 3 & 4 for Treatment A with the combined mean of data in cells 5, 6, 7 & 8 for Treatment B, in order to determine the main effects for the experimental variable. Thus, the main effect for experimental variable, independent of control variable 1 and variable 2 is obtained. Similarly, the main effect for control variable 1, independent of experimental variable and control variable 2, is obtained, if comparison is made between combined mean of data in cells 1, 3, 5 & 7 and combined mean of data in cells 2, 4, 6 & 8 of the 2 × 2 × 2 factorial design. So also, the main effect for control variable 2 independent of experimental variable and control variable 1 can be determined, if the combined mean of data in cells 1, 2, 5 & 6 are compared with the combined mean of data in cells 3, 4, 7 & 8.

To obtain the first order interaction, say, for EV × CV 1, control variable 2 must be ignored, for which purpose 2 × 2 design may be developed from the 2 × 2 × 2 design by comparing the data of the relevant cells of the latter design as shown in Fig below.

		Experimental variables	
		Treatment A	**Treatment B**
Control	Level II	Cells 1, 3	Cells 5, 7
Variable 1	Level II	Cells 2, 4	Cells 6, 8

On similar lines, other first order interactions can be determined too. As described above, the analysis of the first order interaction is essentially a simple factorial analysis as only two variables are considered at a time and the remaining one is ignored. But the analysis of the second order interaction would not ignore one of the three independent variables in case of a 2 × 2 × 2 design. The analysis would be termed as a complex factorial analysis.

It is important to note that the complex factorial design need not necessarily be of 2 × 2 × 2 type design, but can be generalized to any number of independent variables included in a complex factorial design, the higher the order of the interaction analysis possible. But the overall task goes on becoming more and more complicated with the inclusion of more and more independent variables in the design.

Advantages of Factorial Design

1. It provides equivalent accuracy (as happens in case of experiments with only one factor) with less labour and as such is a source of economy.
2. It is possible to determine the main effects of two (in simple factorial design) or more (in complex factorial design) factors (or variables) in one single experiment.
3. It also facilitates various other comparisons of interest e.g., it gives information about such effects which cannot be obtained by treating one single factor at a time. The determination of interaction effects is possible in case of factorial design.

B-5 Split Plot Design (S.P. Design)

Split-plot designs were originally developed by Fisher (1925) for use in agricultural experiments. In simple terms, a split-plot experiment is a blocked experiment, where the blocks themselves serve as experimental units for a subset of the factors. Thus, there are two levels of experimental units. The blocks are referred to as whole plots, while the experimental units within blocks are called as subplots, split plots or split units. There are two levels of randomization corresponding to the two levels of experimental units. One randomization is conducted to determine the assignment of block-level treatments to whole plots. Then, as always in a blocked experiment, a randomization of treatments to split-plot experimental units occurs within each block or whole plot.

As a simple illustration, consider a study of the effects of two irrigation methods (factor A) and two fertilizers (factor B) on yield of a crop, using four available fields as experimental units. In this investigation, it is not possible to apply different irrigation methods (factor A) in areas smaller than a field, although different fertilizer types (factor B) could be applied in relatively small areas. For example, if we subdivide each whole plot (field) into two split plots, each of the two fertilizer types can be applied once within each whole plot, as shown in figure below. In this split-plot design, a first randomization assigns the two irrigation types to the four fields (whole plots); then within each field, a separate randomization is conducted to assign the two fertilizer types to the two split plots within each field.

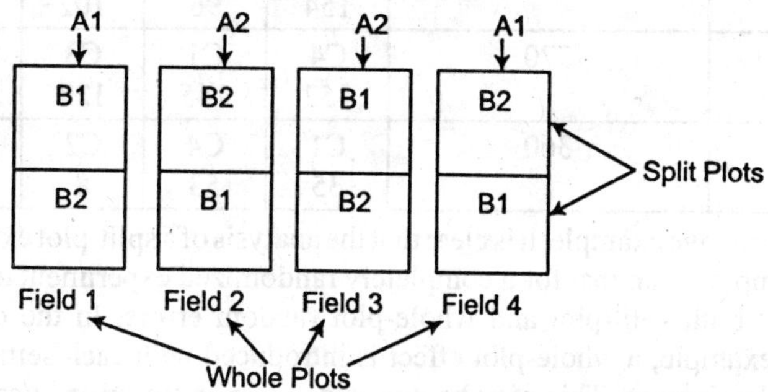

In industrial experiments, factors are often differentiated with respect to the ease with which they can be changed from experimental run to experimental run; that means, with one easy-to-change factor and one hard-to-change factor. This may be due to the fact that a particular treatment is expensive or time-consuming to change, or it may be due to the fact that the experiment is to be run in large batches and the batches can be subdivided later for additional treatments. Take for example an experiment designed to study the corrosion resistance of steel bars treated with four coatings, C1, C2, C3, and C4, at three furnace temperatures, 3600°C, 3700°C, and 3800°C. Furnace temperature is the hard-to-change factor because of the time it takes to reset the furnace and reach a new equilibrium temperature. Once the equilibrium temperature is reached, four steel bars with randomly assigned coatings C1, C2, C3, and C4 are randomly positioned in the furnace and heated. The layout of the experiment as performed is given in the table below. It may be noticed that each whole-plot treatment (temperature) is replicated twice and that there is just one complete replicate of the split-plot treatments (coatings) within each whole plot. Thus, we have six whole plots and four subplots within each whole plot.

Whole Plot	Temperature (°C)	Coating (Randomized order)			
1	360	C2 75	C3 84	C1 68	C4 91
2	370	C1 66	C3 89	C4 88	C2 95
3	380	C3 148	C1 157	C2 130	C4 210

Whole Plot	Temperature (°C)	Coating (Randomized order)			
4	380	C4 154	C3 96	C2 102	C1 108
5	370	C4 152	C1 145	C3 128	C2 140
6	360	C1 35	C4 53	C2 8	C3 47

From the above example, it is clear that the analysis of a split-plot experiment is more complex than that for a completely randomized experiment due to the presence of both split-plot and whole-plot random errors. In the corrosion resistance example, a whole-plot effect is introduced with each setting or re-setting of the furnace. This may be due, e.g., to operator error in setting the temperature, to calibration error in the temperature controls, or to changes in ambient conditions. Split-plot errors might arise due to lack of repeatability of the measurement system, to variation in the distribution of heat within the furnace, to variation in the thickness of the coatings from steel bar to steel bar, and so on.

More Examples on Split Plot Design

1. Many times packaged-food manufacturers wish to develop an optimal formulation of a cake mix. Because cake mixes are made in large batches, the ingredient factors (flour, shortening, and egg powder) are hard to change. Many packages of cake mix are produced from each batch, and the individual packages of cake mix can be baked using different baking times and baking temperatures. Here, the easy-to-change, split-plot factors are time and temperature.
2. Another example is involving large batches and the subsequent processing of sub-batches. In this instance, a company may wish to study the effect of various factors on the swelling of a wood product after it has been saturated with water and allowed to dry. A batch is characterized by the type and size of wood being tested, the amounts of two additives used, and the level of moisture saturation. Batches can later be subdivided for processing, which is characterized by three easy-to-change factors, i.e., process time, pressure, and material density.
3. Take one example, an experiment involving four prototype combustion engine designs and two grades of gasoline. The engineers' preference is to make up the four motor designs and test both of the gasoline

types while a particular motor was on a test stand. Here the hard-to-change factor is motor type and the easy-to-change factor is gasoline grade.

Uses of Split-Plot Design

- Variable natural plot sizes for different factors in a factorial.
- In situations where a second factor is brought into an experiment to increase its scope.
- In situations where it is known that larger differences will occur for some factors than others.
- Where greater precision is desired for some factors than others.

CONCLUSION

As experimental design forms the vital step in any research work, appropriate type of design must be decided prior to collection and analysis of data. Due weightage should be given to various points such as the type universe and its nature, the objective of study, source list of the sampling frame, desired level of accuracy etc when taking decision in respect of the design of the research project.

EXERCISES

1. Explain the meaning of the following:
   (i) Extraneous variable
   (ii) Confounded relationship
   (iii) Research hypothesis
   (iv) Experimental group and control group
   (v) Treatments
2. What is research design? Discuss the basis of stratification to be employed in sampling public opinion on inflation.
3. Explain the principles of Replication, Randomization and Local control used in experimental designs.
4. Explain and illustrate the following research designs:
   (i) Latin square design
   (ii) Random replications design
   (iii) Simple factorial design and Complex factorial design
   (iv) Informal experimental design
   (v) Two group simple randomized design

5. Explain the meaning and significance of Research Design.
6. "Research design in exploratory studies must be flexible but in descriptive studies, it must minimize bias and maximize reliability."- Discuss
7. Explain Split-Plot Design with suitable examples. State its uses.

SAMPLING DISTRIBUTION STATISTICAL INFERENCE

We have already studied how sample data can be used to calculate statistics such as the mean and standard deviation. We also we studied how samples are taken from the population. But the question is- Are the statistics of all samples collected from the population same? In fact, the statistics computed from several samples need not be same. The statistics vary from sample to sample.

Let us say, from a population of 10,00,000 persons living in a city, the mean height of randomly selected ten samples each consisting of say 25 persons are calculated. We will see that the mean and standard deviation of each of these ten samples would be different. *So a probability distribution of all the possible means of the samples is a distribution of the sample means, which is called as sampling distribution of the mean.* Similarly, we may have sampling distribution of median, standard deviation and proportion etc.

Consider only the sample means. It is highly unlikely that all of these sample means would be the same. We see some variability in our observed means. This variability in the sample means results from *sampling error* due to chance. The ten samples as stated above will have ten sample means, which differ from each other characterised by their standard deviation, called as standard deviation of sample means. Statisticians term this standard deviation of sample means as *Standard Error of Mean*. Similarly we can have standard deviation of sample medians called *standard error of median* and standard deviation of sample proportions called as *standard error of proportion.*

The term "standard error" conveys a specific meaning. The standard error indicates not only the size of the chance error that has been made, but also the accuracy we are likely to get if we use a sample statistic to estimate a population parameter. If sample means are less spread, it has a small standard error and is thus a better estimator of the population mean than a distribution of sample means that is widely spread and has a larger standard error.

CENTRAL LIMIT THEOREM

When sampling is from a normal population, the means of samples drawn from such a population are themselves normally distributed. But when sampling is not from a normal population, the size of the sample plays a very crucial role. When n is small, the shape of the distribution will depend largely on the shape of the parent population, but as n becomes large ($n > 30$), the shape of the sampling distribution becomes more and more like a normal population, irrespective of the shape of the parent population. In other words, even if the parent population is not normal, then also the means of the samples collected from this non-normal population are Approximately Normally Distributed (AND). The theorem which explains this sort of relationship between the shape of the population distribution and the sample distribution of the mean is known as Central Limit Theorem. This theorem is by far the most important theorem in statistical inference. It assures that the sampling distribution of the mean approaches normal distribution as the sample size increases. In formal terms, we may say that population having mean μ and finite variance σ^2 approaches the normal distribution with mean μ and variance $\frac{\sigma^2}{n}$ as n goes to infinity.

The significance of the central limit theorem lies in the fact that it permits us to use sample statistic to make inference about population parameter without knowing anything about the shape of the frequency distribution of that population other than what we can get from the sample. This is the basic building block in all statistical analyses. In fact decision makers use normal distribution as an approximation to the sampling distribution whenever the sample size is more than 30.

The following illustration explains the theorem clearly.

Let us say, in a particular soil moisture condition; four varieties of earthworms are found having lengths of 6 cm, 9 cm, 12 cm and 15 cm. Suppose we select earthworm samples of any two lengths from the population (with replacement). We will now calculate the standard error.

In the random sample of two varieties of earthworms (with replacement), the possible samples would be as follows:

	6	9	12	15
6	(6,6)	(6,9)	(6,12)	(6,15)

9	(9,6)	(9,9)	(9,12)	(9,15)
12	(12,6)	(12,9)	(12,12)	(12,15)
15	(15,6)	(15,9)	(15,12)	(15,15)

The means of these samples are respectively:

	6	7.5	9	10.5
	7.5	9	10.5	12
	9	10.5	12	13.5
	10.5	12	13.5	15
Total:	33	39	45	51

The mean of sampling distribution of means is given by:

$$\bar{x} = \frac{33+39+45+51}{16} = 10.5$$

Now let us calculate the values of the various $(x - \bar{x})$, which are given below:

6 - 10.5	7.5 - 10.5	9 - 10.5	10.5 - 10.5
7.5 - 10.5	9 - 10.5	10.5 - 10.5	12 - 10.5
9 - 10.5	10.5 - 10.5	12 - 10.5	13.5 - 10.5
10.5 - 10.5	12 - 10.5	13.5 - 10.5	15 - 10.5

Which on simplification gives:

–4.5	–3	–1.5	0
–3	–1.5	0	1.5
–1.5	0	1.5	3
0	1.5	3	4.5

Now let us calculate their squares, which are given below:

	20.25	9	2.25	0
	9	2.25	0	2.25
	2.25	0	2.25	9
	0	2.25	9	20.25
Total =	31.5	13.5	31.5	13.5

The variance of sampling distribution of mean is given by

$$\frac{31.5+13.5+13.5+31.5}{16} = 5.625$$

The Standard Deviation of Sampling Distribution of Means = Standard Error. Hence,

S.E. = √(Variance) = √(5.625) = 2.372

Now let us see the population figures. Because, population is assumed to be normally distributed, the population mean μ is given by,

$\mu = \frac{6+9+12+15}{4} = 10.5$, it is equal to the mean of sampling distribution of mean.

Similarly, the variance of the population figures σ^2 is given by,

$$\sigma^2 = \frac{(6-10.5)^2 + (9-10.5)^2 + (12-10.5)^2 + (15-10.5)^2}{4} = 11.25$$

Now the standard error is obtained by, S.E.

$$= \frac{\sigma}{\sqrt{n}} = \sqrt{\frac{\sigma^2}{n}} = \sqrt{\frac{11.25}{2}} = \sqrt{5.625} = 2.372$$

= Standard deviation of sampling distribution of the means, found out earlier.

So it is concluded that:

(i) The Mean of the Sampling Distribution of Means is equal to the Mean of the Population.

(ii) Variance of the Sampling distribution of Means is equal to the Variance of the Population divided by the size of sample or σ^2 / n, where σ^2 is the variance of the population.

(iii) The Standard Error of the Sampling Distribution of Means which is equal to the Standard Deviation of the Population divided by the square root of sample size or $\sigma/\sqrt{n}$.

The Standard Errors of some well-known statistics for large samples are given below. Here we consider n as sample size, σ^2 as population variance and P as the population proportion.

Statistic	Standard Error (S.E.)
Sample Mean = $\bar{x}$	$\frac{\sigma}{\sqrt{n}}$

Statistic	Standard Error (S.E.)
Sample Standard Deviation = s	$\frac{\sigma}{\sqrt{2n}}$
Sample variance = s^2	$\frac{\sigma^2}{\sqrt{\frac{2}{n}}}$
Sample Proportion = p	$\sqrt{\frac{PQ}{n}}$
Difference between two sample means ($\overline{x}_1$ & $\overline{x}_2$)	$\sqrt{\frac{\sigma_1^2}{n_2}+\frac{\sigma_2^2}{n_1}}$
Difference between two sample proportions (p_1 & p_2)	$\sqrt{\frac{P_1Q_1}{n_1}+\frac{P_2Q_2}{n_2}}$

The equation for standard error of mean is designed for situations in which the population is infinite or in which we sample from a finite population with replacement (that is, after each item is sampled, it is put back into the population before the next item is chosen, so that the same item can possibly be chosen more than once). But the decision makers face many situations where the population is finite or limited in size, e.g. employees in a company, costumers of a product, number of students in an institution, etc. Hence, the above equation for standard error of mean needs to be modified for finite population or sampling without replacement, which is given by:

$$\text{Standard Error of Mean for Finite Population} = \frac{\sigma}{\sqrt{n}} \times \sqrt{\frac{N-n}{N-1}}$$

Where, N = size of the population

n = size of the sample

The new term on the right-hand side, which we multiply by our original standard error, is called the *finite population multiplier* = $\sqrt{\frac{N-n}{N-1}}$. When the size of the sample is very small in comparison to a very large population, this multiplier has little effect. So the general rule is *when the sampling fraction is less than 0.05, the finite population multiplier need not be used.*

Example 10.1: From a population of 130 items with a mean of 102 and a standard deviation of 16, 60 items are chosen. Calculate the standard error of mean(SEM).

Solution: Here N = 130; μ = 102; σ = 16; n = 60

Fraction of sample to the population

$$= \frac{60}{130} = 0.46, \text{ which is greater than} 0.05$$

So the finite population multiplier is to be used.

$$\text{SEM} = \frac{\sigma}{\sqrt{n}} \times \sqrt{\frac{N-n}{N-1}} = \frac{16}{\sqrt{60}} \times \sqrt{\frac{130-60}{130-1}} = 2.065 \times 0.7366 = 1.521$$

Difference between Standard Deviation and Standard Error

The main difference between the two lies in the fact that the former concerns with original values and the latter with the statistic computed from the samples of original values. The standard deviation of the distribution of sample means is called the standard error of the mean. The standard deviation applies to the distribution of items around their averages, while the Standard Error of the mean applies to the distribution of averages of samples around the true average of the population.

USES OF STANDARD ERROR

1. Standard error provides an idea about the degree of precision of a sample or in other words it is an indication about the extent to which a sample is reliable.
2. It is used in test statistic to know whether a sample is drawn from a known population or not.
3. It is used to determine the size of the sample.
4. Standard error helps in determining the limits within which a parameter value is expected to lie. The sample distribution tends to normal distribution, if the sample size is large. In many experiments 95% confidence is the accepted level of accuracy. In normal distribution 95% of the items have a value within Sample Mean ± 1.96 Standard deviation (or standard error). So the chance of population mean going outside this range is only 5% or in other words only 5 times out of 100, the population mean can cross these limits. Similarly various

levels of significance can be attached for any calculation purposes or testing of hypothesis.

STATISTIC AND PARAMETER

Anything that pertains to sample is called as a "statistic" and that pertaining to population is called a "parameter". For example: sample mean $(\overline{x})$, sample variance (s^2) and sample standard deviation(s) are each a statistic, whereas, population mean (μ), population variance (σ) and population standard deviation (σ) are each a parameter. It may be noted that a *statistic* pertains to the sample value, whereas the subject being studied is *Statistics*.

Difference between Large and Small Samples

Though no hard and fast line can be drawn between large and small samples, yet conventionally if a sample consists of more than 30 items it is considered to be large. A sample consisting items up to 30 (30 inclusive) would be deemed to be small sample.

The tests of significance used in large samples, which we will study in subsequent chapters, are different from those used in small samples because small samples fail to satisfy the assumptions under which the large sample analyses are done. The assumptions under which significance tests are applied in case of large samples are:

1. The random sampling distribution of statistic in case of large samples has the properties of the normal distribution. In case of small samples this may not hold good.
2. In large samples, sample values are close to parameter values and can be used for analyses in the absence of parameter values. For example- for calculation of standard error, if the population standard deviation is not known, it can be substituted by the sample standard deviation, which is called ëstimated standard deviation. This is not possible in case of small samples unless necessary adjustments are made.

STATISTICAL INFERENCE

The topics covered in the chapters 8, 9 and 10 form the foundation for statistical inference, which deals in decision making processes using probability concepts. Managers usually want to make quick estimates for their organizations' growth. As it has been already discussed that the population study or census survey is quite expensive and time consuming, samples are drawn and studied, which are then regarded as estimates of the population parameters. For example- the head of a degree college attempts to estimate enrolments next year from current enrolment status. A financial manager attempts to estimate the creditworthiness of prospective customers from a sample of their past repayment habits. In all such cases, inferences are made about the population from information taken from a sample. So statistical inference is broadly divided into two areas of inferential procedures:

- Theory of Estimation
- Theory of Testing of Hypothesis

In the first category of estimation, two types of estimates are made: *point estimate* and *interval estimate*. A point estimate is a single number that is used to estimate an unknown population parameter, whereas an interval estimate is a range of values used to estimate a population parameter. Consider an example- after analysing previous years' enrolment figures, the head of the degree college might say that next year this course will have 200 students. This is a case of point estimate. On the other hand, if the head gives a statement like- it is very likely that the next year's the enrolment figure will lie between 180 to 225. This is a case of interval estimate.

CRITERIA OF A GOOD ESTIMATOR

1. **Unbiasedness:** If average value of all possible values of sample statistics is equal to population parameter, then this is an unbiased estimator. This is a desirable property for a good estimator to have. The term unbiasedness refers to the fact that, a sample mean $\bar{x}\bar{x}$ is an unbiased estimator of a population mean μ, because the mean of the sampling distribution of sample means taken from the same population is equal to the population mean itself. It may be noted that the sample variance s^2, when divided by (n-1) instead of n is an unbiased estimator of population variance σ^2. That is $\hat{\sigma}^2 = \frac{1}{n-1}\sum_{i=1}^{n}(x_i - \bar{x})^2$ is an unbiased estimate of σ^2. But if μ is known in advance, then the unbiased estimator of population variance becomes equal to $\frac{1}{n}\sum_{i=1}^{n}(x_i - \mu)^2$.

2. **Consistency:** The estimate should move and concentrate more around the true value as the sample size increases. That means, if the sample size increases, it becomes almost certain that the value of the statistic comes very close to the value of the population parameter. There might be large number of consistent estimators and in practice one should choose the one which approaches the parameter value more rapidly than others as the sample size increases.

3. **Efficiency:** Efficiency is somewhat related to variance. From among several unbiased estimators, the one which has the minimum possible variance is considered to be efficient. If we compare two statistics from a sample of the same size and try to decide which one is a more efficient estimator, we would pick the statistic that has the smaller standard error, or standard deviation of the sampling distribution.
4. **Sufficiency**: A sufficient estimator extracts more information from the sample than any other estimator.

MAXIMUM LIKELIHOOD ESTIMATOR

Once the criteria for a good estimator is stated, then how to find the estimators with desirable properties? Out of several methods like, Maximum Likelihood Method, Method of Least Square and Method of Moments, the first one is more popular, which was developed by R.A.Fisher. The method of maximum likelihood provides estimators that have both a reasonable intuitive basis and many desirable statistical properties. The method is very broadly applicable and is simple to apply. Once a maximum-likelihood estimator is derived, the general theory of maximum-likelihood estimation provides standard errors, statistical tests, and other results useful for statistical inference. A disadvantage of the method is that it frequently requires strong assumptions about the structure of the data.

Example: We want to estimate the probability π of getting a head upon flipping a particular coin.

- We flip the coin 'independently' 10 times (i.e. we sample n = 10 flips), obtaining the following result: H H T H H H T T H H
- The probability of obtaining this sequence— in advance of collecting the data — is a function of the unknown parameter :
  Pr(data|parameter) = Pr(H H T H H H T T H H | π) = $\pi\ \pi\ \pi\ (1 - \pi)$ $(1 - \pi)\ (1 - \pi) = \pi^7(1 - \pi)^3$
- But the data for our particular sample are fixed: We have already collected them.
- The parameter π also has a fixed value, but this value is unknown, and so we can let it vary in our imagination between 0 and 1, treating the probability of the observed data as a function of π.
- This function is called the likelihood function: L (parameter|data) = L(π | H H T H H H T T H H) = $\pi^7(1 - \pi)^3$
- The probability function and the likelihood function are given by the same equation, but the probability function is a function of the data

with the value of the parameter fixed, while the likelihood function is a function of the parameter with the data fixed.

Here are some representative values of the likelihood for different values of π:

π	$L(\pi \mid data) = \pi^7(1-\pi)^3$
0.0	0.0
0.1	0.0000000729
0.2	0.00000655
0.3	0.0000750
0.4	0.000354
0.5	0.000977
0.6	0.00179
0.7	0.00222
0.8	0.00168
0.9	0.000478
1.0	0.0

- The probability of obtaining the sample of data that we have in hand, H H T H H H T T H H, is small, regardless of the true value of π. This is usually the case: Any specific sample result — including the one that is realized — will have low probability.
- Nevertheless, the likelihood contains useful information about the unknown parameter π. For example, π cannot be 0 or 1, and is 'unlikely' to be close to 0 or 1.
- Reversing this reasoning, the value of that is most supported by the data is the one for which the likelihood is largest.
- This value is the maximum-likelihood estimate (MLE), denoted by $\hat{\pi}$. Here, $\hat{\pi} = 0.7$, which is the sample proportion of heads, 7/10.

More generally, for n independent flips of the coin, producing a particular sequence that includes x heads and n − x tails,

- $L(\pi|data) = Pr(data|\ \pi) = \pi^x(1-\pi)^{n-x}$
- We want the value of π that maximizes $L(\pi \mid data)$, which we often abbreviate $L(\pi)$.
- It is simpler — and equivalent — to find the value of π that maximizes the log of the likelihood

- $\log_e L(\pi) = x \log_e \pi + (n - x) \log_e(1 - \pi)$
- Differentiating $\log_e L(\pi)$ with respect to π produces

$$= \frac{d \log_e L(\pi)}{d\pi} = \frac{x}{\pi} + (n-x)\frac{1}{1-\pi}(-1) = \frac{x}{\pi} - \frac{n-x}{1-\pi}$$

- Setting the derivative to 0 and solving produces the MLE which, as before, is the sample proportion $\frac{x}{n}$. The maximum-likelihood estimator is $\hat{\pi} = \frac{X}{n}$

Properties of Maximum-Likelihood Estimator

Under very broad conditions, maximum-likelihood estimators have the following general properties:

- Maximum-likelihood estimators are consistent.
- They are asymptotically unbiased, although they may be biased in finite samples.
- They are asymptotically efficient—no asymptotically unbiased estimator has a smaller asymptotic variance.
- They are asymptotically normally distributed.
- If there is a sufficient statistic for a parameter, then the maximum likelihood estimator of the parameter is a function of a sufficient statistic.
- It is expected to give a very reliable estimator in large samples, say n > 30.
- A sufficient statistic is a statistic that exhausts all of the information in the sample about the parameter of interest.

POINT ESTIMATION

An estimator is a sample statistic used to estimate a population parameter. Sample mean $\bar{x}$ can be considered as an estimator of population mean μ, and sample proportion can be used as an estimator of population proportion. When a specific numerical value of the estimator is observed, that value is called as an estimate. *A point estimate is a specific observed value of a statistic.*

Example 10.2: The following are the attendances (in thousands) at nine randomly selected sporting events. Find the point estimates of the mean and the variance of the population from which the sample was taken. Also calculate the standard error.

8.8 14.0 21.3 7.9 12.5 20.6 16.3 14.1 13.0

Solution: Observe the following table:

x	8.8	14	21.3	7.9	12.5	20.6	16.3	14.1	13
$(x-\bar{x})^2$	30.006	0.077	49.311	40.676	3.16	39.97	4.089	0.031	1.632

$$\sum x = 128.5$$

$$\text{Mean} = \bar{x} = \frac{128.5}{9} = 14.2778$$

$$\hat{\sigma}^2 = \frac{\sum(x-\bar{x})^2}{n-1} = \frac{30.006+0.007+49.311+.....+1.632}{9-1} = 21.119$$

$$\text{Standard Error} = \frac{\hat{\sigma}}{\sqrt{n}} = \sqrt{\frac{21.119}{9}} = 1.5318$$

Example 10.3: In the above example, if the mean of the population is known to be 15000, calculate the point estimate of the population variance.

Solution: Given that μ = 15 (thousand of people)

Observe the following table:

x	8.8	14	21.3	7.9	12.5	20.6	16.3	14.1	13
$(x-\mu)^2$	38.44	1.000	39.69	50.41	6.25	31.36	1.69	0.81	4

$$\hat{\sigma}^2 = \frac{\sum(x-\mu)^2}{n-1} = \frac{38.44+1.00+39.69+......+4.00}{9} = 19.294$$

Example 10.4: Independent samples of size 10 and 12 from urban and rural housewives showed the following body weights (in kg).

Urban	64	54	71	72	69	57	59	56	62	67		
Rural	59	53	58	59	61	52	61	63	57	56	64	67

Estimate the difference between body weights of urban and rural housewives and find its standard error.

Solution: The following table is prepared, which is self-explanatory:

Urban (x)	$(x-\bar{x})^2$	Rural (y)	$(y-\bar{y})^2$
64	0.81	59	0.0278
54	82.81	53	38.028

	Urban (x)	$(x-\bar{x})^2$		Rural (y)	$(y-\bar{y})^2$
	71	62.41		58	1.3611
	72	79.21		59	0.0278
	69	34.81		61	3.3611
	57	37.21		52	51.361
	59	16.81		61	3.3611
	56	50.41		63	14.694
	62	1.21		57	4.6944
	67	15.21		56	10.028
Total:	631	380.90		64	23.361
Mean =	63.10			67	61.361
			Total:	710	211.67
			Mean =	59.167	

So the difference between the means = 63.10 – 59.167 = 3.933

$$\hat{\sigma}_1 = \sqrt{\frac{\sum(x-\bar{x})^2}{n_1 - 1}} = \sqrt{\frac{380.90}{10-1}} = 6.505$$

$$= \hat{\sigma}_2 = \sqrt{\frac{\sum(y-\bar{y})^2}{n_2 - 1}} = \sqrt{\frac{211.67}{12-1}} = 4.386$$

$$\text{Standard Error} = \sqrt{\frac{\hat{\sigma}_1^2}{n_1} + \frac{\hat{\sigma}_2^2}{n_2}} = \sqrt{\frac{6.505^2}{10} + \frac{4.386^2}{12}} = 2.4155$$

Example 10.5: Out of a random sample of 200 factories in a certain district, 105 were found to be emitting exhaust gases in the atmosphere beyond permissible limit. Estimate the proportion of such industries in the entire district and also estimate its standard error.

Solution:

$$\text{p} = \text{sample proportion} = \frac{105}{200} = 0.525 = \text{estimate of population proportion}$$

Standard Error of Population Proportion

$$= \sqrt{\frac{p(1-p)}{n}} = \sqrt{\frac{0.525(1-0.525)}{200}} = 0.0353$$

Example 10.6: In a survey conducted in two cities regarding the people's choice on a Government's declaration of a particular savings scheme, it was found that 300 out of 500 people in city-A and 180 out of 400 people in city-B favoured it. Compute the estimate of the difference between proportions of people in favour of the Government's scheme. Also estimate its standard error.

Solution: Let p_1 and p_2 denote the proportions of people favouring the scheme in two cities respectively.

$$\text{So } p_1 = \frac{300}{500} = 0.6 \text{ and } p_2 = \frac{180}{400} = 0.45$$

So the estimate of difference between proportions $= p_1 - p_2 = 0.60 - 0.45 = 0.15$

Estimated Standard Error of difference of proportions

$$= \sqrt{\frac{p_1(1-p_1)}{n_1} + \frac{p_2(1-p_2)}{n_2}} = \sqrt{\frac{0.6\times(1-0.6)}{500} + \frac{0.45\times(1-0.45)}{400}} = 0.0331$$

INTERVAL ESTIMATION

An interval estimate describes a range of values within which a population parameter is expected to lie. Before delving deep into the matter, let us understand the basic concept of how normal probability distribution curve is used in interval estimation. We already know that specific portions of the area under normal probability curve is defined by plus and minus any given number of standard deviations from the mean. This property of normal distribution is now applied to standard error of mean for making statements about range of values.

Let us take an example- a randomly selected sample of 200 shop owners revealed that the mean daily profit is 36. This can be the point estimate of the mean daily profit of all shop owners in the region of study. But the Commissioner, Commercial Tax wants to know the uncertainty that is likely to be attached with this statement, that is, a statement about the range within which the unknown population mean is likely to lie. To know this, we need to find the standard error of the mean (SEM). If it is already known from previous studies that the population standard deviation is 10, then the SEM = $\frac{\sigma}{\sqrt{n}} = \frac{10}{\sqrt{200}} = 0.707$. Now the Commissioner can be apprised of the fact that the population mean may lie within the range of $\pm 1SEM$ i.e. between 36 – 0.707 and 36 + 0.707 i.e. 35.293 and 36.707.

From the standard normal distribution table, we know that the area under the normal curve with $\pm 1\sigma$ on either side is 68.3%. So we can say that the probability is 68.3% or 0.683 that the mean of a sample size of 200 will be within ± 1 standard errors of population mean μ. In other words, 68.3% of all sample means are within ± 1 standard errors from μ and hence, μ lies within .293 and 36.070. ±1 standard errors of 68.3% of all possible sample means i.e.within 35 So we can say that we are 68.3% confident that the population mean will lie in this range. If we select 1000 samples at random from the population and construct an interval of ± 1 standard errors around each sample mean, then about 683 of these intervals out of 1000 intervals, so constructed, will contain the true population mean value. The figure below illustrates the concept graphically, which shows such intervals for four samples. Please note that sample 3 does not contain the population mean. Statisticians usually say-the population mean μ will lie within ± 1 standard errors from the sample mean 68.3% of the time.

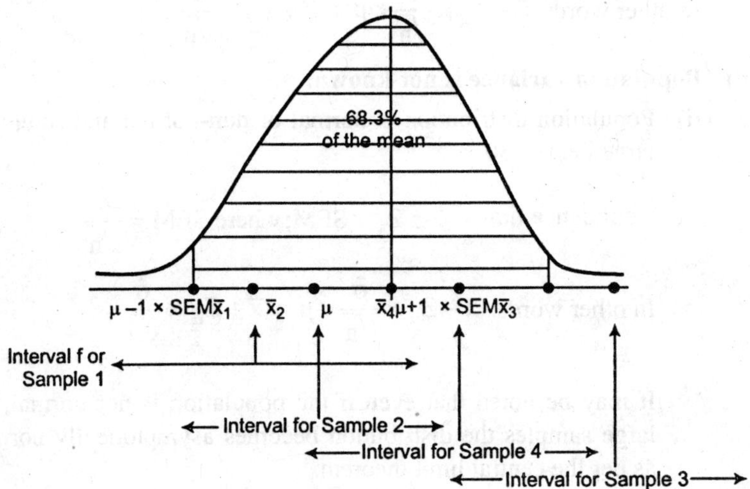

Similarly, the probability is 95.5% or 0.955 that the population mean will be within ± 2 standard error of the mean i.e. 36 – 2 0.707 and 36 + 2 0.707 i.e. within 34.586 and 37.414. So also we can be 99.7% confident that the population mean will lie within ± 3 standard error of mean i.e. within 36 . To become 95% confident, we have to find the range within 1.96 times the standard error at both sides of the sample mean i.e. $\overline{x} \pm 1.96\text{SEM}$. For higher confidence levels like 99%, this range will be $\overline{x} \pm 2.58 \times \text{SEM}$.

Hence, going by the above discussions, the sample statistic is used to find the two limits of the population parameter known as ***lower confidence limit and upper confidence limit*** at any desired *level of confidence* (1-α), where α is called as the *significance level*. It is both ways correct to say 95% confidence level or 5% significance level (α = 0.05). For 95% confidence level, the area under two tails of normal curve will be obviously 5% or 0.05(=α), thereby making each tail area of 0.025(=α/2). Similarly, for 99% confidence level, area under each tail will be 0.005.

In the following let us construct confidence intervals of certain well-known parameters under different pre-conditions commonly used in statistical analyses.

A. Confidence Interval for Single Mean

(a) Population variance (s^2) is known: Sample is of any size, population distribution is either normal or non-normal

Confidence limits: $\overline{x} \pm Z_{\frac{\text{å}}{2}} \times \text{SEM}$; where $\text{SEM} = \frac{\acute{o}}{\sqrt{n}}$;

In other words $\overline{x} - Z_{\frac{\alpha}{2}} \times \frac{\sigma}{\sqrt{n}} < \mu < \overline{x} + Z_{\frac{\alpha}{2}} \times \frac{\sigma}{\sqrt{n}}$:

(b) Population variance is not-known:

(i) Population distribution is normal or non-normal and sample is large i.e. n > 30

Confidence limits: $\overline{x} \pm Z_{\frac{\text{å}}{2}} \times \text{SEM}$; where $\text{SEM} = \frac{\hat{o}}{\sqrt{n}}$

In other words: $\overline{x} - Z_{\frac{\alpha}{2}} \times \frac{\hat{\sigma}}{\sqrt{n}} < \mu < \overline{x} + Z_{\frac{\alpha}{2}} \times \frac{\hat{\sigma}}{\sqrt{n}}$

It may be noted that even if the population is not normal, for large samples the distribution becomes asymptotically normal, as per the central limit theorem.

(ii) Population distribution is normal and sample size is small i.e. $\mu \leq 30$

Confidence limits: $\overline{x} \pm t_{\frac{\alpha}{2}} \times \text{SEM}$; where $\text{SEM} = \frac{\hat{\sigma}}{\sqrt{n}}$; is the estimated population standard deviation $= \sqrt{\frac{\sum (x_i - \overline{x})^2}{n-1}}$; t-value is to be found for n-1 degrees of freedom.

In other words: $\overline{x} - t_{\frac{\alpha}{2}} \times \frac{\hat{\sigma}}{\sqrt{n}} < \mu < \overline{x} + t_{\frac{\alpha}{2}} \times \frac{\hat{\sigma}}{\sqrt{n}}$

t-distribution: The conditions for use of t-distribution are (1) samples must come from normal population, (2) sample size is less than or equal to 30 and (3) population variance is not known. t-distribution curve is symmetrical and is flatter than normal curve. As sample size increases, the flatness at top is gradually lost and the curve more or less attains shape of a normal curve. In fact, for sample sizes of more than 30, the t-distribution is close to the normal distribution, so that normal approximation to t can be made. t-distribution is lower at the mean and higher at the tails than a normal distribution (see the figure below).

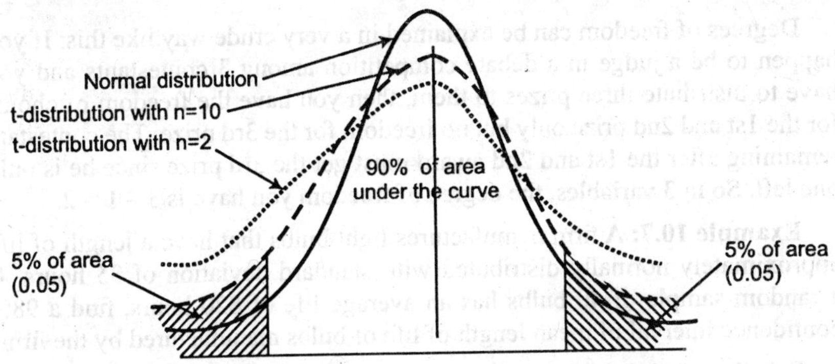

Unlike Z-distribution, where the figures in the table represent the area (or the probabilities) under the curve on any one side of the central axis (mean), the figures in t-distribution table represents the area under the curve at any one tail corresponding to a particular degree of freedom. If it is required to find the confidence interval of 10 items at 95% confidence level or 5% significance level, then the figure corresponding to α=0.025 (area at one tail end; area at two tails will be 0.025 2 = 0.05) for 10 - 1=9 degrees of freedom (df) is to be taken.

Degrees of Freedom

Since the sample is a part of the population with N number of items, it is desirable to get an estimate of the population parameter from the sample statistic. It becomes possible to draw an estimate of the various population parameters from the sample statistic by attaching a restriction to the n number of sample observations. Thus, the formula for sample variance $s^2 = \dfrac{\sum(x-\bar{x})^2}{n-1}$

which is an appropriate estimate of the population variance σ^2. Hence in all calculations of the estimates of population variance one should divide the appropriate value by n-1 instead of n. However, in the samples of large size where n is (say 1000) estimate of population parameter value will show little or negligible difference if the division is made either by n or by n-1. Hence, in all calculations of sample variance or sample standard deviation where sample size is less than or equal to 30, the division should be by n-1. When the number of restrictions is subtracted from the total number of items, we get the number of degree of freedom. In subsequent chapters many problems will show restrictions more than one. In these cases degree of freedom (df) is to be calculated by deducting the number of restrictions, say 2 or 3 etc. from the total number of observations.

Degrees of freedom can be explained in a very crude way like this: If you happen to be a judge in a debate competition among 3 contestants and you have to distribute three prizes to them, then you have the freedom of choice for the 1st and 2nd prize only but no freedom for the 3rd prize. The contestant remaining after the 1st and 2nd award must get the 3rd prize since he is only one left. So in 3 variables, the degree of freedom you have is $3 - 1 = 2$.

Example 10.7: A firm manufactures light bulbs that have a length of life approximately normally distributed with standard deviation of 35 hours. If a random sample of 40 bulbs has an average life of 820 hours, find a 98% confidence interval of mean length of life of bulbs manufactured by the firm.

Solution: Point estimate of population mean = $\overline{x} = 820$

Given that population standard deviation is known $\sigma = 35$ and $n = 40$

$$= \text{SEM} = \frac{\sigma}{\sqrt{n}} = \frac{35}{\sqrt{40}} = 5.534$$

From the normal distribution table, it is found that = $Z_{\frac{\alpha}{2}}$ 2.33 for 98% area under normal curve

Confidence limits are: $\overline{x} \pm Z_{\frac{\alpha}{2}} \times \text{SEM} = 820 \pm 2.33 \times 5.534 = 807.106 \,\&\, 832.894$

That means $807.106 < \mu < 832.894$

Example 10.8: The mean and standard deviation of the grade-points of a randomly selected 30 MBA students was found to be 2.7 and 0.4 respectively. Find 96% confidence limits of mean for the entire students.

Solution: Although the sample size n 30, let us make normal approximation to the distribution.

Population standard deviation is not known.

Given that sample standard deviation s = 0.4 = estimated population standard deviation $\hat{\sigma}$; n = 30 (Large sample considered)

Sample mean = $\overline{x} = 2.7$ = Point estimate of population mean

$$= \text{SEM} = \frac{\hat{\sigma}}{\sqrt{n}} = \frac{0.4}{\sqrt{30}} = 0.073$$

From the normal distribution table, it is found that = $Z_{\frac{\alpha}{2}}$ 2.05 for 96% area under normal curve

Confidencelimitsare: $\bar{x} \pm Z_{\frac{\alpha}{2}} \times SEM = 2.7 \pm 2.05 \times 0.073 = 2.550$ & 2.849

That means $2.550 < \mu < 2.849$

Example 10.9: The heights of a random sample of 8 college students have a mean of 158 cm with a standard deviation of 2.6 cm. Construct a 99% confidence interval for the mean height of all college students assuming an approximate normal distribution.

Solution: Population is approximately normally distributed. Population standard deviation is not known. Sample size is small ($n \leq 30$). So t-distribution would apply.

Given that sample standard deviation s = 2.6 = estimated population standard deviation $\hat{\sigma}$; n = 8 (Small sample)

Sample mean = $\bar{x} = 158$ = Point estimate of population mean

$$SEM = \frac{\hat{\sigma}}{\sqrt{n}} = \frac{2.6}{\sqrt{8}} = 0.919$$

From the t-distribution table, it is found that

$t_{\frac{\alpha}{2}}$ (for 8-1=7 degrees of freedom at α=0.005) = 3.499 ≅ 3.5 for 99% area under for 99% area under t-curve or $\frac{100-99}{2}$% area at one end of tail i.e. 0.005.

Confidence limits are:

$$\bar{x} \pm t_{\frac{\alpha}{2}} \times SEM = 158 \pm 3.5 \times 0.919 = 154.784 \text{ \& } 161.217$$

That means $154.784 < \mu < 161.217$.

Example 10.10: The owner of a tyre repairing shop wants to ascertain the average time that it takes to repair a tyre. How many number of mechanics he should observe to be 95% sure that his mean will be within 4 minutes within true mean? Assume that it is known from previous studies that $\sigma = 12$ minutes.

Solution: The difference between sample mean and population mean would be = 4 minutes = $\bar{x} - \mu$.

Presume that the sample would be large, so that Z = 1.96 read from normal distribution table for 95% area under the curve. We know that,

$$\bar{x} - Z_{\frac{\alpha}{2}} \times \frac{\sigma}{\sqrt{n}} < \mu < \bar{x} + Z_{\frac{\alpha}{2}} \times \frac{\sigma}{\sqrt{n}}$$

Considering only one side of the inequality: $\overline{x} - Z_{\frac{\alpha}{2}} \times \frac{\sigma}{\sqrt{n}} < \mu$

i.e., $(\overline{x} - \mu) < Z_{\frac{\alpha}{2}} \times \frac{\sigma}{\sqrt{n}}$

i.e., $4 < 1.96 \times \frac{12}{\sqrt{n}}$

i.e., $\sqrt{n} > \frac{1.96 \times 12}{4}$

i.e., $n > \left(\frac{1.96 \times 12}{4}\right)^2 = 34.57 \cong 35$

So the owner of the factory should observe at least 35 mechanics in order to arrive at the desired results.

Example 10.11: Data collected from a health clinic regarding the time taken to dispose of 8 patients were recorded as 12, 8, 16, 14, 9, 9, 11, 17 minutes respectively. Find a 95% confidence interval for the mean, assuming an approximate normal distribution.

Solution:

x	$(x - \overline{x})^2$
12	0
8	16
16	16
14	4
9	9
9	9
11	1
17	25
Total: 96	80

Mean $= \overline{x} = \frac{96}{8} = 12$

Estimated population standard deviation $\hat{\sigma} = \sqrt{\frac{(x - \overline{x})^2}{n-1}} = \sqrt{\frac{80}{8-1}} = 3.380$

From the t-table $t_{0.025} = 2.365$ for 8-1=7 degree of freedom

$$SEM = \frac{\hat{\sigma}}{\sqrt{n}} = \frac{3.380}{\sqrt{8}} = 1.195$$

Confidencelimitsare: $\bar{x} \pm t_{\frac{\alpha}{2}} \times SEM = 12 \pm 2.365 \times 1.195 = 9.174$ & 14.826

That means $9.174 < \mu < 14.826$.

B. Confidence Interval for Difference between two Means

(a) Population variances (σ_1^2 & σ_2^2) are both known:

Sample is of any size, population distribution is either normal or non-normal

Confidence limits: $(\bar{x}_1 - \bar{x}_2) \pm Z_{\frac{\alpha}{2}} \times SEM$; where $SEM = \sqrt{\frac{\sigma_1^2}{n_1} + \frac{\sigma_2^2}{n_2}}$

In other words:

$$(\bar{x}_1 - \bar{x}_2) - Z_{\frac{\alpha}{2}} \times \sqrt{\frac{\sigma_1^2}{n_1} + \frac{\sigma_2^2}{n_2}} < (\mu_1 - \mu_2) < (\overline{x_1 - x_2}) + Z_{\frac{\alpha}{2}} \times \sqrt{\frac{\sigma_1^2}{n_1} + \frac{\sigma_2^2}{n_2}}$$

(b) Population variances (any one or both) are not-known:

(i) Population distribution is normal or non-normal and both n_1 & n_2 >30

Confidence limits: $(\bar{x}_1 - \bar{x}_2) \pm Z_{\frac{\alpha}{2}} \times SEM$; where $SEM = \sqrt{\frac{\hat{\sigma}_1^2}{n_1} + \frac{\hat{\sigma}_2^2}{n_2}}$

In other words:

$$(\bar{x}_1 - \bar{x}_2) - Z_{\frac{\alpha}{2}} \times \sqrt{\frac{\hat{\sigma}_1^2}{n_1} + \frac{\hat{\sigma}_2^2}{n_2}} < (\mu_1 - \mu_2) < (\overline{x_1 - x_2}) + Z_{\frac{\alpha}{2}} \times \sqrt{\frac{\hat{\sigma}_1^2}{n_1} + \frac{\hat{\sigma}_2^2}{n_2}}$$

(ii) Populations are normal and either of n_1 & $n_2 < 30$

Confidence limits: $(\bar{x}_1 - \bar{x}_2) \pm t_{\frac{\alpha}{2}} \times SEM$; where $SEM = \sqrt{\frac{\hat{\sigma}_1^2}{n_1} + \frac{\hat{\sigma}_2^2}{n_2}}$

In other words:

In other words $(\bar{x}_1 - \bar{x}_2) - t_{\frac{\alpha}{2}} \times \sqrt{\frac{\hat{\sigma}_1^2}{n_1} + \frac{\hat{\sigma}_2^2}{n_2}} < (\mu_1 - \mu_2) < (\overline{x_1 - x_2}) + t_{\frac{\alpha}{2}} \times \sqrt{\frac{\hat{\sigma}_1^2}{n_1} + \frac{\hat{\sigma}_2^2}{n_2}}$

where $\hat{\sigma}_1$ & $\hat{\sigma}_2$ are the estimated population standard deviations using formula as $\sqrt{\frac{\sum(x_i - \bar{x})^2}{n-1}}$;

t-value is to be found for $n_1 - 1 + n_2 - 1 = n_1 + n_2 - 2$ degrees of freedom.

(iii) Population distribution is normal, population variances are equal ($\sigma_1^2 = \sigma_2^2$) and either n_1 or n_2 or both ≤ 30

So, we have to find the pooled standard deviation for both the samples, which is given by,

$$\hat{\sigma}^2 = \frac{\hat{\sigma}_1^2(n_1 - 1) + \hat{\sigma}_2^2(n_2 - 1)}{n_1 + n_2 - 2}$$

Confidence limits: $(\overline{x}_1 - \overline{x}_2) \pm t_{\frac{\alpha}{2}} \times \text{SEM}$; where $\text{SEM} = \hat{\sigma}\sqrt{\frac{1}{n_1} + \frac{1}{n_2}}$

In other words:

$$(\overline{x}_1 - \overline{x}_2) - t_{\frac{\alpha}{2}} \times \hat{\sigma}\sqrt{\frac{1}{n_1} + \frac{1}{n_2}} < (\overline{x}_1 - \overline{x}_2) < (\overline{x}_1 - \overline{x}_2) + t_{\frac{\alpha}{2}} \times \hat{\sigma}\sqrt{\frac{1}{n_1} + \frac{1}{n_2}}$$

t-value is to be found for $n_1 + n_2 - 2$ degrees of freedom.

Example 10.12: It is known that the weekly overtime worked by the workers of two factories are normally distributed with means μ_1 and μ_2 and standard deviations of 3 and 5 hours respectively. Random samples of size 30 and 40 selected from both the groups of workers showed sample averages to be equal to 12 and 15 respectively. Find 95% confidence interval for $\mu_1 - \mu_2$.

Solution: Normal population, variances known. So Z-table would be applied.

Point estimate of $(\mu_1 - \mu_2) = (\overline{x}_1 - \overline{x}_2) = 15 - 12 = 3$ hours

$$\text{SEM} = \sqrt{\frac{\sigma_1^2}{n_1} + \frac{\sigma_2^2}{n_2}} = \sqrt{\frac{3^2}{30} + \frac{5^2}{40}} = 0.9617$$

From the Z-table

$Z_{0.025} = 1.96$ for 95% means area under curve at one side = 95 2 = 0.475

Confidence limits:

$$(\overline{x}_1 - \overline{x}_2) \pm Z_{\frac{\alpha}{2}} \times \text{SEM} = 3 \pm 1.96 \times 0.9617 = 1.115 \text{ and} 4.885$$

That means $1.115 < (\mu_1 - \mu_2) < 4.885$

Example 10.13:

A study undertaken for 40 male workers and 50 female workers in a factory revealed that the average hours in a week they work are 56 hours and 35 hours with standard deviations 3 hours and 4 hours respectively. Find

96% confidence interval for the difference between the mean weekly working hours of all the male workers (μ1) and all the female workers (μ2) in the factory.

Solution: Here both n_1 and n_2 are large (more than 30). Population variances are not known. So Z-table would apply.

Given that: $\overline{x}_1 = 56$ & $\overline{x}_2 = 35$; $s_1 = 3$ & $s_2 = 4$; $n_1 = 40$ & $n_2 = 50$

$$\text{Where SEM} = \sqrt{\frac{\hat{\sigma}_1^2}{n_1}+\frac{\hat{\sigma}_2^2}{n_2}} = \sqrt{\frac{s_1^2}{n_1}+\frac{s_2^2}{n_2}} = \sqrt{\frac{3^2}{40}+\frac{4^2}{50}} = 0.738$$

From Z-table for 96%, Z = 2.05 (corresponding to area at one side under the curve = 0.48)

Confidence limits:

$$\left(\overline{x}_1 - \overline{x}_2\right) \pm Z_{\frac{\alpha}{2}} \times \text{SEM} = (56-35) \pm 2.05 \times 0.738 = 19.487 \text{ and} 22.513$$

That means $19.487 < (\mu_1 - \mu_2) < 22.513$

Example 10.14: Samples of two types of heating filaments were tested for their length of life and the following data were obtained:

Type-I	**Type-II**
$n_1 = 6$	$n_2 = 8$
$\overline{x}_1 = 1100$ hours	$\overline{x}_2 = 1040$ hours
$s_1 = 34$ hours	$s_2 = 40$ hours

Assuming normality, compute the 95% confidence interval for the difference between population mean length of life of the two types of filaments (Assume that the standard deviation of length of life for the two types are the same).

Solution: As population standard deviations are equal, we have to find the pooled standard deviation, which is given by:

$$\hat{\sigma}^2 = \frac{\hat{\sigma}_1^2(n_1-1)+\hat{\sigma}_2^2(n_2-1)}{n_1+n_2-2} = \frac{34^2(6-1)+40^2(8-1)}{6+8-2} = 1415$$

$$\text{SEM} = \hat{\sigma}\sqrt{\frac{1}{n_1}+\frac{1}{n_2}} = \sqrt{1415} \times \sqrt{\frac{1}{6}+\frac{1}{8}} = 20.315$$

Here t-distribution would apply, as samples are small and from normal populations. From the table, for 6+8-2= 12 degrees of freedom and α=0.025, t = 2.179

Confidence limits:

$$(\overline{x}_1 - \overline{x}_2) \pm t_{\frac{\alpha}{2}} \times SEM = (1100 - 1040) \pm 2.179 \times 20.315 = 15.734 \;\&\; 104.267$$

That means $15.734 < (\mu_1 - \mu_2) < 104.267$

CONFIDENCE INTERVAL FOR PAIRED OBSERVATIONS

There can be situations when the observations in two samples occur in pairs. Two observations on a single subject are taken at two different conditions, so they are related with each other. Thus, the samples are not independent, but the variances of the two populations may not necessarily be equal. For example, the effectiveness of a drug is tested taking observations on body weights before administration of the drug and again after administration of the drug to n number of animals. We can have n pairs of observations. There will be n number of values of difference between each pair of observation, which is assumed to be normal with mean μ_d and variance σ_d^2. The unknown σ_d^2 is estimated by s_d^2, which is the sample variance of the differences.

$\mu_d = \mu_1 - \mu_2$ and is given by mean of d's

i.e. $\overline{d} = \frac{1}{n}\sum_{i=1}^{n} d_i$ and $SE = \frac{s_d}{\sqrt{n}}$; $s_d^2 = \frac{\sum d^2 - n\overline{d}^2}{n-1}$

Confidence limits: $\overline{d} - t_{\frac{\alpha}{2}} \frac{s_d}{\sqrt{n}} < \mu_d < \overline{d} + t_{\frac{\alpha}{2}} \frac{s_d}{\sqrt{n}}$

t-value is to be found for n-1 degrees of freedom.

Example 10.15: All the medium scale industries in a district were renovated for better safety standards and practices. The following are the number of accidents occurred in 10 different industries before and after renovation during past one year.

Before	6	4	8	5	3	1	7	6	5	2
After	5	2	5	3	4	3	4	3	2	3

Compute a 95% confidence interval for the true difference between average number of accidents in the district.

Solution:

											Summation
Before	6	4	8	5	3	1	7	6	5	2	
After	5	2	5	3	4	3	4	3	2	3	
Difference(d)	1	2	3	2	–1	–2	3	3	3	–1	13
d^2	1	4	9	4	1	4	9	9	9	1	51

$$\bar{d} = \frac{\sum d}{n} = \frac{13}{10} = 1.3$$

and $s_d^2 = \frac{\sum d^2 - n\bar{d}^2}{n-1} = \frac{51 - 10 \times 1.3^2}{10-1} = 3.789$ and $SE = \frac{s_d}{\sqrt{n}} = \sqrt{\frac{3.789}{10}} = 0.6155$

t-value for $10 - 1 = 9$ degrees of freedom at $\alpha = 0.05$ is 2.262

Confidence limits:

$\bar{d} - t_{\frac{\alpha}{2}} \frac{s_d}{\sqrt{n}} < \mu_d < \bar{d} + t_{\frac{\alpha}{2}} \frac{s_d}{\sqrt{n}}$ i.e. $1.3 - 2.262 \times 0.6155 < \mu_d < 1.3 + 2.262 \times 0.6155$

i.e. $-0.092 < \mu_d < 2.692$

C. Confidence Interval for a Proportion

Confidence limits: $p \pm Z_{\frac{\alpha}{2}} \times SE$; where $SE = \sqrt{\frac{p(1-p)}{n}}$; where p is the sample proportion.

In other words: $p - Z_{\frac{\alpha}{2}} \times \sqrt{\frac{p(1-p)}{n}} < P < p + Z_{\frac{\alpha}{2}} \times \sqrt{\frac{p(1-p)}{n}}$

D. Confidence Interval for Difference between two Proportions

Consider two independent large samples of sizes n_1 and n_2 with p_1 and p_2 successes respectively.

Confidence limits: $(p_1 - p_2) \pm Z_{\frac{\alpha}{2}} \times SE$; where $SE = \sqrt{\frac{p_1(1-p_1)}{n_1} + \frac{p_2(1-p_2)}{n_2}}$

i.e. $(p_1 - p_2) - Z_{\frac{\alpha}{2}} \times \sqrt{\frac{p_1(1-p_1)}{n_1} + \frac{p_2(1-p_2)}{n_2}} < (P_1 - P_2)$

$$< (p_1 - p_2) + Z_{\frac{\alpha}{2}} \times \sqrt{\frac{p_1(1-p_1)}{n_1} + \frac{p_2(1-p_2)}{n_2}}$$

Now it is time to take some illustrative examples to fully understand the theory.

Example 10.16: From a random sample of 200 cement bags in one consignment of a cement manufacturing industry, 15 were found to be underweight. Find a 90% confidence interval for proportion of underweight cement bags in the consignment.

Solution: If P be the proportion of all underweight cement bags in the consignment, the point estimate is given by, $p = \frac{15}{200} = 0.075$

$1-\alpha=0.90 \Rightarrow \alpha=0.10$

From Z-table, for area under the curve at one side 0.45, Z = 1.645

$$SE=\sqrt{\frac{p(1-p)}{n}}=\sqrt{\frac{0.075(1-0.075)}{200}}=0.0186$$

Confidence Interval:

$$=p-Z_{\frac{\alpha}{2}}\times\sqrt{\frac{p(1-p)}{n}}<P<p+Z_{\frac{\alpha}{2}}\times\sqrt{\frac{p(1-p)}{n}}$$

i.e. $0.075-1.645\times0.0186<P<0.075+1.645\times0.0186$

i.e. $0.0444<P<0.1056$

Example 10.17: In random samples of 300 and 500 students from two city colleges showed that 120 and 150 students respectively were smokers. Find a 90% confidence interval for the true difference in the proportions.

Solution: $p_1=\frac{120}{300}=0.40$ and $p_2=\frac{150}{500}=0.30$, $p_1-p_2=0.4-0.3=0.1$

$$SE=\sqrt{\frac{p_1(1-p_1)}{n_1}+\frac{p_2(1-p_2)}{n_2}}=\sqrt{\frac{0.4(1-0.4)}{300}+\frac{0.3(1-0.3)}{500}}=0.0349$$

As shown in solution of Example 10.16, for $1-\alpha=90\%$,

Confidence limits: $(p_1-p_2)\pm Z_{\frac{\alpha}{2}}\times SE$

i.e. $0.1\pm1.645\times0.0349=0.0426$ and 0.1574

CONFIDENCE INTERVAL FOR VARIANCE OF A NORMAL DISTRIBUTION

As has already been discussed, if population mean μ is not known, the point estimate of population variance is: $\hat{\sigma}^2=\frac{\sum(x-\bar{x})^2}{n-1}$ with $n-1$ degrees of freedom. But if the population mean μ is known, then the point estimate of population variance is: $\hat{\sigma}^2=\frac{\sum(x-\mu)^2}{n}$. We can establish the confidence interval of population variance with the aid of a statistic called Chi square (χ^2). In fact, the square of normal variate (Z^2) follows χ^2-distribution.

Confidence interval is: $\frac{(n-1)\hat{\sigma}^2}{\chi^2_{\frac{\alpha}{2}}} < \sigma^2 < \frac{(n-1)\hat{\sigma}^2}{\chi^2_{1-\frac{\alpha}{2}}}$ with (n-1) df (for unknown population mean μ)

Confidence interval is: $\frac{n\hat{\sigma}^2}{\chi^2_{\frac{\alpha}{2}}} < \sigma^2 < \frac{n\hat{\sigma}^2}{\chi^2_{1-\frac{\alpha}{2}}}$ with n df (for known population mean μ)

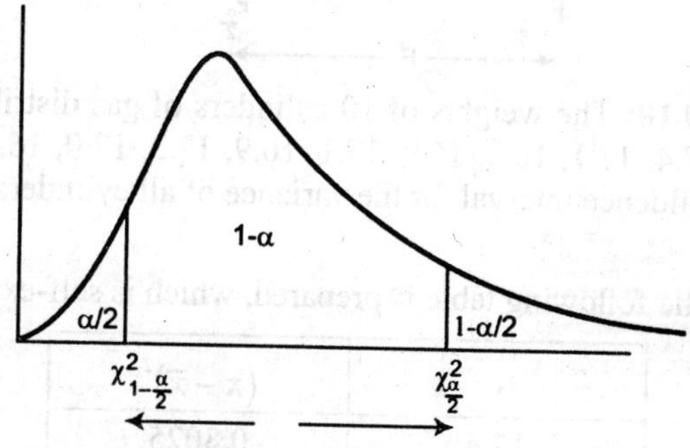

E. Confidence Interval for Ratio of Variances from two Independent Normal Distributions

If σ_1^2 and σ_2^2 are the population variances from 1st and 2nd normal populations, from which two independent samples of sizes n_1 and n_2 are drawn. If both the population variances (μ_1 & μ_2) are known, the point estimate of $\frac{\sigma_1^2}{\sigma_2^2}$ is: $\frac{\hat{\sigma}_1^2}{\hat{\sigma}_2^2}$ and a statistic F is defined by:

$$F = \left(\frac{s_1^2}{\sigma_1^2}\right) / \left(\frac{s_2^2}{\sigma_2^2}\right); \text{ with } v_1 = n_1 - 1 \text{ and } v_2 = n_2 - 1 \text{ df}$$

The variance with higher value is taken in the numerator.

Confidence interval is: $\frac{\hat{\sigma}_1^2}{\hat{\sigma}_2^2} \times \frac{1}{F_{\frac{\alpha}{2}(v_1,v_2)}} < \frac{\sigma_1^2}{\sigma_2^2} < \frac{\hat{\sigma}_1^2}{\hat{\sigma}_2^2} \times F_{\frac{\alpha}{2}(v_2,v_1)}$

It is verified that: $F_{1-\frac{\acute{a}}{2}(v_1,v_2)} = \frac{1}{F_{\frac{\acute{a}}{2}(v_2,v_1)}}$

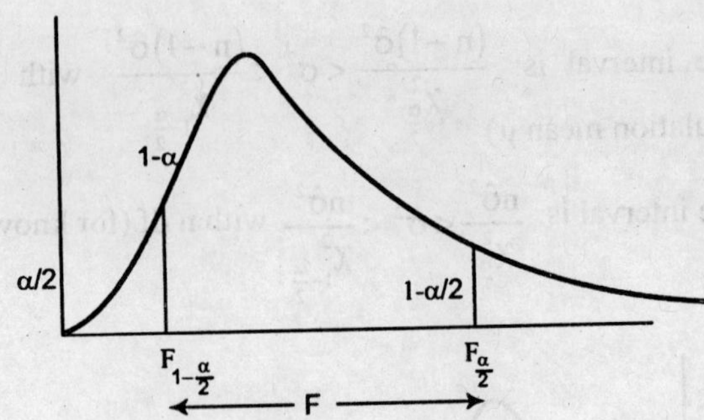

Example 10.18: The weights of 10 cylinders of gas distributed by a gas company are 17.4, 17.1, 16.5, 16.9, 17.1, 16.9, 17.3, 17.0, 16.2 and 16.1 kg. Find a 95% confidence interval for the variance of all cylinders distributed by this company.

Solution: The following table is prepared, which is self-explanatory:

x	$(x-\bar{x})^2$
17.4	0.3025
17.1	0.0625
16.5	0.1225
16.9	0.0025
17.1	0.0625
16.9	0.0025
17.3	0.2025
17.0	0.0225
16.2	0.4225
16.1	0.5625
Total = 168.5	Total = 1.765
Mean = 16.85	

Estimated Population variance = $\hat{\sigma}^2 = \dfrac{\sum(x-\bar{x})^2}{n-1} = \dfrac{1.765}{10-1} = 0.1961$

95% confidence interval means, $\alpha = 0.05$ and $\alpha/2 = 0.025$. So we have to find the values of $\chi^2_{\frac{\alpha}{2}}$ and $\chi^2_{1-\frac{\alpha}{2}}$ from the χ^2 – Table, we have,

$\chi^2_{0.025} = 19.023$ and $\chi^2_{0.975} = 2.70$

Hence, Confidence interval is: $\frac{(n-1)\hat{\sigma}^2}{\chi^2_{\frac{\alpha}{2}}} < \sigma^2 < \frac{(n-1)\hat{\sigma}^2}{\chi^2_{1-\frac{\alpha}{2}}}$

i.e. $\frac{9 \times 0.1961}{19.023} < \sigma^2 < \frac{9 \times 0.1961}{2.70}$

i.e. $0.093 < \sigma^2 < 0.653$

Example 10.19: A sample of 16 boys and 13 girls was selected from the students appearing at the Mathematics Olympiad. The boys made an average grade of 80 with a S.D. of 8, while the girls made average grade of 76 with a S.D. of 7. Find a 98% confidence interval for $\frac{a_1}{a_2}$.

Solution: Estimated Population variance for boys = $\hat{\sigma}_1^2 = s_1^2 = 8 \times 8 = 64$

Estimated Population variance for girls = $\hat{\sigma}_2^2 = s_2^2 = 7 \times 7 = 49$

$\hat{\sigma}_2^2 = s_2^2 = 7 \times 7 = 49$

$s_1^{\,2}$ is greater than $s_2^{\,2}$. Hence we will compute $\frac{\hat{\sigma}_1^2}{\hat{\sigma}_2^2} = \frac{64}{49} = 1.306$

$v_1 = 16 - 1 = 15$

$v_2 = 13 - 1 = 12$

For 98% confidence interval, $\alpha = 2\% = 0.02$; $\alpha/2 = 0.01$

From the F-table, we find the value of $F_{\frac{\alpha}{2}(v_1, v_2)} = F_{0.01(15,12)} = 4.01$ (between 3.97 and 4.05) $F_{\frac{\alpha}{2}(v_2, v_1)} = F_{0.01(12,15)} = 3.67$

Confidence interval is: $\frac{\hat{\sigma}_1^2}{\hat{\sigma}_2^2} \times \frac{1}{F_{\frac{\alpha}{2}(v_1, v_2)}} < \frac{\sigma_1^2}{\sigma_2^2} < \frac{\hat{\sigma}_1^2}{\hat{\sigma}_2^2} \times F_{\frac{\alpha}{2}(v_2, v_1)}$

i.e. $\frac{64}{49} \times \frac{1}{4.01} < \frac{\sigma_1^2}{\sigma_2^2} < \frac{64}{49} \times 3.67$

i.e. $0.326 < \frac{\sigma_1^2}{\sigma_2^2} < 4.793$

So, $\sqrt{0.326} < \frac{\sigma_1}{\sigma_2} < \sqrt{4.793}$

Or, $0.570 < \frac{\sigma_1}{\sigma_2} < 2.189$

For convenience a table is presented below showing various conditions which govern the formula to be used in finding out the confidence intervals.

SINGLE MEAN

S. No.	Population Variance (σ^2)	n	Population Type	Test Statistic	
1	Known	$n \leq 30$ or $n > 30$	Normal or Non-Normal	$\frac{\lvert \bar{x} - \mu \rvert}{\sigma / \sqrt{n}}$	
2 (a)	Unknown	$n \leq 30$	Normal	$t = \frac{\lvert \bar{x} - \mu \rvert}{\hat{\sigma} / \sqrt{n}}$ with (n-1) d.f.	$\hat{\sigma} = \frac{\Sigma\left(x - \bar{x}\right)^2}{n-1}$
2 (b)		$n > 30$	Normal or Non-Normal	$z = \frac{\lvert \bar{x} - \mu \rvert}{\hat{\sigma} / \sqrt{n}}$	

TWO MEANS

S. No	Population Variances (σ_1^2, σ_2^2)	n_1, n_2	Population Type	Test Statistic
1	Known Equal or Unequal ($s_1^2 = s_2^2$ or $s_1^2 \neq s_2^2$)	$n_1, n_2 \leq 30$ or $n_1, n_2 > 30$	Normal or Non-Normal	$z = \frac{\lvert \bar{x}_1 - \bar{x}_2 \rvert}{\sqrt{\frac{\sigma_1^2}{n_1} + \frac{\sigma_2^2}{n_2}}}$

S. No	Population Variances (σ_1^2, σ_2^2)	n_1, n_2	Population Type	Test Statistic
2 (a)	Unknown Equal ($s_1^2 = s_2^2$)	Either n_1 or n_2 or both ≤ 30	Normal	$t = \dfrac{\lvert\overline{x_1} - \overline{x_2}\rvert}{\sqrt{\hat{\sigma}^2\left(\dfrac{1}{n_1}+\dfrac{1}{n_2}\right)}}$ *with* (n_1+n_2-2) *d.f.* $\hat{\sigma}^2 = \dfrac{(n_1-1)s_1^2+(n_2-1)s_2^2}{n_1+n_2-2}$ [*Pooled Variance*]
2 (b)	Unknown Unequal ($s_1^2 \neq s_2^2$)	$n_1 > 30$ & $n_2 > 30$	Normal or Non-Normal	$z = \dfrac{\lvert\overline{x_1} - \overline{x_2}\rvert}{\sqrt{\dfrac{\hat{\sigma}_1^2}{n_1}+\dfrac{\hat{\sigma}_2^2}{n_2}}}$

PAIRED OBSERVATIONS

(Population Variances may not be equal)

$$t = \frac{\lvert\bar{d}\rvert}{\dfrac{s_d}{\sqrt{n}}} \qquad s_d^2 = \frac{\sum d^2 - n\bar{d}^2}{n-1}$$

Single Proportion two Proportions

Binomial Experiment

(Samples are generally large)

$$z = \frac{\lvert\hat{p} - p\rvert}{\sqrt{\dfrac{pq}{n}}} \qquad z = \frac{\lvert\hat{p}_1 - \hat{p}_2\rvert}{\sqrt{\dfrac{\hat{p}_1\hat{q}_1}{n_1}+\dfrac{\hat{p}_2\hat{q}_2}{n_2}}}$$

Confidence Interval for Variance

Unknown Population mean (μ)

(n – 1) d.f.

$$\frac{(n-1)s^2}{\chi^2_{\alpha/2}} < \sigma^2 < \frac{(n-1)s^2}{\chi^2_{1-\alpha/2}}$$

Known Population Mean (μ)

n d.f.

$$\frac{n\sigma^2}{\chi^2_{\alpha/2}} < \sigma^2 < \frac{n\sigma^2}{\chi^2_{1-\alpha/2}}$$

Confidence Interval for Ratio of two Variances

Unknown Population mean (μ)

(n – 1) d.f.

Unknown Population

Means (μ_1 & μ_2)

(n_1 – 1, n_2 – 1) d.f.

$$\frac{s_1^2}{s_2^2}\frac{1}{F_{(n_1-1,n_2-1)}} < \frac{\sigma_1^2}{\sigma_2^2} < \frac{s_1^2}{s_2^2}F_{(n_2-1,n_1-1)}$$

Known Population Mean (μ)

n d.f.

Known Population

Means (μ_1 & μ_2)

(n_1, n_2) d.f.

$$\frac{s_1^2}{s_2^2}\frac{1}{F_{(n_1,n_2)}} < \frac{\sigma_1^2}{\sigma_2^2} < \frac{s_1^2}{s_2^2}F_{(n_2,n_1)}$$

EXERCISES

1. 48 students were selected from a school having total student strength of 260 with their annual higher secondary examination marks of 443 and a standard deviation of 9. Calculate the standard error of mean(SEM).
2. The following are the time in minutes to clear traffic congestions at rush hours at ten randomly selected market squares of a big city. Find the point estimates of the mean and the variance of the time taken to clear the traffic congestion at all the traffic squares (population) existing in the city area. Also calculate the standard error.

   6.5 12.5 20.0 8.5 12.5 18.5 16.0 14.0 13.0 7.5
3. In the above example, if the mean time taken for all the traffic squares in the city is known to be 14.5, calculate the point estimate of the population variance.
4. Independent samples of 9 and 11 garment shops from urban and rural areas showed the following sales figures in thousand rupees.

Urban	62	52	69	70	67	55	57	54	60		
Rural	57	51	56	57	59	50	59	61	55	54	62

Estimate the difference between sales of urban and rural garment shops and find its standard error.

5. In a sample of 380 textile workers, 165 expressed extreme dissatisfaction regarding a prospective plan to modify working conditions. Because the dissatisfaction was strong enough to allow management to interpret plan reaction as being highly negative, they were curious about the proportion of total workers harbouring this sentiment. Estimate the proportion of such dissatisfied workers in the entire textile industry and also estimate its standard error.
6. A survey was conducted in two districts regarding the students' drop-out in elementary schools, after Government's introduction of mid-day meal scheme. It was found that 118 out of 500 students in district A and 106 out of 400 students in city B dropped out of school on a particular day. Compute the estimate of the difference between proportions of students' drop out. Also estimate its standard error.
7. Owner of a recently opened restaurant had difficulty in estimating the number of pizzas to be prepared each evening. He observed a sample of 35 nights and found that the mean number was 115. The population standard deviation is known to be 8. Find a 98% confidence interval of mean number of pizzas to be prepared by the restaurant owner.
8. A small-scale industry manufactures 4 inches size nails and the mean and standard deviation of the lengths of a randomly selected 30 nails were found to be 3.88 inches and 0.04 inches respectively. Find 96% confidence limits for the mean of the nails manufactured in the industry.
9. The heights of a random sample of 6 college students have a mean of 162 cm with a standard deviation of 1.8 cm. Construct a 99% confidence interval for the mean height of all college students assuming an approximate normal distribution.
10. The owner of a C&F agency wants to ascertain the average time that it takes to unload the goods from a truck. How many number of trucks he should observe to be 95% sure that the mean will be within 3 minutes within true mean? Assume that it is known from previous studies that $\sigma = 12$ minutes.
11. The manager wants to find the proportion of its employees who are in favour of the company's changing policy. How many employees the manager should interview so that he becomes sure that the true proportion remains ± 0.05 at 95% confidence level. How would

the sample size change if the manager believes that about 75% of employees favour the proposal?

12. A restaurant owner wanted to purchase new furniture. Seven randomly selected custumers were investigated regarding the revenue earnings and the amounts being recorded as 12, 8, 16, 14, 9, 9, 11 thousand respectively. Construct a 95% confidence interval for the mean revenue earnings, assuming an approximate normal distribution.

13. A two-lane road was surveyed and it was found that the number of passengers per kilometre on both the lanes are normally distributed with means and and standard deviations of 3 and 5 hours respectively. Random samples of size 30 and 40 number of stretches of 1 kilometre length selected from both the lanes showed sample averages to be equal to 12 and 15 respectively. Find 95% confidence interval for .

14. A study undertaken for 45 male workers and 50 female workers in a factory revealed that the average hours in a week they work are 55 hours and 32 hours with standard deviations 2 hours and 3 hours respectively. Find 96% confidence interval for the difference between the mean weekly working hours of all the male workers and all the female workers in the factory.

14. MBA colleges at two different districts revealed the following as regards the mean number of semester hours taught and the standard deviations.

District I	District II
$n_1 = 6$	$n_2 = 8$
Mean of x_1 = 648 hours	Mean of x_2 = 652 hours
$s_1 = 6$ hours	$s_2 = 8$ hours

Assuming normality, compute the 95% confidence interval for the difference between mean semester hours taught in the two districts (Assume that the standard deviation of mean semester hours taught in two districts are the same).

16. A costumer wanted to buy a modem for transmission of data from one computer to another. She tried with two types of modems available in the market and did a test run of the modems by sending 6 randomly chosen files with both modems and recorded the following rates (in thousand baud)

File	1	2	3	4	5	6	
Ultima	28.8	9.53	10.16	10.32	10.01	10.70	9.65

Max Pro	28.8	10.90	11.45	11.16	12.22	10.40	11.35

Compute a 95% confidence interval for the true difference between average rates of data transfer by both modems.

17. All the medium-scale industries in a district were renovated for better safety standards and practices. The following are the number of accidents occurred in 10 different industries before and after renovation during past one year.

Before	6	4	8	5	3	1	7	6	5	2
After	5	2	5	3	4	3	4	3	2	3

Compare a 95% confidence interval for the true difference between average number of accidents in the district.

18. From a random sample of 250 cement bags in one consignment of a cement manufacturing industry, 18 were found to be underweight. Find a 90% confidence interval for the proportion of underweight cement bags in the consignment.
19. In random samples of 400 and 550 students from two city colleges showed that 105 and 135 students respectively were late comers. Find a 90% confidence interval for the true difference in the proportions.
20. The weights of 10 cylinders of gas distributed by a gas company are 17.4, 17.1, 16.5, 16.9, 17.1, 16.9, 17.3, 17.0, 16.2 and 16.1 kg. Find a 95% confidence interval for the variance of all cylinders distributed by this company.
21. A sample of 16 boys and 13 girls was selected from the students appearing at the Mathematics Olympiad. The boys made an average grade of 80 with an S.D. of 8, while the girls made average grade of 76 with an S.D. of 7. Find a 98% confidence interval for $\frac{\sigma_1}{\sigma_2}$.

HYPOTHESIS TESTING

Test of Hypothesis forms the all-important section in statistics, which has wide applications in industrial production, drug manufacturing companie, finance department of companies, share markets and in many decision making situations. The decision makers may be interested in the relation between the values of the two populations rather than their actual values. For example- a drug manufacturer may need to know whether a new drug so manufactured is effective in curing the disease or not. A company may want to ascertain whether its female employees receive lower salaries than its male counterparts for the same work. A training director may wish to know whether the efficiency of the employees has increased after the training imparted to them.

In computing the confidence interval, we use sample data to find the range within which the population parameter is expected to lie with certain degree of confidence. But in hypothesis testing, we use sample statistic to find with certain degree of confidence that the sample has been drawn from a given population. In other words, if there is any difference between sample statistic and population parameter, then it has arisen due to sampling fluctuations. This concept holds good in all types of hypothesis testing, e.g. test for single mean, single proportion; difference between two means, difference between two proportions, paired data, variance and ratio of variances etc.

A *hypothesis* is a conclusion, which is drawn on a logical basis. The *test of hypothesis* is a process of testing of significance, which concerns with the testing of some hypothesis regarding a population parameter on the basis of a sample. At the outset a sample is drawn from a population, its statistic are found and on the basis of such statistic it is seen whether the sample so drawn has come from the parent population with certain specified characteristics or not. This computed sample statistic may differ from the hypothetical value of the population parameter. If the difference is small, it is considered that the difference has arisen due to sampling fluctuations and hence is not significant and the hypothesis that has been set up is accepted. If the difference is large, it is concluded that the sample might have come from the same population, but there is no evidence to prove it and hence is significant and the hypothesis that has been set up is rejected. So the test of hypothesis discloses the fact whether the difference between, (i) the computed statistic and the hypothetical parameter and (ii) two sample statistics, is significant or not significant. That is why, it is, in other words, known as *test of significance.*

HYPOTHESIS TESTING

We draw samples from the population to obtain an estimate of the population parameter values. Estimates of the population parameters thus obtained may not exactly match with the true values. So there is an element of risk involved in it. That is why the standard error of the estimate is calculated. This enables us to know the limits within which the parameter value is expected to lie. So the idea lies in finding out whether the difference in the estimated value of the parameter and the true value, if known, is significant or it could have arisen due to fluctuations in sampling. For this reason a hypothesis is formulated and then is tested for its validity.

A hypothesis is a quantitative statement about a population. It may or may not be true. By testing a hypothesis we can find out whether it deserves acceptance or rejection.

Some important terms used in testing of hypothesis are discussed below before going into details of the testing procedure.

1. **Test Of Significance:** Test of hypothesis is based on test of significance. The test of significance enables us to decide on the basis of sample results, if (i) the deviation between the observed sample statistic and hypothetical parameter value or (ii) the deviation between two sample statistics is significant or might be attributed due to sampling fluctuations.

   It is well known that if the sample size is large almost all distributions can be approximated by normal distribution. So for large sample one can safely use the normal test of significance and for small samples t-test, F-test etc.

2. **Null Hypothesis:** For applying the test of significance one has to set up a hypothesis i.e. a definite statement about the population parameter. Such hypothesis is a hypothesis of no difference. This hypothesis is known as Null hypothesis and denoted by H_0

   Definition: Null Hypothesis is the hypothesis, which is tested for possible rejection under the assumption that it is true. It is usually the case that the null hypothesis is either accepted or rejected basing on the test of significance.

3. **Alternative Hypothesis:** Any hypothesis which is complementary to null hypothesis is called alternative hypothesis. This is usually denoted by H_1. Suppose we are interested to verify whether a population has mean μ_0 or not. Then we set up the null hypothesis:

   $$H_0 : \mu = \mu_0$$

   Then the alternative hypothesis could be any one of the statements:

   (i) $H_1 : \mu \neq \mu_0$ (i.e. $\mu > \mu_0$ or, $\mu < \mu_0$)

   (ii) $H_1 : \mu > \mu_0$

   (iii) $H_1 : \mu < \mu_0$

   The alternative hypothesis in (i) is known as a two-tailed alternative, whereas the other two (ii) & (iii) are known as right tailed and left tailed alternatives respectively. In general (ii) and (iii) are known as one tailed alternatives. The setting of alternative hypothesis is very important since on the basis of alternative hypothesis only, we decide whether a one tailed (right or left) or two tailed test is to be used.

4. **Errors in Testing of Hypothesis:** After applying a test, a decision is taken about the acceptance or rejection of null hypothesis vis-à-vis the alternative hypothesis. There is always some possibility of committing an error in taking a decision about the hypothesis. These errors can be of two types.

   Now we write:

   Prob.[Rejecting H_0 when it is true] = α (called as Producer's risk)

   And Prob.[Accepting H_0 when it is wrong] = β (called as Consumer's risk)

   Here the α and β are called the size of type I and type II errors respectively. $1 - \beta$ is called the Power of a test.

5. **Level of Significance:** It is the quantity of risk of type-I error, which can be readily tolerated in making a decision about the null hypothesis H_0. In other words, it is the probability of type I error (α) which is tolerable. The level of significance is denoted by α and conveniently chosen as 0.05 or 0.01. $\alpha = 0.01$ is used for high precision and $\alpha =$ 0.05 for moderate precision.

6. **Critical Region:** A statistic is used to test the hypothesis H_0. The test statistic follows a known distribution. In a test, the area under the probability density curve is divided into two regions i.e. the region of acceptance and the region of rejection. The region of rejection is the region in which H_0 is rejected. It indicates that if the value of test statistic lies in this region, H_0 will be rejected. This region is called *critical region.* The area of the critical region is equal to the

level of significance α. The critical region is always on the tail of the distribution curve. It may be on both sides or on one tail (right or left), depending upon the alternative hypothesis.

Procedure for Hypothesis Testing

1. The first step is to set up a hypothesis called *Null Hypothesis* (H_0). For example, we may say that the average age of workers in a factory is 32 i.e. population mean $\mu = 32$. This is an assumption about the population mean of age of workers. Let us denote the sample mean by $\overline{x}$. Then the null hypothesis $H_0 : \overline{x} = 32$.
2. The complement of the null hypothesis is *Alternative Hypothesis* (H_1). As against the null hypothesis, the alternative hypothesis assumes that the difference is not due to sampling fluctuations, but they are real. So the alternative hypothesis in the present case would be the age of workers is not 32.

   Here three situations can arise.

   (i) The average age is either more than 32 or less than 32. In this case *Two-Tailed Test* is to be applied i.e. $H_1 : \neq 32$

   (ii) The average age is more than 32. Apply *Right-Tailed Test* i.e. H_1: $\overline{x} > 32$.

   (iii) The average age is less than 32. *Apply Left-Tailed Test* i.e. H_1: $\overline{x} < 32$.
3. The third step is to set up a level of significance α depending on reliability of the estimates and permissible risks at which the hypothesis is to be tested. This is to be decided before sample is drawn i.e. α is fixed in advance. It means the level of confidence with which a particular hypothesis is to be accepted or rejected. The level of confidence determines the probability of being right or wrong in accepting or rejecting the hypothesis. Usually α value is fixed in 0.01 or 0.05 or 0.1 i.e. same thing as setting confidence coefficient in 99% or 95% or 90% respectively.
4. The fourth step is to decide the test statistic. It is a statistic computed from a sample. It is generally based on some probability distribution like Z, t, χ^2 and F etc. Considering a large sample the test statistic for testing sample mean is:

$$Z = \frac{|\overline{x} - \mu|}{\text{SEM}} = \frac{|\overline{x} - 32|}{\frac{\sigma}{\sqrt{n}}}$$ where σ is the population standard deviation.

$\hat{\sigma}$ is the estimated population standard deviation $= \sqrt{\dfrac{\sum(x_i - \bar{x})^2}{n-1}}$

t-value is to be found for n-1 degrees of freedom.

Example 11.1: Is it likely that a sample of 16 items with mean 10.6 mm could have been drawn from a population of mean 10.4 mm and standard deviation of 0.4 mm?

Solution: Population standard deviation (σ) = 0.4 and μ = 10.4

Given that $\bar{x} = 10.6$ and $n = 16$

Let the sample has come from the population with mean 10.4 and sd 0.4

So, $H_0 : \bar{x} =$ (That means there is no difference between sample mean and population mean)

$H_1 : \bar{x} \neq$ (Sample mean and population mean are different) $\alpha = 0.05$

$$\text{SEM} = \frac{\sigma}{\sqrt{n}} = \frac{0.4}{\sqrt{16}} = 0.1$$

$$Z_c = \frac{|\bar{x} - \mu|}{\text{SEM}} = \frac{|10.6 - 10.4|}{0.1} = 2 \text{ (Calculated value of test statistic Z)}$$

From normal distribution table, for 5% significance level ($\alpha = 0.05$), Z_t = 1.96 (Table value of Z). Here $Z_c > Z_t$.

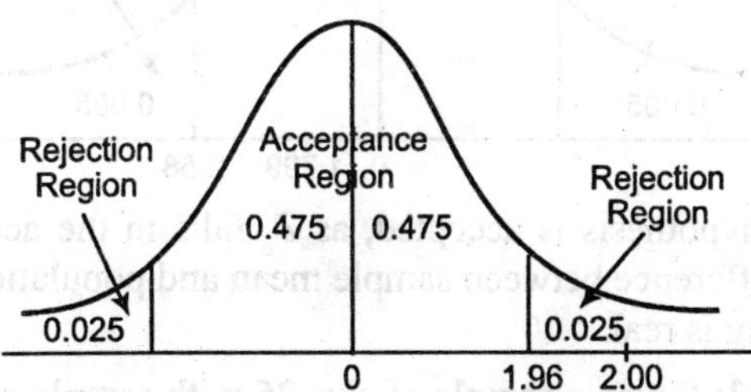

So the null hypothesis is rejected, as Z_c falls in the rejection region. There is significant difference between sample mean and population mean. Hence, there is no evidence to prove that the sample has come from the population having mean 10.4 mm and sd 0.4 mm.

Example 11.2: A sales manager feels that the monthly demand of a product is normally distributed with mean 3600 units and standard deviation of 250 units. Test the hypothesis regarding the manager's feeling about the mean with a sample size of 28, which gives a sample mean of 3515 units at 1% significance level.

Solution: Based on the sales manager's feeling we have $\mu = 3600$ and $\sigma = 250$

Given that $\overline{x} = 3515$ and $n = 28$

Let there is no difference between sample mean and population mean.

So, $H_0 : \overline{x} =$ (That means there is no difference between sample mean and population mean)

$H_1 : \overline{x} \neq$ (Sample mean and population mean are different) $\alpha = 0.01$

$$SEM = \frac{\sigma}{\sqrt{n}} = \frac{250}{\sqrt{28}} = 47.245$$

$$Z_c = \frac{|\overline{x} - \mu|}{SEM} = \frac{|3515 - 3600|}{47.245} = 1.799 \text{ (Calculated value of test statistic Z)}$$

From normal distribution table, for 1% significance level (α=0.05), Z_t = 2.58 (Table value of Z). Here $Z_c < Z_t$.

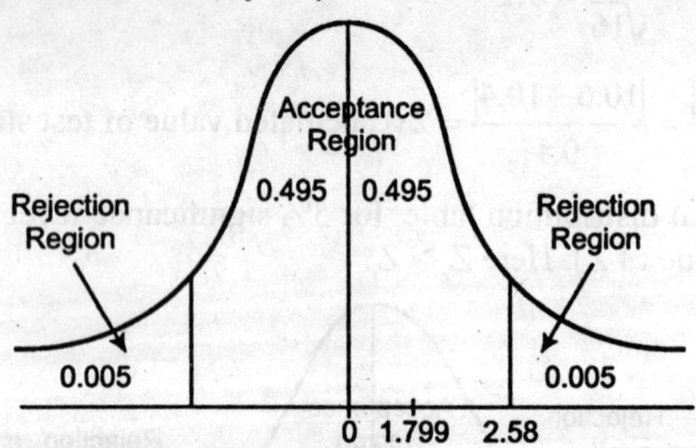

So the null hypothesis is accepted, as Z_c falls in the acceptance region. So there is no difference between sample mean and population mean. Hence, manager's feeling is real.

Example 11.3: Given a sample of size 25 with sample mean of 19.2 and sample standard deviation 4.5. Test the hypothesis H_0: population mean = 16, against the alternative H_1: that the population mean is greater than 16, at 1% significance level.

Solution: Here the population standard deviation (σ) is not known. Sample size is small (n = 25). So we have to take the sample sd as estimated population sd, $\hat{\sigma} = 4.5$. Given that $\overline{x} = 19.2$

H_0: $\mu = 16$

H_1: $\mu > 16$

It is a one-tailed test.

$\alpha = 0.01$

$$SEM = \frac{\hat{\sigma}}{\sqrt{n}} = \frac{4.5}{\sqrt{25}} = 0.9$$

$$Z_c = \frac{|\bar{x} - \mu|}{SEM} = \frac{|19.2 - 16|}{0.9} = 3.556 \text{ (Calculated value of test statistic Z)}$$

From normal distribution table, for one tailed test, at 1% significance level ($\alpha = 0.01$), $Z_t = 2.325$ (Table value of Z), shown in the figure below. Here $Z_c > Z_t$.

So the null hypothesis is rejected, as Z_c falls in the rejection region. There is significant difference between sample mean and population mean. Hence, null hypothesis is rejected and alternative hypothesis is accepted that the population mean is greater than 16.

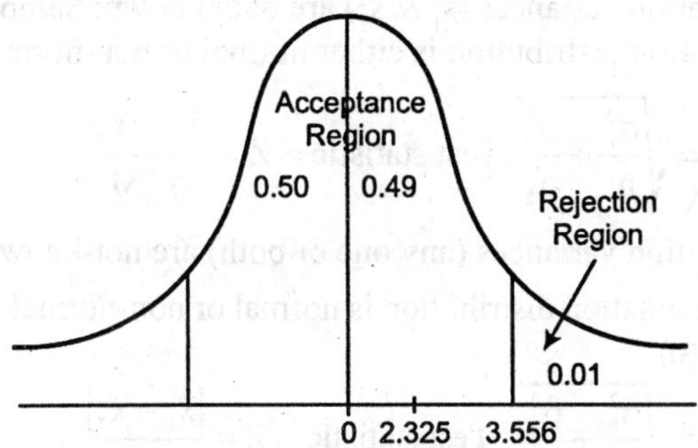

Example 11.4: Is it likely that the following sample came from a population of mean 1.4? Sample values are 1.6, 1.8, 1.8, 1.7, 1.9, 1.5.

Solution: Given that $\mu = 1.4$, sample size is small ($n = 6$). Population standard deviation σ is not known. Hence t-test would apply. We have to find the estimated population standard deviation by using the formula:

$$\hat{\sigma} = \sqrt{\frac{\sum(x_i - \bar{x})^2}{n-1}} = 1.567 \text{ (students are advised to do the calculations) and}$$

$\bar{x} = 1.717$

Let the sample come from the population with mean 1.4. That means there is no difference between sample mean and population mean.

So, $H_0 : \bar{x} = \mu$

$H_1 : \bar{x} \neq \mu$

$$SEM = \frac{\hat{\sigma}}{\sqrt{n}} = \frac{1.567}{\sqrt{6}} = 0.6397$$

Test statistic = $t = \frac{|\bar{x} - \mu|}{SEM} = \frac{|1.717 - 1.4|}{0.6397} = 0.4956$ (Calculated value of test statistic Z) $\alpha = 0.05$

t-value is to be found for n – 1 degrees of freedom. n – 1 = 6 – 1 = 5

From t- distribution table, for two tailed test, at 5% significance level (α=0.05), $t_t = 2.57$ (Table value of t). Here $t_c < t_t$.

So the null hypothesis is accepted, as t_c falls in the acceptance region. There is no significant difference between sample mean and population mean. The sample might have come from the population whose mean is 1.4.

B. Test of Difference between two Means

(a) Population variances (s_1^2 & s_2^2) are both known: Sample is of any size, population distribution is either normal or non-normal

$$SEM = \sqrt{\frac{\sigma_1^2}{n_1} + \frac{\sigma_2^2}{n_2}}; \text{Test statistic} = Z = \frac{|\bar{x}_1 - \bar{x}_2|}{SEM}$$

(b) Population variances (any one or both) are not-known:

(i) Population distribution is normal or non-normal and both n_1 & n_2 > 30

$$SEM = \sqrt{\frac{\hat{\sigma}_1^2}{n_1} + \frac{\hat{\sigma}_2^2}{n_2}}; \text{ Test statistic} = Z = \frac{|\bar{x}_1 - \bar{x}_2|}{SEM}$$

(ii) Both populations are normal and either of n_1 & n_2 < 30

$$SEM = \sqrt{\frac{\hat{\sigma}_1^2}{n_1} + \frac{\hat{\sigma}_2^2}{n_2}}; \text{ Test statistic} = t = \frac{|\bar{x}_1 - \bar{x}_2|}{SEM}$$

where $\hat{\sigma}_1, \hat{\sigma}_2$ are the estimated population standard deviations using formula as $\sqrt{\frac{\sum(x_i - \bar{x})^2}{n-1}}$; t-value is to be found for n_1-1 + n_2-1 (= n_1 + n_2 – 2) degrees of freedom.

(iii) Population distribution is normal or non-normal, population variances are equal

$\sigma_1^2 = \sigma_2^2$ and both n_1 & n_2 > 30

So, we have to find the pooled standard deviation for both the samples, which is given by, $\hat{\sigma}^2 = \dfrac{\hat{\sigma}_1^2(n_1-1)+\hat{\sigma}_2^2(n_2-1)}{n_1+n_2-2}$

$$SEM = \hat{\sigma}\sqrt{\frac{1}{n_1}+\frac{1}{n_2}}; \text{Test statistic} = Z = \frac{|\overline{x}_1-\overline{x}_2|}{SEM}$$

Example 11.5: Independent random samples of 32 boys and 35 girls in a certain college have the average body weights 52 kgs and 48 kgs respectively. If it is known that the standard deviations of all boys and girls in the college are 2 kgs and 1.8 kgs respectively, test whether there is any significant difference between average body weights of boys and girls at 1% level of significance.

Solution:

Population standard deviations are known; $\sigma_1 = 2.0$ & $\sigma_2 = 1.8$

Given that, $\overline{x}_1 = 52$ & $\overline{x}_2 = 48$; $n_1 = 32$ & $n_2 = 35$

$H_0: \overline{x}_1 = \overline{x}_2$

$H_1: \overline{x}_1 \neq \overline{x}_2$

$\alpha = 0.01$

$$SEM = \sqrt{\frac{\sigma_1^2}{n_1}+\frac{\sigma_2^2}{n_2}} = \sqrt{\frac{2^2}{32}+\frac{1.8^2}{35}} = 0.466$$

$$\text{Test statistic} = Z = \frac{|\overline{x}_1-\overline{x}_2|}{SEM} = \frac{|52-48|}{0.466} = 8.584 = Z_c$$

Table value of Z (Z_t) for $\alpha = 0.01$ is 2.58. So, $Z_c > Z_t$. Z_c falls in the rejection region. Hence null hypothesis is rejected. That means there is significant difference between average body weights of boys and girls.

Example 11.6: The average monthly income of 40 male workers in a factory is Rs 5600.00 with sd of Rs.12.00 in comparison to that of 35 female workers as Rs 5565.00 with sd of Rs 10.00 respectively. Test at 5% level of significance, whether there is any significant difference between average incomes of all male and female workers in the factory.

Solution: Population s.d. are not known and sample sizes are large.

Given that, $\overline{x}_1 = 5600$ & $\overline{x}_2 = 5565$; $n_1 = 40$ & $n_2 = 35$ (both > 30)

$\hat{\sigma}_1 = s_1 = 12$; $\hat{\sigma}_2 = s_2 = 10$

$H_0: \overline{x}_1 = \overline{x}_2$

$H_1: \overline{x}_1 \neq \overline{x}_2$

$\alpha = 0.05$

$$SEM = \sqrt{\frac{\hat{\sigma}_1^2}{n_1} + \frac{\hat{\sigma}_2^2}{n_2}} = \sqrt{\frac{12^2}{40} + \frac{10^2}{35}} = 2.541$$

$$\text{Test statistic} = Z = \frac{|\bar{x}_1 - \bar{x}_2|}{SEM} = \frac{|5600 - 5565|}{2.514} = 13.774 = Z_c$$

Table value of Z (Z_t) for α=0.05 is 1.96. So, $Z_c > Z_t$. Z_c falls in the rejection region. Hence null hypothesis is rejected. That means there is significant difference between average monthly incomes of male and female workers.

Example 11.7: Two independent samples of 8 and 7 items respectively had the values of the variable

Sample 1:	9	10	12	16	14	24	11	7
Sample 2:	9	11	17	13	15	14	15	

Assuming that the samples have come from populations of approximate normal distributions and having the same variances, test whether the means of two populations are equal.

Solution: Here population distributions are normal and sd's are not known.

Sample sizes are small. So t-test would apply.

x_1	$(x_1 - \bar{x}_1)^2$	x_2	$(x_2 - \bar{x}_2)^2$
9	15.016	9	19.612
10	8.266	11	5.898
12	0.766	17	12.755
16	9.766	13	0.184
14	1.266	15	2.469
24	123.766	14	0.327
11	3.516	15	2.469
7	34.516	Sum = 94	43.71429
103	196.875	Mean = 13.43 = $\bar{x}_2$	
Mean = 12.875 = $\bar{x}_1$			

$$= s_1 = \sqrt{\frac{\sum (x_i - \bar{x})^2}{n-1}} = \sqrt{\frac{196.875}{8-1}} = 5.303 = \hat{\sigma}_1$$

$$\text{and } s_2 = \sqrt{\frac{\sum (x_i - \bar{x})^2}{n-1}} = \sqrt{\frac{43.714}{7-1}} = 2.699 = \hat{\sigma}_2$$

Pooled Variance

$$= \hat{\sigma}^2 = \frac{\hat{\sigma}_1^2(n_1-1)+\hat{\sigma}_2^2(n_2-1)}{n_1+n_2-2} = \frac{5.303^2 \times (8-1) + 2.699^2 \times (7-1)}{8+7-2} = 18.504$$

$$= \hat{\sigma} = \sqrt{18.504} = 4.3017$$

H_0: $\bar{x}_1 = \bar{x}_2$

H_1: $\bar{x}_1 \neq \bar{x}_2$

$\alpha = 0.05$

$$= \text{SEM} = \hat{\sigma} \times \sqrt{\frac{1}{n_1}+\frac{1}{n_2}} = 4.3017 \times \sqrt{\frac{1}{8}+\frac{1}{7}} = 2.226$$

$$= \text{Test statistic} = t = \frac{|\bar{x}_1 - \bar{x}_2|}{\text{SEM}} = \frac{|12.875 - 13.43|}{2.226} = 0.249 = t_c$$

Table value of t (t_t) for α=0.05 at 13 df is 2.16 So, $t_c < t_t$. t_c falls in the acceptance region. Hence null hypothesis is accepted. That means there is no significant difference between the means of two populations.

Example 11.8: Test the above hypothesis, if it is known that the two population variances are not equal.

Solution: In such case,

$$= \text{SEM} = \sqrt{\frac{\hat{\sigma}_1^2}{n_1}+\frac{\hat{\sigma}_2^2}{n_2}} = \sqrt{\frac{5.303^2}{8}+\frac{2.699^2}{7}} = 2.134$$

$$= \text{est statistic} = t = \frac{|\bar{x}_1 - \bar{x}_2|}{\text{SEM}} = \frac{|12.875 - 13.43|}{2.134} = 0.260 = t_c$$

Table value of t (t_t) for α = 0.05 at 13 df is 2.16 So, $t_c < t_t$. t_c falls in the acceptance region. Hence null hypothesis is accepted. That means there is no significant difference between the means of two populations.

C) Paired t-Test

When observations are taken pairwise, paired t-test is applied. We can have n pairs of observations. There will be n number of values of difference between each pair of observation, which is assumed to be normal with mean m_d and variance σ_d^2. The unknown σ_d^2 is estimated by σ_d^2, which is the sample variance of the differences.

$$\bar{d} = \frac{1}{n} \times \sum_{i=1}^{n} d_i \text{ and SE} = \frac{s_d}{\sqrt{n}};\ s_d^2 = \frac{\sum d^2 - n\bar{d}^2}{n-1}$$

$$\text{Test statistic} = t = \frac{|\bar{d}|}{SE}$$

t-value is to be found for n-1 degrees of freedom.

Example 11.9: 12 workers of a company were imparted a skill development training and the following are their IQs measured before and after the training respectively. Was the training effective?

Before:	35	48	38	52	47	50	45	51	40	46	49	52
After:	38	48	39	51	50	52	48	51	42	47	50	50

Solution: As the test is conducted on the same set of samples, paired t-test would apply.

Before:	35	48	38	52	47	50	45	51	40	46	49	52
After:	38	48	39	51	50	52	48	51	42	47	50	50
Difference:	3	0	1	-1	3	2	3	0	2	1	1	-2
d^2:	9	0	1	1	9	4	9	0	4	1	1	4

$$\text{Sum of difference} = \sum d = 13; \bar{d} = \frac{13}{12} = 1.083; \sum d^2 = 43$$

$$s_d^2 = \frac{\sum d^2 - n\bar{d}^2}{n-1} = \frac{43 - 12 \times 1.083^2}{12-1} = 2.629$$

H_0: There is no difference between the mean IQ before and after training.

H_1: There is difference between the mean IQ before and after training.

Let $\alpha = 0.05$

$$= SE = \frac{s_d}{\sqrt{n}} = \sqrt{\frac{2.629}{12}} = 0.468$$

$$\text{Test statistic} = t = \frac{|\bar{d}|}{SE} = \frac{1.083}{0.468} = 2.314 = t_c$$

Table value of t (t_t) for α=0.05 at 11 df is 2.20. So, $t_c > t_t$. t_c falls in the rejection region. Hence null hypothesis is rejected. That means there is significant difference between the mean IQ of workers before and after training. Training was effective.

D. Test for a Single Proportion

$SE = \sqrt{\frac{p(1-p)}{n}}$; p is the sample proportion. If P is the hypothesised population proportion,

$$\text{Test statistic} = Z = \frac{|p-P|}{SE}$$

Example 11.10: A study conducted on 120 employees of a company revealed that 18 employees work over-time every week. But the personnel manager of the company hypothesised this proportion to be 0.13. Test whether we can accept personnel manager's view at 5% level of significance ?

Solution: We have,

$$p = \frac{18}{120} = 0.15 \text{ and } P = 0.13$$

H_0 : There is no difference between sample proportion and hypothesised proportion.

H_1 : There is difference between sample proportion and hypothesised proportion.

$\alpha = 0.05$

$$= SE = \sqrt{\frac{p(1-p)}{n}} = \sqrt{\frac{0.15 \times (1-0.15)}{120}} = 0.0326$$

$$\text{Test statistic} = Z = \frac{|p-P|}{SE} = \frac{|0.15-0.13|}{0.0326} = 0.6135 = Zc$$

Table value of Z (Z_t) for α=0.05 is 1.96 So, $Z_c < Z_t$. Z_c falls in the acceptance region. Hence null hypothesis is accepted. That means there is no significant difference between the hypothesised value and the sample proportion. Hence, personnel manager's views is rational.

E. Test for Difference between Two Proportions

$$SE = \sqrt{\frac{p_1(1-p_1)}{n_1} + \frac{p_2(1-p_2)}{n_2}}$$

$$\text{Test statistic} = Z = \frac{|p_1 - p_2|}{SE}$$

Example 11.11: 200 patients were divided into two halves and were administered two drugs A & B. The recovery rates were 76% and 83% respectively. Test whether the difference is significant at 1% level of significance.

Solution: Given data: $p_1 = 0.76$ and $p_2 = 0.83$; $n_1 = 100$ and $n_2 = 100$; $\alpha =$ 0.05

$$SE = \sqrt{\frac{p_1(1-p_1)}{n_1} + \frac{p_2(1-p_2)}{n_2}} = \sqrt{\frac{0.76(1-0.76)}{100} + \frac{0.83(1-0.83)}{100}} = 0.0569$$

$$\text{Test statistic} = Z = \frac{|p_1 - p_2|}{SE} = \frac{|0.76 - 0.83|}{0.05689} = 1.2302 = Z_c$$

Table value of Z (Z_t) for $\alpha = 0.05$ is 1.96 So, $Z_c < Z_t$. Z_c falls in the acceptance region. Hence null hypothesis is accepted. That means there is no significant difference.

F. Test for Variance of a Normal Distribution

(i) If population mean (μ) is known:

$$= \chi^2 = \sum_{i=1}^{n} \frac{(x_i - \mu)^2}{\sigma^2}$$

The calculated value of chi-square should be tested against table value of chi-square with n degrees of freedom.

(ii) If population mean (μ) is not known:

$$= \chi^2 = \sum_{i=1}^{n} \frac{(x_i - \overline{x})^2}{\sigma^2}$$

This calculated value of chi-square should be tested against table value of chi-square with (n-1) degrees of freedom.

Example 11.12: A random sample of 8 cigars of a certain brand showed nicotine content having standard deviation of 2.5 milligram. What can you conclude about the hypothesis that the standard deviation is not 3 milligram?

Solution: Given:

$n = 8; s = 2.5; \sigma = 3$

$H_0: \sigma = 3$

$H_1: \sigma \neq 3$

Let, $\alpha = 0.01$

$s^2 = 2.5^2 = 6.25$

$$\sum(x - \overline{x})^2 = (n-1)s^2 = (8-1) \times 6.25 = 43.75$$

$$\div^2 = \sum_{i=1}^{n} \frac{(x_i - \overline{x})^2}{\acute{o}^2} = \frac{43.75}{3^2} = 4.861 \text{ (calculated value)}$$

Table value of $\chi^2_{0.01}$ at $\alpha = 0.01$ for 7 df is 18.475

Table value of $\chi^2_{0.09}$ at $\alpha = 0.99$ for 7 df is 1.239

Since calculated value of lies between $\chi^2_{0.01}$ and $\chi^2_{0.99}$, there is no reason to reject the null hypothesis.

Example 11.13: The weights of 12 containers are 10.4, 9.8, 10.2, 10.4, 10.8, 9.9, 10.6, 10.5, 9.8, 9.6, 10.2 and 9.8 kg. Test the hypothesis that $\sigma^2 = 0.08$ against alternative hypothesis $\sigma^2 > 0.08$ at 1% level of significance.

Solution: On working out we will get $\sum(x-\bar{x})^2 = 1.607$

$$\chi^2 = \sum_{i=1}^{n}\frac{(x_i-\bar{x})^2}{\sigma^2} = \frac{1.607}{0.08} = 20.0875\,(\text{calculated value})$$

df = 12 – 1 = 11

Table value of χ^2 at α=0.01 for 11 df = 24.725; which is greater than the calculated value. So there is no reason to doubt the hypothesis that $\sigma^2 = 0.08$.

χ^2-TEST FOR GOODNESS OF FIT

This test is used to know how well any particular theoretical distribution (such as Binomial, Poisson or Normal distribution) fit to the observed field data. Chi-square test gives a solution to this. If the calculated value of χ^2 is less than the table value of χ^2 at a given level of significance, the fit is considered to be a good one, which in other words means that the divergence between the observed and expected frequencies is due to sampling fluctuations. But if the calculated value of χ^2 is greater than the table value, the fit is not considered to be a good one.

$$\chi^2 = \sum\frac{(O_i - E_i)^2}{E_i}$$; where O_i and E_i are observed and expected frequencies respectively.

Example 11.14: In an experiment on breeding of Pea, Mendel obtained the following frequencies of seeds. Round & Yellow 315, Wrinkled & Yellow 101, Round & Green 108, Wrinkled & Green 32. Theory predicts that the frequencies should be in the proportion 9:3:3:1. Are the experimental results compatible with genetic theory?

Solution: Total number of seeds under test = 315 + 101 + 108 + 32 = 556

As per theory they should remain in proportion of 9:3:3:1. So the numbers should be

Round & Yellow: $556 \times \frac{9}{16} = 312.75$

Wrinkled & Yellow: $556 \times \frac{3}{16} = 104.25$

Round & Green: $556 \times \frac{3}{16} = 104.25$

Wrinkled & Green: $556 \times \frac{1}{16} = 34.75$

Class	Observed Frequency (O)	Expected Frequency (E)	$\frac{(O-E)^2}{E}$
Round & Yellow	315	312.75	0.01618
Wrinkled & Yellow	101	104.25	0.101318
Round & Green	108	104.25	0.13489
Wrinkled & Green	32	34.75	0.21762
		Total:	0.470008
		$\chi^2 = \sum \frac{(O_i - E_i)^2}{E_i} = 0.470008$	

Critical value of χ^2 for 3 df at $\alpha = 0.05$ is 7.815. Calculated value is much less than table value. So there is no reason to suspect the hypothesis and the experimental results seem to confirm the genetic theory.

Example 11.15: Fit a binomial distribution with the parameters n = 4 and p = 0.2 to the following frequency distribution and test the goodness of fit. Assume $\alpha = 0.05$.

Number of successes:	0	1	2	3	4
Frequency:	38	42	16	3	1

Solution: For a binomial distribution with n = 0 and p = 0.2 & q = 0.8, we have the following expected frequencies: (using the formula ${}^nC_r p^r q^{n-r}$). Total frequency = 38 + 42 + 16 + 3 + 1 = 100

0 success: ${}^4C_0 p^0 q^{4-0} = 0.4096 \times 100 = 40.96$

1 success: ${}^4C_1 p^1 q^{4-1} = 0.4096 \times 100 = 40.96$

2 successes: ${}^4C_2 p^2 q^{4-2} = 0.1536 \times 100 = 15.36$

3 successes: ${}^4C_3 p^3 q^{4-3} = 0.0256 \times 100 = 2.56$

4 successes: ${}^4C_4p^4q^{4-4} = 0.0016 \times 100 = 0.16$

Successes	Observed Frequency (O)	Expected Frequency (E)	$\frac{(O-E)^2}{E}$
0	38	40.96	0.2139
1	42	40.96	0.0264
2	16	15.36	0.0267
3	3	2.56	0.0756
4	1	0.16	4.41

It may be noted that for categories of 3 and 4 successes, the expected values are 2.56 and 0.16 respectively, which are less than 5. If the cell frequencies become less than 5, then they should be clubbed into one, so as to make the expected frequency more than 5. So, the values for successes 2, 3 and 4 are clubbed together and the revised table is presented below.

Successes	Observed Frequency(O)	Expected Frequency(E)	$\frac{(O-E)^2}{E}$
0	38	40.96	0.2139
1	42	40.96	0.0264
2	20	18.08	0.2039
		Total:	0.4442
		$\chi^2 = \sum \frac{(O_i - E_i)^2}{E_i} = 0.4442$	

Critical value of χ^2 for 2 df at $\alpha = 0.05$ is 5.991. Calculated value is much less than table value. So there is no reason to suspect the null hypothesis and hence, it is concluded that the binomial distribution with $p = 0.2$ provides a good description of the observed frequencies.

χ^2- TEST OF INDEPENDENCE

This test enables to ascertain the degree of association. For example, we may be interested to know, whether a particular training on the workers is effective or not. So the attributes of workers before and after training are recorded and the chi-square test is conducted. Null hypothesis is constructed by stating that the training is ineffective i.e. the attributes before and after training are same. In other words, the training and attributes of workers are independent of each other and not associated. Calculated value and table value of χ^2 decide whether to accept the null hypothesis or alternative hypothesis. It is to be noted that χ^2 is not a measure of the degree of relationship or the form of relationship between two attributes, but is simply a technique of judging the significance of such association or relationship between two attributes.

$\chi^2 = \sum \frac{(O_i - E_i)^2}{E_i}$; where O_i and E_i are observed and expected frequencies respectively.

Certain conditions are necessary for its validity. These are:

1. The sample size n should be large.
2. Observations recorded and used are collected on random basis.
3. None of the expected frequencies should be less than 5. If it happens, one or more classes should be combined together.
4. Sum of observed frequencies = Sum of expected frequencies = n
5. The constraints must be linear.
6. When the expected frequency is less than 5, Yates has proposed a correction, known as Yates' correction of continuity. This correction consists in replacing the observed frequency d by $\left(d+\frac{1}{2}\right)$ if ad > bc and by $\left(d-\frac{1}{2}\right)$ if ad < bc and adjusting the remaining frequencies accordingly in such a way that the marginal totals remain unaltered. The effect of Yates' correction is to replace $(ad - bc)^2$ in the formula by $\{|ad - bc| - n/2\}^2$. Thus, the corrected statistic becomes:

$$= \div^2 = \frac{n \times \left[|ad - bc| - \frac{n}{2}\right]^2}{(a+b)\times(c+d)\times(a+c)\times(b+d)};$$

where cell frequencies are like,

Class	Condition-I	Condition-II	Total
X	a	b	a + b
Y	c	d	c + d
Total:	a + c	b + d	N

Degree of freedom plays an important role here. If there are 10 frequency classes and there is one independent constraint, then the df is 10 – 1 = 9.In case of a contingency table (a table with 2 columns and 2 rows or, 2 columns and more than 2 rows or, 2 rows and more than 2 columns or, more than 2 columns and more than 2 rows), the d.f. is worked out as d.f. = (c-1) X (r-1); where c is number of columns and r is the number of rows.

Example 11.16: A new method of treating a disease was tried in a hospital on randomly selected patients admitted to a hospital with the following results:

	Survived	Died	Patients Admitted
Old treatment:	115	25	140
New treatment:	260	40	300

Test the validity of the claim that the new treatment reduced the fatality of the disease.

Solution: H_0: The null hypothesis is constructed that the new treatment and survival rate are independent of each other i.e. new treatment has no effect on survival rate.

H_1: Alternative hypothesis will be new treatment and survival rate are dependent on each other i.e. new treatment is effective.

In the given data: a = 115; b = 25; c = 260; d = 40; n = 140 + 300 = 440

$$= \chi^2 = \frac{n \times \left[|ad - bc| - \frac{n}{2}\right]^2}{(a+b)\times(c+d)\times(a+c)\times(b+d)}$$

$$= \frac{440 \times \left[|115 \times 40 - 25 \times 260| - \frac{440}{2}\right]^2}{(115+25)\times(260+40)\times(115+260)\times(25+40)} = 1.213$$

This value can also be obtained by using the formula $\chi^2 = \sum \frac{(O_i - E_i)^2}{E_i}$ and same result will occur. Students are advised to work out this method for check.

Degree of freedom = (r – 1) × (c – 1) = (2 – 1) × (2 – 1) = 1

The table value of for 1 df at $\alpha = 0.01$ is equal to 6.635. So the calculated value is well within the tabulated value. Hence, null hypothesis is accepted that the new treatment and survival rate are independent of each other. So the claim that the new treatment reduced the fatality of disease is false.

Example 11.17: In a study undertaken by family planning organization, a random sample of 700 families were classified according to their income and the number of children.

Income Group	Number of children		
	0-1	2-3	more than 3
Higher	43	25	12
Middle	65	102	83
Lower	72	138	160

Test the hypothesis at 1% significance level that the family income is independent of the number of children.

Solution: The observed values (O_i) are as below:

Income	**Number of children**		**Row Total**	
Group	**0-1**	**2-3**	**more than 3**	
Higher	43	25	12	80
Middle	65	102	83	250
Lower	72	138	160	370
Column Total:	180	265	255	700

The expected cell frequencies (E_i) are calculated as below:

Income Group	**Number of children**		
	0-12-3	**more than 3**	
Higher	$\frac{80\times180}{700}=20.57$	$\frac{80\times265}{700}=30.285$	$\frac{80\times255}{700}=29.143$
Middle	$\frac{250\times180}{700}=64.28$	$\frac{250\times265}{700}=94.64$	$\frac{250\times255}{700}=91.07$
Lower	$\frac{370\times180}{700}=95.14$	$\frac{370\times265}{700}=140.07$	$\frac{370\times255}{700}=134.78$

$$\div^2=\sum\frac{(O_i-E_i)^2}{E_i}$$

$$=\frac{(43-20.57)^2}{20.57}+\frac{(25-30.285)^2}{30.285}+\frac{(12-29.143)^2}{29.143}+\frac{(65-64.28)^2}{64.28}$$

$$+\frac{(102-94.64)^2}{94.64}+\frac{(83-91.07)^2}{91.07}+\frac{(72-95.14)^2}{95.14}+\frac{(138-140.07)^2}{140.07}$$

$$+\frac{(160-134.78)^2}{134.78}=46.778$$

Degree of freedom = (r – 1) × (c – 1) = (3 – 1) × (3 – 1) = 2 × 2 = 4

Table value of χ^2 at $\alpha = 0.01$ for 4 df = 13.277. Calculated value is greater than table value. So the null hypothesis is rejected. That means number of children in a family is dependent upon the family income.

χ^2-TEST OF HOMOGENITY

Here in test of homogeneity, unlike in test of independence, 2 or more samples are drawn. The test is designed to determine whether two or more independent samples are drawn from the same population or from different populations. For example, we may be interested to know, people in various age groups say 21-30, 31-40, 41-50 and so on of a region have the same level of satisfaction on functioning of public representatives or not. So we collect samples from each age group and hypothesise that there is no difference in their level of satisfaction. That is to say the groups are homogeneous. The principle that governs the decision is same as that in test of independence.

Example 11.18: Two candidates A and B obtained the following votes in a college election. Examine whether the students have any voting preference.

Students	Vote for		Total
	A	B	
Undergraduate:	132	268	400
Post Graduate:	83	67	150
Total:	215	335	550

Solution: Null Hypothesis H_0: Voting preference of under graduate and post graduate students is same.

Alternative Hypothesis H_1: Voting preference of under graduate and post graduate students is not same.

Let $\alpha = 0.05$

The cell frequencies are calculated as below. (Expected Values)

Students	Vote for	
	A	B
Undergraduate	$\frac{400 \times 215}{550} = 156.36$	$\frac{400 \times 335}{550} = 243.64$
Graduate	$\frac{150 \times 215}{550} = 58.64$	$\frac{150 \times 335}{550} = 91.36$

$$\chi^2 = \sum \frac{(O_i - E_i)^2}{E_i} = \frac{(132-156.36)^2}{156.36} + \frac{(268-243.64)^2}{243.64}$$

$$+ \frac{(83-58.64)^2}{58.64} + \frac{(67-91.36)^2}{91.36} +$$

$= 3.795 + 2.436 + 10.120 + 6.495 = 22.846$

Degree of freedom = $(2 - 1) \times (2 - 1) = 1$

Table value of χ^2 at $\alpha = 0.05$ for 1df = 3.841. Calculated value is more than the table value. Null hypothesis is rejected. So the preference of two groups of students is not same.

G. Test for Equality of Variances from Normal Populations (F-Test)

This test is carried out to know if two samples have come from populations having same variances or not. As defined earlier a statistic F is found using the formula $F = \frac{\hat{\sigma}_1^2}{\hat{\sigma}_2^2}$; where $\hat{\sigma}_1^2 = \frac{1}{n_1 - 1}\sum(x_{1i} - \bar{x}_1)^2$ and $\hat{\sigma}_2^2 = \frac{1}{n_2 - 1}\sum(x_{2i} - \bar{x}_2)^2$. The larger variance is taken in the numerator to facilitate finding the F-value from F-table, where the values pertain to right tail only. Degrees of freedom are $v_1 = n_1 - 1$ and $v_2 = n_2 - 1$. Null hypothesis is established as H_0: $\sigma_1^2 = \sigma_2^2$ with the alternative hypothesis H_1: (i) $\sigma_1^2 \neq \sigma_2^2$ or (ii) $\sigma_1^2 > \sigma_2^2$. In the former case (i), if calculated value of F remains within the region confined between $F_{\alpha/2}$ and $F_{1-\alpha/2}$, then null hypothesis is accepted and alternative hypothesis is rejected. In the latter case (ii), it becomes one tailed test and if calculated value of F remains within F_α, then null hypothesis is accepted and alternative hypothesis is rejected.

Example 11.19: Two random samples from two independent populations are:

Sample-I:	26	32	40	18	26	44	30	38	40	36
Sample-II:	22	18	24	26	32	16	24	18		

Find the estimates of variances of the populations and test whether the two samples have come from populations having same variance. Test at 95% confidence level.

Solution: By using the formulae, $\hat{\sigma}_1^2 = \frac{1}{n_1 - 1}\sum(x_{1i} - \bar{x}_1)^2$ and $\hat{\sigma}_2^2 = \frac{1}{n_2 - 1}\sum(x_{2i} - \bar{x}_2)^2$, we can find the estimated population variances as $\hat{\sigma}_1^2 = 65.111$ and $\hat{\sigma}_2^2 = 27.143$. Students are advised to work out these values.

Hence, $F = \frac{\hat{\sigma}_1^2}{\hat{\sigma}_2^2} = \frac{65.111}{27.143} = 2.399$

Construct Null Hypothesis H_0: Two samples have come from populations having same variances i.e. $\sigma_1^2 = \sigma_2^2$.

$H_1: \sigma_1^2 \neq \sigma_2^2$

Let $\alpha = 0.05$

$v_1 = 10-1=9$ and $v_2 = 8-1=7$

It is a two-tailed test.

Table value of $F_{0.05(9,7)}$ at $\alpha = 0.05$ for 9,7 df is 3.68.

Table value of $F_{0.95(9,7)} = 1 \div F_{0.05(7,9)} = 1 \div 3.29 = 0.304$

The calculated value remains within the range and hence there is no reason to reject the null hypothesis. The samples might have come from the populations having same variances.

Example 11.20: Given $n_1 = 13$; $s_1 = 8.4$ and $n_2 = 10$; $s_2 = 4.6$. Test the null hypothesis H_0: $\sigma_1^2 = \sigma_2^2$ against H_1: $\sigma_1^2 > \sigma_2^2$. Use 5% level of significance.

Solution: It is one tailed test.

$$F = \frac{s_1^2}{s_2^2} = \frac{8.4^2}{4.6^2} = 3.335$$

Table value of $F_{0.05(12,9)} = 3.07$

As the calculated value is more than the table value, null hypothesis is rejected and alternative hypothesis is accepted at 5% significance level.

Producer's Risk (Type I Error = α) & Consumer's Risk (Type II Error = β)

In a typical business scenario, buyers (or consumers) are generally willing to accept lots that contain small percentages of defectives as "good," especially if the cost related to a few defects is low. Hence, mistakes will be made in acceptance of bad lots and rejection of good lots. The probability that a bad lot containing tolerable defectives (Lot Tolerance Percent Defective- LTPD) will be accepted by the consumer is known as the consumer's risk, or beta (β), or the probability of making a Type II error. On the other hand, the probability that a good lot containing acceptable defectives equal (Accepted Quality Level- AQL) will be rejected by the consumer is known as the producer's risk, or alpha (α), or the probability of making a Type-I error. Many sampling plans are designed to have a producer's risk of 5 percent and a consumer's risk of 10 percent, although other combinations are also used. The probability of accepting a lot with AQL quality is $1 - \alpha$.

		State of Nature	
		Product is Good	**Product is Bad**
Outcome	Consumer Accepts Goods	OK	Consumer's Risk **(Type-Ii Error = B)**
	Consumer Rejects Goods	Producer's Risk **(Type-I Error = A)**	OK

EXERCISES

1. What statistical parameters are to be considered in designing laboratory and field experiments.
2. Write short notes on Null Hypothesis within 150 words.
3. Write within 150 words on tests of significance.
4. Describe principles and applications of S.E.M.
5. Write brief accounts of Standard Deviation, Standard Error and methods of sampling in about 200 words.
6. Write brief account of t-test and its significance in about 200 words.
7. Explain statistical terms: (i) Null Hypothesis (ii) Degrees of freedom.
8. What is t-test? Discuss its application.
9. Describe 't'-test to test the significant difference between two sample means when samples are drawn independently.
10. What is paired 't' test. Describe the test. Mention the condition when it is used.
11. Write a brief note on test of significance.
12. What is Student t-test? Explain its application with examples.
13. A wanted to buy a house and learnt from his colleague that houses in Green Villa are sold at a price Rs.15,000 less than the house sold at Turbo Villa. So he enquired about 10 houses each at both the places and found their costs as follows. Does the data support his claim?

Cost of House at Green Villa (Rs Lakh)	14	13	18	19	20	15	16	17	15	18
Cost of House at Turbo Villa (Rs Lakh)	8	9	14	15	12	18	16	12	15	10

14. In 2001, a survey of 40 municipal hospitals revealed an average occupancy rate of 84.6% and the sample standard deviation was 9.4%. Another survey of 65 municipal hospitals in 2004 found an average occupancy rate of 68.5% and the sample standard deviation was 18.2%. At 10% level of significance, can we conclude that the average occupancy rate changed significantly during the 3 years between surveys?

15. Eight computer- dealers in a city were asked for their prices on two similar laser printers. The results of this survey are given below. At 5% significance level, is it reasonable to assert that, on average, the Mapple price is less expensive than the Accord price?

Mapple price (in thousand rupees)	2.5	2.2	1.7	1.9	1.8	1.7	2.1	1.9
Accord price (in thousand rupees)	2.0	2.7	2.4	1.9	2.1	2.3	2.6	1.8

16. An aerobics club claims to curb body weights after certain newly developed exercise technique. Ten persons were put under observation for one month and the following data were recorded regarding their body weights before and after the exercise. Does it support the club's claim? Test at α=0.05.

Initial Body weight	105	108	110	98	96	106	112	90	101	107
Body weight after 1 month	103	105	104	95	92	101	107	88	98	102

17. Eight suppliers of certain glass ware item were interviewed regarding the number of broken items received by them per shipment, before and after adopting a special kind of packaging method. State whether the new method of packaging has lowered the number of broken items per shipment. Test at α = 0.05.

Supplier:-	1	2	3	4	5	6	7	8
Before	16	11	17	6	13	18	5	16
After	13	12	11	5	8	14	7	14

18. Albino rats (n = 10) were administered with an Ayurvedic mercurial drug @ 10 mg/kg/day for 7 days. Initial and final body weights of the rats were recorded as shown in the table. Determine whether administration of the drug has any significant effect on the gain or loss in the body weight using paired t-test.

Rat No.	1	2	3	4	5	6	7	8	9	10
Initial Body Weight in gm.	109	114	101	97	111	109	96	119	101	109
Final Body Weight after 7 days of drug administration	108	115	99	94	107	111	97	114	97	110

19. Two varieties of potato plants (X and Y) yield tubers as following. Does the mean number of tubers of the variety A significantly differ from that of Variety B ?

Tuber Yield, kg / plant

Variety X	2.5	2.7	1.9	2.7	1.4	1.9	2.1	2.2	2.5	2.6
Variety Y	2.8	2.5	2.7	2.9	3.1	2.3	2.3	3.2	2.5	2.9

20. Samples of two types of pencil batteries were tested for length of life and the following data were obtained:

Type I	**Type II**
$n_1 = 5$	$n_2 = 7$
$\bar{x}_1 = 1115$ hour	$\bar{x}_2 = 1030$ hour
$sd_1 = 32$ hour	$sd_2 = 36$ hour

Is the difference in the means sufficient to conclude that the type – I is superior to type – II in respect of length of battery-life?

21. The sale of chocolates by a company on randomly selected 7 days are given below. Can we conclude that the average sale of chocolates of the company is 23.67 kg?

Days	1	2	3	4	5	6	7
Sale (kg)	80.89	8.81	18.01	22.13	15.62	18.26	6.4

22. From 2001 to 2005, the mean price / earnings (P/E) ratio of approximately 100 stocks listed on a Stock Exchange was 14.38 and the standard deviation was 8.63. In a sample of 28 randomly chosen Stock Exchanges, the mean P/E ratio in 2006 was 11.78. Does this sample present sufficient evidence to conclude at 5% significance level that in 2006 the mean P/E ratio had changed from its earlier value?

23. Some financial theoreticians say that the stock market's daily prices constitute a "random walk with positive drift". If this is accurate, then the Wankhade Industrial Average should show a gain on more than 50% of all trading days. If the average increased on 105 of

180 randomly chosen days, what do you think about the suggested theory? Use a 0.01 level of significance.

24. MicroSwaft estimated last year that 36% of potential software buyers were planning to wait to purchase the new operating system until an upgrade had been released. After an advertisement campaign to reassure the public, MicroSwaft surveyed 2800 people and found 960 who were still skeptical. At 5% significance level, can the company conclude the proportion of skeptical people has decreased?
25. 52 of 2010 randomly sampled corporations had their 2005 income tax returns audited. In another sample of 2400 corporations, 62 had their 2004 returns audited. Was the fraction of corporate returns audited by 2005 significantly different from the 2004 fraction? Test the appropriate hypothesis at α=0.01
26. Mr. Brian, the new manager of cosmetic make-up saloon, is interested in the percentage of costumers who are totally satisfied with the store. The previous manager had 84% of the costumers totally satisfied and Brian claims the same is true today. Brian sampled 185 costumers and found 156 were totally satisfied. At 1% significance level, is there evidence that Brian's claim is valid?
27. In a survey of injection drug users in a city, it was found that 12 out of 405 were HIV positive. Can we conclude that fewer than 5% of the injection drug users in the sampled population are HIV positive?
28. The mean yield of wheat from a district A was 208 kg with standard deviation 6 kg per acre from a sample of 125 plots. In another district B, the mean yield was 210 kg with standard deviation of 7 kg from a sample of 130 plots. Test whether there is any significant difference between the mean yield of crops in the two districts.
29. A random sample of 8 customers at a banking institution were studied as to the time of disposal by the bank authorities and showed the average time of disposal to be 2.5 minutes. What can you conclude about the hypothesis that the standard deviation is not 3 seconds ?
30. The time taken by 12 graduate students for reading and writing e-mails are 10.4, 9.8, 10.2, 10.4, 10.8, 9.9, 10.6, 10.5, 9.8, 9.6, 10.2 and 9.8 minutes. Test the hypothesis that $\sigma^2 = 0.08$ against alternative hypothesis $\sigma^2 > 0.08$ at 1% level of significance.
31. In an experiment on breeding of Pea, a biologist obtained the following frequencies of seeds. Pale & Black 305, Fresh & Black 98, Pale & White 106, Fresh & White 30. Theory predicts that the frequencies should be in the proportion 9:3:3:1. Are the experimental results compatible with genetic theory?

32. Fit a binomial distribution with the parameters n = 3 and p = 0.15 to the following frequency distribution and test the goodness of fit. Assume $\alpha = 0.05$.

Number of successes:	0	1	2	3
Frequency:	32	45	15	4

33. A brand manager is concerned that his brand's share may be unevenly distributed throughout the state. In a survey in which the state was divided into four zones, a random sampling of 120 costumers in each zone was surveyed with the following results.

	ZONE				
	NE	**NW**	**SE**	**SW**	**TOTAL**
Purchase the brand:	45	60	50	55	210
Do not purchase:	65	50	60	55	230
Total:	110	110	110	110	440

Develop a table of observed and expected frequencies for this problem. Calculate the sample χ^2 value. State the null and alternative hypothesis. At $\alpha = 0.05$, test whether brand share is the same across the four zones.

34. In a study undertaken by an advertising farm the following results were obtained regarding the demographics of a new product.

Future	**Age Group**				
Activity	14-16	17-19	20-22	23-25	26-29
Purchase frequently:	12	18	17	22	32
Seldom purchase	18	25	29	24	30
Never purchase	45	32	29	29	13

Test the hypothesis at 1% significance level that purchase is independent of the age group.

35. Two candidates A and B obtained the following votes in a college election. Examine whether the students have any voting preference.

Students	**Vote for**		**Total**
	A	**B**	
Undergraduate:	135	265	400
Post Graduate:	85	65	150
Total:	220	330	550

36. Two random samples from two independent populations are:

Sample-I:	25	31	39	17	27	45	31	39	41	37
Sample-II:	21	17	23	25	33	17	25	19		

Find the estimates of variances of the populations and test whether the two samples have come from populations having same variance. Test at 90% confidence level.

37. Given $n_1 = 18$; $s_1 = 8.6$ and $n_2 = 12$; $s_2 = 4.5$. Test the null hypothesis H_0: $\sigma_1^2 = \sigma_2^2$ against H_1: $\sigma_1^2 < \sigma_2^2$. Use 5% level of significance.
38. Write brief account of chi-square and t-test within 200 words.
39. Write notes on 2 × 2 contingency table.
40. What is chi-square test? Describe its uses in research.
41. Define chi-square test and discuss its uses in testing of hypothesis.
42. What is chi-square test? What are its uses? How do you calculate χ^2 in a 2 × 2 contingency table, where the frequencies are given?
43. Describe χ^2 test of significance and state the various uses of this test.
44. Write brief account of Chi-square and its importance in about 200 words.

ANALYSIS OF VARIANCE (ANOVA)

In the previous chapter we discussed the comparison of two normal populations on the basis of two sample means (Fisher's t-test). The analysis of variance is an effective statistical technique that is widely used for comparison of three or more population means. The test of significance based on t statistic is only a procedure of testing two population means, whereas this analysis of variance procedure helps us to compare several population means at a time. For example, the rates of production of different industries of same locality may be compared with the help of analysis of variance. So the basic purpose of the analysis of variance is to test the homogeneity of several means. Using analysis of variance, it is possible to make inferences about whether the samples are drawn from populations having the same mean.

ASSUMPTIONS IN F-TEST

1. Samples are drawn from normal populations and that each of these populations has the same variance,σ^2. However, if the sample sizes are large enough, assumption of normality may not be required.
2. Samples are drawn independently.
3. The various effects (treatment effects and error effects) are additive in nature.

The technique of Analysis of variance was developed by Prof. R.A. Fisher. The total variations in any set of numerical data may be due to different causes. The causes may be classified into groups (i) Assignable causes (ii) Chance or random causes. The variation due to assignable factor may be detected and measured but the variation due chance factor or random factor is not possible to detect. Hence the sum of variations due to assignable factor and random factor is the total variations. Hence analysis of variance is the technique where the total variation in the sample data is expressed as the sum of variations arising due to some specific independent source.

Total Variations = Variations due to Assignable Factors + Variations due to Random Factors

Let us take up an example to understand the technique fully. Suppose in one investigation, the haemoglobin percentage(mg) of patients after administration of drugs of different concentrations were obtained. Randomly selected patients were taken for investigation. The total variation is measured by the sum of the squares of the deviations of haemoglobin percentage values from mean value. But the total variation is composed of variations from two sources. Firstly, the haemoglobin percentage values within each type of concentration will show some variation. Secondly, the mean volume of haemoglobin percentage for the four different types of concentrations may show some variation. Hence the total variation is partitioned into two components: (i) between concentration types (between groups) (ii) within concentration (within groups) types. Or in other words (i) variance between samples and (ii) variance within samples.

The actual analysis of variance is carried out on the basis of a ratio between the variances rather than the difference between the variances. The variance ratio is obtained by dividing the variance between samples by the variance within samples. This ratio forms the F-statistic.

$$\text{Test Statistic } F = \frac{\text{Variance Between Samples}}{\text{Variance Within Samples}}$$

$$= \frac{\text{Variance due to Assignable Factor}}{\text{Variance due to Chance or random Factor}}$$

Now we have to set up a table of variance analysis and find out if their means differ significantly among the wards. Let us say, we obtain the following results after the investigation.

Haemoglobin percentage of Patients

Milder	Mild	Strong	Heavy
9	13	19	14
11	12	13	9
13	10	17	13
9	15	7	18
8	10	9	8

Let the values in this table are indexed in the following manner.

Group I	Group II	Group III	Group IV
x_{11}	x_{21}	x_{31}	x_{41}
x_{12}	x_{22}	x_{32}	x_{42}
x_{13}	x_{23}	x_{33}	x_{43}

	x_{14}	x_{24}	x_{34}	x_{44}
	x_{15}	x_{25}	x_{35}	x_{45}
Total:	T_1	T_2	T_3	T_4

Here, $n_1 = 5$, $n_2 = 5$, $n_3 = 5$ and $n_4 = 5$. If we calculate the means of each sample, we get, $\bar{x}_1 = 10; \bar{x}_2 = 12; \bar{x}_3 = 13; \bar{x}_4 = 12.4$. The grand mean(using all data) will be

$$\frac{9+11+13+.....+18+8}{20} = 11.85$$

Statement of Hypothesis

Now it is to be concluded whether the four samples were drawn from populations having the same mean μ, by constructing the hypotheses as:

Null Hypothesis H_0: $\mu_1 = \mu_2 = \mu_3 = \mu_4$

Alternative Hypothesis H_1: μ_1, μ_2, μ_3 & μ_4 are not all equal.

If the sample means do not differ significantly, we can infer that the different concentrations of the drug do not have any impact in the haemoglobin percentage of patients. On the other hand, if differences exist among sample means, then we can conclude that the different concentrations of the drug have effect among the patients. Accordingly, appropriate concentration of the drug is to be prescribed.

ANALYSIS FROM FIRST PRINCIPLE

In order to calculate the F statistic, we need to find the values of variance of between groups and that within groups. In the first case, variance among sample means i.e. 10, 12, 13 and 12.4 are determined. For the second case, variation within four samples themselves i.e. (9, 11, 13, 9, 8); (13, 12, 10, 15, 10); (19, 13, 17, 7, 9); (14, 9, 13, 18, 8). Then these two estimates are compared. The three steps are:

1. Determine one estimate of the population variance among the sample means.
2. Determine a second estimate of the population variance from the variance within the samples.
3. Compare these two estimates. If they are approximately equal in value, accept the null hypothesis.

Lets us now carry out the above operation step wise.

Step 1: As we know that the sample variance $s^2 = \frac{\sum(x - \bar{x})^2}{n-1}$

There are four sample means: 10, 12, 13 & 12.4 and one grand mean $\overline{\overline{x}} = 11.85$.

Following table shows various calculations for finding out the estimate of the population variance among the sample means or between column variance.

Drug concentrations	Sample No (k)	Sample size (n)	Sample mean $(\overline{x})$	Grand mean $(\overline{\overline{x}})$	$(\overline{x}-\overline{\overline{x}})$	$(\overline{x}-\overline{\overline{x}})^2$	$n(\overline{x}-\overline{\overline{x}})^2$
Milder	1	5	10	11.85	–1.85	3.4225	17.1125
Mild	2	5	12	11.85	0.15	0.0225	0.1125
Strong	3	5	13	11.85	1.15	1.3225	6.6125
Heavy	4	5	12.4	11.85	0.55	0.3025	1.5125
						Total =	25.35

$$\text{So between column variance} = \frac{\sum n(\overline{x}-\overline{\overline{x}})^2}{k-1} = \frac{25.35}{4-1} = 8.45$$

Step 2: The following table illustrates various calculations involving determination of estimate of the population variance from the variance within the samples otherwise called the within column variance.

S. No.	Milder concentration		Mild concentration		Strong concentration		Heavy concentration	
	x	$(x-\overline{x})^2$	x	$(x-\overline{x})^2$	x	$(x-\overline{x})^2$	x	$(x-\overline{x})^2$
1	9	1	13	1	19	36	14	2.56
2	11	1	12	0	13	0	9	11.56
3	13	9	10	4	17	16	13	0.36
4	9	1	15	9	7	36	18	31.36
5	8	4	10	4	9	16	8	19.36
Sum = T	50	16	60	18	65	104	62	65.2
Mean $\overline{x}$ =	10		12		13		12.4	
$\frac{\sum(x-\overline{x})^2}{n-1}$	$\frac{16}{5-1} = s_1^2 = 4$		$\frac{18}{5-1} = s_2^2 = 4.5$		$\frac{104}{5-1} = s_3^2 = 26$		$\frac{65.2}{5-1} = s_4^2 =$ 16.3	

Taking weighted mean, within column variance

$$= \frac{4}{16}\times s_1^2 + \frac{4}{16}\times s_2^2 + \frac{4}{16}\times s_3^2 + \frac{4}{16}\times s_4^2 = 0.25$$

$0.25 \times 4.5 + 0.25 \times 26 + 0.25 \times 16.3 = 12.70$

$$F(\text{or } F_c) = \frac{\text{Between col variance}}{\text{Within col variance}} = \frac{8.45}{12.70} = 0.665$$

SHORT-CUT METHOD

For analysis of variance table we split up the total sum of squares into sum of two sum of squares i.e. due to sum of square between groups and due to sum of squares within groups. For preparation of Analysis of Variance table we do the following calculations:

Gr I	Gr II	Gr III	Gr IV
9	13	19	14
11	12	13	9
13	10	17	13
9	15	7	18
8	10	9	8
50	60	65	62

1. Total number of observations = N = 20 and number of groups = 4
2. Grand Total = $G = x_{11} + x_{12} + \ldots x_{44} + x_{45}$
   = Sum of all observations.
   = Sum of group totals = $T_1 + T_2 + T_3 + T_4 = 50 + 60 + 65 + 62 = 237$
3. Correction Factor = CF = $G^2 \div$ Total No. of Observations = $G^2 \div N$ = $237^2 \div 20 = 2808.45$
4. Total Sum of Squares = TSS = Sum of squares of each observation $- CF = x_{11}^2 + x_{12}^2 + \ldots + x_{44}^2 + x_{45}^2 - CF = 9^2 + 11^2 + \ldots + 18^2 + 8^2 - 2808.45 = 3037 - 2808.45 = 228.55$
5. Between Sum of Squares = BSS =

$$\frac{(\text{Gr} - \text{I total})^2}{\text{No.of obs in Gr} - \text{I}} + \frac{(\text{Gr} - \text{II total})^2}{\text{No.of obs in Gr} - \text{II}}$$

$$\frac{(\text{Gr} - \text{III total})^2}{\text{No.of obs in Gr} - \text{III}} + \frac{(\text{Gr} - \text{IV total})^2}{\text{No.of obs in Gr} - \text{IV}} - CF$$

$$= \frac{T_1^2}{5} + \frac{T_2^2}{5} + \frac{T_3^2}{5} + \frac{T_4^2}{5} - CF = \frac{50^2}{5} + \frac{60^2}{5} + \frac{65^2}{5} + \frac{62^2}{5} - 2808.45$$

$= 2833.80 - 2808.45 = 25.35$

6. Within Sum of Squares WSS = TSS – BSS = 228.55-25.35 = 203.20
7. Calculation of degrees of freedom:

   Total degrees of freedom = Total items – 1 = N – 1 = 20 – 1 = 19

   Between group degrees of freedom = No of groups – 1 = c – 1 = 4 – 1 = 3

   Within group degrees of freedom = Total df – between group df = (N – 1) – (c – 1) = 19 – 3 = 16
8. Calculation of mean squares:- The respective mean squares can be obtained by dividing sum of squares by degrees of freedom.

   Between Mean Square BMS = BSS ÷ (c-1) = 25.35 ÷ 3 = 8.45

   Within Mean Square WMS = WSS ÷ (N-c) = 203.20 ÷ 16 = 12.7

ANOVA Table

(One-Way Classification)

Source of Variation	Degrees of Freedom	Sum of Squares	Mean Squares	F
Between Groups	4 – 1 = 3	BSS = 25.35	BMS = 25.35 ÷ 3 = 8.45	BMS ÷ WMS = F_c (Calculated F) = 8.45 ÷ 12.7 = 0.665
Within Groups (or Error)	(By Subtraction) 19 – 3 = 16	WSS = TSS – BSS = 203.20	WMS = 203.20 ÷ 16 = 12.7	$F_{0.05(3.16)}$ = 3.24
Total	20 – 1 = 19	TSS = 228.55		

The table value of F at α = 0.05 for 3,16 df is 3.24. Calculated value 0.665 is less than the table value 3.24. So there is no evidence to reject the null hypothesis. Hence, the mean values of haemoglobin percentage of all the patients in the entire population treated with drugs having all the four levels of concentrations are same.

It may be noted that whether the null hypothesis is true or false, the denominator is a good estimator of population variance, because it is based on the variance within the samples.

Precautions in using F-test: As it can be seen from the above example that the sample sizes are too small in order to be able to draw valid conclusions about the effectiveness of various levels of concentration of the drug. We chose small samples so that the logic of analysis of variance can be explained with ease. However, in actual practice, the same methodology is adopted, but with larger samples only.

In this example, the test was carried out with only one factor, i.e. levels of concentration of the drug. So it is called "*One way Test*". However, many other factors that might contribute to our results have not been taken into consideration such as, age of patients, body weights of patients, various chemical compositions of the drug and so on. To be able to make significant decisions based on analysis of variance, we need to be certain that all such factors are effectively controlled. Had we wished to measure the effect of two factors, such as the levels of concentration of the drug and ages of the patients, we would need to go for more advanced method of analysis of variance, called "*Two way Test*", which is explained subsequently.

Example 12.1: The following data show the number of claims processed per day for a group of four insurance company employees observed for a number of days. Test the hypothesis that the employees' mean claims per day are all the same. Use the 0.05 level of significance.

Employee 1: 15 17 14 12

Employee 2: 12 10 13 17

Employee 3: 11 14 13 15 12

Employee 4: 13 12 12 14 10 9

Solution:

	Employee 1	**Employee 2**	**Employee 3**	**Employee 4**
	15	12	11	13
	17	10	14	12
	14	13	13	12
	12	17	15	14
			12	10
				9
Total:	58	52	65	70

Total number of observations: N = 19 and number of groups = 4

Grand Total: G = Sum of group totals = $T_1 + T_2 + T_3 + T_4$ = 58 + 52 + 65 + 70 = 245

Correction Factor: CF = $G^2 \div$ Total No. of Observations = $G^2 \div N = 245^2$ 19 = 3159.210

Total Sum of Squares: TSS = Sum of squares of each observation – CF = $x_{11}^2 + x_{12}^2 + \ldots + x_{45}^2 + x_{46}^2 - CF = 15^2 + 12^2 + \ldots + 10^2 + 9^2 - 3159.21 = 3245 - 3159.21 = 85.79$

Between Sum of Squares BSS

$$= \frac{(\text{Gr}-\text{I total})^2}{\text{No.of obs in Gr}-\text{I}} + \frac{(\text{Gr}-\text{II total})^2}{\text{No.of obs in Gr}-\text{II}}$$

$$+ \frac{(\text{Gr}-\text{III total})^2}{\text{No.of obs in Gr}-\text{III}} + \frac{(\text{Gr}-\text{IV total})^2}{\text{No.of obs in Gr}-\text{IV}} - \text{CF}$$

$$= \frac{T_1^2}{4} + \frac{T_2^2}{4} + \frac{T_3^2}{5} + \frac{T_4^2}{6} - \text{CF} = \frac{58^2}{4} + \frac{52^2}{4} + \frac{65^2}{5} + \frac{70^2}{6} - 3159.21$$

$= 3178.67 - 3159.21 = 19.46$

Within Sum of Squares WSS = TSS – BSS = 85.79 – 19.46 = 66.33

Calculation of degrees of freedom:

Total degrees of freedom: Total items – 1 = N – 1 = 19 – 1 = 18

Between group degrees of freedom: No of groups – 1 = c – 1 = 4 – 1 = 3

Within group degrees of freedom: Total df – between group df

= (N – 1) – (c – 1) = 18 – 3 = 15

9. **Calculation of mean squares:** The respective mean squares can be obtained by dividing sum of squares by degrees of freedom.

   Between Mean Square: BMS = BSS ÷ (c – 1) = 19.46 3 = 6.487

   Within Mean Square: WMS = WSS ÷ (N – c) = 66.33 15 = 4.422

ANOVA Table

(One-Way Classification)

Source of Variation	Degrees of Freedom	Sum of Squares	Mean Squares	F
Between Groups	4 – 1 = 3	BSS = 19.46	BMS = 19.46 ÷ 3 = 6.487	BMS/ WMS = F_c (Calculated F) = 6.487 ÷ 4.422 = 1.47
Within Groups (or Error)	(By Subtraction) 18 – 3 = 15	WSS = TSS – BSS = 66.33	WMS = 66.33 ÷ 15 = 4.422	$F_{0.05(3.15)} = 3.29$
Total	19 – 1 = 18	TSS = 85.79		

The table value of F at α = 0.05 for 3,15 df is 3.29. Calculated value falls within the table value(1.47 < 3.29). So there is no evidence to reject the null hypothesis. Hence, there is no difference among the mean claims of employees of the four companies.

DATA CODING

As can be seen in the above example, higher valued digits are to be dealt with and even squaring them makes them still larger, which may be difficult to manage in calculations. Hence data coding helps in reducing large numbers to smaller numbers, which then becomes easy to deal with. Such operation does not change the value of F_c. Any rational number can be added to or subtracted from the numbers in the given table and a new table is prepared after addition or subtraction. Same procedure is then followed with this newly prepared table and F_c is calculated. Please remember that only addition or subtraction is allowed not multiplication or division.

Let us deduct 12 from each of the numbers and the following table is prepared.

CODED DATA

	Employee 1	Employee 2	Employee 3	Employee 4
	3	0	–1	1
	5	–2	2	0
	2	1	1	0
	0	5	3	2
			0	–2
				–3
Total:	10	4	5	–2

Total number of observations: N = 19 and number of groups = 4

Grand Total G = Sum of group totals = $T_1 + T_2 + T_3 + T_4 = 10 + 4 + 5 - 2 = 17$

Correction Factor CF = $G^2 \div$ Total No. of Observations = $G^2 \div N = 17^2\ 19 = 15.210$

Total Sum of Squares TSS = Sum of squares of each observation – CF = $x_{11}{}^2 + x_{12}{}^2 + \ldots + x_{45}{}^2 + x_{46}{}^2 - CF = 3^2 + 5^2 + \ldots + (-2)^2 + (-3)^2 - 15.210 = 101 - 15.210 = 85.79$

Between Sum of Squares BSS

$$= \frac{(\text{Gr} - \text{I total})^2}{\text{No. of obs in Gr} - \text{I}} + \frac{(\text{Gr} - \text{II total})^2}{\text{No. of obs in Gr} - \text{II}}$$

$$+ \frac{(\text{Gr} - \text{III total})^2}{\text{No. of obs in Gr} - \text{III}} + \frac{(\text{Gr} - \text{IV total})^2}{\text{No. of obs in Gr} - \text{IV}} - \text{CF}$$

$$= \frac{T_1^2}{4} + \frac{T_2^2}{4} + \frac{T_3^2}{5} + \frac{T_4^2}{6} - CF$$

$$= \frac{10^2}{4} + \frac{4^2}{4} + \frac{5^2}{5} + \frac{(-2)^2}{6} - 15.210 = 34.67 - 15.210 = 19.46$$

Within Sum of Squares: WSS = TSS – BSS = 85.79 - 19.46 = 66.33

Calculation of degrees of freedom:

Total degrees of freedom: Total items – 1 = N – 1 = 19 – 1 = 18

Between group degrees of freedom: No of groups – 1 = c – 1 = 4 – 1 = 3

Within group degrees of freedom: Total df – between group df

= (N – 1) – (c – 1) = 18 – 3 = 15

10. **Calculation of mean squares:** The respective mean squares can be obtained by dividing sum of squares by degrees of freedom.

    Between Mean Square = BMS = BSS ÷ (c – 1) = 19.46 ÷ 3 = 6.487

    Within Mean Square = WMS = WSS ÷ (N – c) = 66.33 ÷ 15 = 4.422

    $F_c = \frac{6.487}{4.422} = 1.47$ (Same as earlier obtained with the original data)

VARIANCE ANALYSIS IN TWO-WAY CLASSIFICATION

In one-way classification we compared variation due to one factor, i.e. we compared variation between the groups with respect to mean. But in many cases there are two factors which may affect the variable values. In that case we consider one factor in rows and another factor in columns. Further we compare the row means i.e. homogeneity of rows and simultaneously compare column means i.e. homogeneity of columns. Here the total variation is due to three variations i.e. row variation, column variation and chance or random variation.

Total Variation = Row Variation + Column Variation + Chance or Random Variation

Let us consider the same example, but with addition of another factor i.e. age group of the patients. The notations are shown below:

	Level of Concentration			
	C_1	C_2	C_3	**Total**
R_1	x_{11}	x_{12}	x_{13}	TR_1
R_2	x_{21}	x_{22}	x_{23}	TR_2

Age Group	R_3	x_{31}	x_{32}	x_{33}	TR_3
	R_4	x_{41}	x_{42}	x_{43}	TR_4
	Total	TC_1	TC_2	TC_3	G = Grand Total

Under the null hypothesis we consider:

1. Let there be no difference of hemoglobin percentages in the entire population due to different levels of concentration of the drug i.e. H_0: $\mu_1 = \mu_2 = \mu_3$
2. Also let there be no difference of hemoglobin percentages in the entire population due to different age groups i.e. H_0: $\mu_1' = \mu_2' = \mu_3' = \mu_4'$

For analysis of variance table we split up the total sum of square into three sums of squares i.e.

(i) Sum of squares due to row (RSS)

(ii) Sum of squares due to column (CSS)

(iii) Sum of squares due to chance or sum of squares due to error (ESS)

For preparation of analysis of variance table we calculate the following (considering the above data table):

1. The number of rows (r) = 4 & number of columns = c = 3

   Total number of observations $N = r \times c = 4 \times 3 = 12$
2. Grand Total (G) = Sum of all observations.
3. Correction Factor (CF) = $G^2 \div N$
4. Total Sum of Squares (TSS) = Sum of squares of individual observations – CF.
5. Row sum of squares (RSS) = $\dfrac{TR_1^2 + TR_2^2 + TR_3^2 + TR_4^2}{c} - CF$
6. Column Sum of Squares (CSS) = $\dfrac{TC_1^2 + TC_2^2 + TC_3^2}{r} - CF$
7. Error Sum of Squares (ESS) = TSS – RSS – CSS
8. Calculation of degrees of freedom

   Total degrees of freedom = rc – 1

   Row degrees of freedom = r – 1

   Column degrees of freedom = c – 1

   Error degrees of freedom = Total degrees of freedom – Row degrees of freedom – Column degrees of freedom = $(r - 1) \times (c - 1)$

9. The respective mean squares are obtained by dividing sum of squares by the degrees of freedom.

   Row mean square = RMS = RSS ÷ (r – 1)

   Column mean square = CMS = CSS ÷ (c – 1)

   Error mean square = EMS = ESS ÷ [(r – 1) × (c – 1)]

ANOVA TABLE

(Two-Way Classification)

Source of Variation	df	Sum of Squares	Mean Square	F
Row	r – 1 = 3	RSS	RMS = RSS ÷ (r – 1)	RMS ÷ EMS= F_{CR}
Column	c – 1 = 2	CSS	CMS = CSS ÷ (c – 1)	CMS ÷ EMS = F_{CC}
Error	(r – 1)(c – 1) = 6	ESS	EMS = ESS ÷ [(r – 1)(c – 1)]	
Total	rc – 1 = 11	TSS		

Here F_{CR} is distributed with {(r – 1), (r – 1)(c – 1)} df_n i.e. in this particular example F_{CR} is distributed with (3,6) df. Now to test the homogeneity of rows, we compare F_{CR} with the tabulated Fα with (3,6) df where α (0.01 or 0.05) is the desired level of significance. If F_{CR} > Fα then we reject the null hypothesis about row i.e. we say there is a significant difference between at least two rows. If F_{CR} ≤ Fα then we accept the null hypothesis about rows i.e. no difference exists between the means of rows. Again F_{CC} is distributed with {(c – 1), (r – 1)(c – 1)} df, i.e. in this case F_{CC} is distributed with (2,6) df. To test the homogeneity of columns we compare Fα′ with (2,6) df where α' (0.01 or 0.05) is the desired level of significance. If F_{CC} > Fα′ then the null hypothesis about columns is rejected, otherwise it is accepted.

Note: Here we get 3 cases (1) row significant and column significant (2) row significant but column not significant (3) row not significant but column significant.

Example:12.2: Soil pH of 3 sites at 3 depths was measured and the results are shown in the table below. Perform an analysis of variance and comment whether the 3 sites and 3 depths have any significant differences in their pH values.

	Soil pH		
Depth	**Site-I**	**Site-II**	**Site-III**
Surface	7.8	6.8	5.4
20 cm	7.5	6.9	5.5
40 cm	7.3	6.4	5.3

Solution:

Step 1: Null Hypothesis:

(i) There is no difference in soil pH at the three sites.

(ii) There is no difference in soil pH at three depths.

Alternative Hypothesis: Differences exist.

Step 2: Select $\alpha = 0.05$

Step 3: Let us subtract 6.5 from each figure to code the data. The new coded data and the sum of squares table are given below (*Note: - As any constant subtraction from or addition with each data will not change the analysis of variance)*

Depth	Site I	Site II	Site III	Total
Surface	1.3	0.3	–1.1	**0.5**
20 cm	1.0	0.4	–1.0	**0.4**
40 cm	0.8	–0.1	–1.2	**–0.5**
Total	**3.1**	**0.6**	**–3.3**	**0.4**

Grand Total (G) = 0.4

Correction Factor (CF) = $(0.4)^2 \div 9 = 0.018$

TSS = Sum of squares of observations – CF = 7.24 – 0.018 = 7.222

$$CSS = (3.1)^2 \div 3 + (0.6)^2 \div 3 + (-3.3)^2 \div 3 - 0.018 = 9.61 \div 3 + 0.36 \div 3 + 10.89 \div 3 - 0.018 = 3.203 + 0.12 + 3.63 - 0.018 = 6.953 - 0.018 = 6.935$$

$$RSS = (0.5)^2 \div 3 + (0.4)^2 \div 3 + (-0.5)^2 \div 3 - 0.018 = 0.25 \div 3 + 0.16 \div 3 + 0.25 \div 3 - 0.018 = 0.083 + 0.053 + 0.083 - 0.018$$

$$= 0.22 - 0.018 = 0.201$$

ESS = TSS – RSS – CSS = 7.222 – 6.935- 0.202 = 0.085

Analysis of Variance Table (Two-Way Classification)

Source of Variation	Degree of freedom	Sum of Squares	Mean Squares	Variance Ratio
Between Sites:	2	6.935	CMS = 6.935 ÷ 2 = 3.4675	3.4675 ÷ 0.0213 = 162.793= F_{CC}
Between Depths:	2	0.202	RMS = 0.201 ÷ 2 = 0.101	0.101 ÷ 0.0213 = 4.742= F_{CR}
Residual or Error:	4	0.085	EMS = 0.086 ÷ 4 = 0.0213	
Total:	8	7.222		

Step 4: Decision: Let us make another table showing the degrees of freedom and the F-values

Source of Variation	Degrees of Freedom	Computed F	Critical F	Probability P	Significance Type
Between Sites:	$d_1 = 2$ (num) $d_2 = 4$ (den)	162.793 > *(Null Hypothesis is rejected)*	6.94	α = 0.05	Highly Significant
Between Depths	$d_1 = 2$ (num) $d_2 = 4$ (den)	4.742 < *(Null Hypothesis is accepted)*	6.94	α = 0.05	Not Significant

Note: num = numerator; den = denominator

***Conclusion*:** From the above table we find between sites is highly significant. It indicates the soil quality with respect to pH of three different sites are completely different. Further as the between depth is not significant there is no effect of depth with regard to pH values.

LEAST SIGNIFICANCE DIFFERENCE (LSD)

By using analysis of variance test we may find the treatment (Group or Row or Column) effects are significant. From this we may conclude that there exists a significant difference between at least two treatments. Then we may be interested to find out which pairs of treatments significantly differ. This can be known by the help of t-test. But instead of calculating t for each pair, we calculate the least significant difference at the desired level of significance.

$$LSD = \sqrt{EMS} \times \sqrt{\frac{1}{n_1} + \frac{1}{n_2}} \times t_\alpha \text{ at error degrees of freedom.}$$

EMS = Error mean square = Within sum of square in case of one way classification.

N_1 and n_2 be the observations present in the first group and second group which will be compared.

T_α = Table value of t at α (0.01 or 0.05) level of significance at error degrees of freedom.

If $n_1 = n_2 = n$, then,

$$LSD = \sqrt{\frac{2EMS}{n}} \cdot t_\alpha \textit{ at error degrees of freedom.}$$

Note: This is also known as critical difference (C.D.).

Example 12.3: Four training methods were compared to see whether they led to greater productivity after training. The following are productivity measures for individuals trained by each method.

Individuals	Training I	Training II	Training III	Training IV
1	72	80	79	90
2	75	75	85	95
3	68	82	88	98
4	79	85	86	85
5	74	82	90	92

Prepare an ANOVA table. Calculate LSD and give your views regarding the training methods.

Solution: Here, we are interested to know effect of four different training methods. Four treatments are considered. Hence, one-way classification is applicable.

Step I:

Null Hypothesis: There is no effect on productivity by different training methods.

Step II: A = 0.05

Step III

Individuals	Productivity measures			
	Training I	Training II	Training III	Training IV
1	72	80	79	90
2	75	75	85	95
3	68	82	88	98
4	79	85	86	85
5	74	82	90	92
Total	**368**	**404**	**428**	**460**
Mean	**73.6**	**80.8**	**85.6**	**92.0**

Sum of all observations G = 1660

Correction factor $G^2 \div 20 = (1660)^2 \div 20 = 137780$

Total Sum of Squares TSS = $72^2 + 75^2 + \ldots.. + 85^2 + 92^2 - CF$ = 138972 – 137880 = 1192

Treatment Sum of Squares $T_rSS = [\, 368^2 + 404^2 + 428^2 + 460^2 \,] \div 5 - CF$ = 138684.8 – 137780 = 904.8

Error Sum of Squares ESS = TSS – T_rSS = 1192 – 904.8 = 287.2

Total Degrees of Freedom 20 – 1 = 19

Treatment Degrees of Freedom 4 – 1 = 3

Error Degrees of freedom 19 – 3 = 16

Treatment Mean Square 904.8 ÷ 3 = 301.6

Error Mean Square EMS = 287.2 ÷ 16 = 17.95

ANOVA Table

Source of Variation	Degrees of freedom	Sum of Squares	Mean Squares	F
Treatment	4 – 1 = 3	904.8	$\frac{904.8}{3} = 301.6$	$\frac{301.6}{17.95} = 16.8(F_c)$**
Error	19 – 3 = 16	287.2	$\frac{287.2}{16} = 17.95$	
Total	:	19	1192	

The critical value of $F_{0.05}$ at (3, 16) d.f. = 3.24, as the F_c value > 3.24

**F value is significant at 5% level.

Critical value of t for 16 df is 2.12 for α=0.05

$$LSD = \sqrt{\frac{2EMS}{n}}\, t_{0.05} \text{ at 16 df} = \sqrt{\frac{2\times 17.95}{5}} \times 2.12 = 5.680$$

Decision: Let us now see the difference of means between productivity measures due to different training programs.

	Difference of Mean Value	**LSD Value**
Difference of Means between Tr-I & Tr-II =	80.8 – 73.6 = 7.2 >	5.68
Difference of Means between Tr-I & Tr-III=	85.6 – 73.6 = 12.0 >	5.68
Difference of Means between Tr-I & Tr-IV =	92.0 – 73.6 = 18.4 >	5.68
Difference of Means between Tr-II & Tr-III =	85.6 – 80.8 = 4.8 <	5.68
Difference of Means between Tr-II & Tr-IV =	92.0 – 80.8 = 11.2 >	5.68
Difference of Means between Tr-III & Tr-IV =	92.0 – 85.6 = 6.4 >	5.68

Hence, it can be concluded that there is no significant difference of productivity measures for training methods- II and III. But all other means in the example differ by more than the LSD and hence their differences are statistically significant.

EXERCISES

1. Compare one-way and two-way F-test with suitable examples.
2. What is 'F' test? Describe 'F' test giving its assumptions and null hypothesis to be tested.
3. What is 'Analysis of Variance' and where it is used. Write the assumptions of this technique.
4. Explain clearly the technique of analysis of variance for data with one way classification.
5. Describe the technique of analysis of variance for a two-way classification data.
6. Explain 'F' test. How is the one way F-test applied?
7. In an investigation relating to effect of mercuric chloride of three different concentrations on the germination of rice grain the following results were obtained. Perform an analysis of variance to show whether there are significant differences in the germination percentages between four concentrations of $HgCl_2$.

   Percentage of germination $HgCl_2$ concentration (mg / l)

Replicates:	0 (Control)	10	20	30
1	97	78	76	60
2	85	85	70	65
3	90	81	65	70
4	96	89	75	55
5	92	85	70	56

8. Soil pH of 3 sites at 3 depths was measured and the results are shown in the table below. Perform an analysis of variance and comment whether the 3 sites and 3 depths have any significant differences as regards their pH values.

		Soil pH	
Depth	**Site I**	**Site II**	**Site III**
Surface	7.7	6.7	5.3
15 cm	7.4	6.8	5.4
35 cm	7.2	6.3	5.2

9. Seeds soaked in conc. H_2SO_4 for 4, 6, 8 and 10 hours were washed thoroughly and then germinated. Germination percentages of the treated seeds are presented in the following table. Calculate the LSD and give your conclusions.

Germination Percentage Period soaked in H_2SO_4 in hours

Replicates	4	6	8	10
1	70	81	80	89
2	74	74	84	94
3	67	81	87	97
4	78	84	85	84
5	75	81	89	91

10. Calculate LSD (Least significant difference) in the following problem. Draw your conclusion on the differences in the number of tillers/plant of the 4 varieties of rice plant.

No. of tillers per plant of the four varieties of rice

Plot Nos.	A	B	C	D
1	8	6	5	14
2	9	7	7	13
3	10	6	6	12
4	11	7	8	15
5	9	8	5	11

SIMPLE REGRESSION ANALYSIS

INTRODUCTION

The term "*Regression*" was first used as a statistical concept in 1877 by Sir Francis Galton. The literal meaning of regression is "stepping back towards average". The study of Galton showed that the height of children born to tall or short parents would tend to move back or 'regress' to mean or average height of population. However, the above concepts came from following observations made by Galton.

1. Tall fathers had tall sons and short fathers had short sons.
2. The mean height of the sons of tall fathers is less than the mean height of their fathers.
3. The mean height of the sons of short fathers is more than the mean height of their fathers.
4. The deviations in the mean height of the sons from the mean height of race were less than the deviations in the mean height of the fathers from the mean height of the race. When the fathers move above the mean or below the mean, the sons tend to move back or regress towards the mean.

DEFINITION

Regression analysis is a mathematical measure of the average relationship between two or more variables in terms of the original units of the data.

Regression analysis attempts to establish the nature of the relationship between variables that is to study the functional relationship between the variables and thereby provide a mechanism for predicting or forecasting.

Regression and correlation analysis are based on the relationship or association between two or more variables. The regression analysis, which is restricted to two variables only, is known as simple regression and the regression analysis, which is made considering more than two variables at a time, is called multiple regression. When two variables are well associated i.e. correlation coefficient between them is reasonably high, one can form a mathematical equation. This mathematical equation is known as regression equation. With the help of regression equation one can predict or estimate the effect of one variable or the other variables. The variable whose value is influenced or is to be predicted is called the dependent or regressed or explained variable. The variable, which influences the values or used for prediction is called independent or regressor or predictor or explanatory variable.

SCATTER DIAGRAM

For a bivariate data (X, Y), the plotting of points can be made on a graph paper. This graph is known as scatter diagram. A scatter diagram is able to produce two types of information. Visually, the pattern of points indicates presence or absence of relation between two variables. Secondly, joining the points smoothly shows the mathematical relationship (straight line or curve) between two variables approximately. If the value in one axis increases or decreases in the same linear ratio as that in the other axis throughout, then the relationship is said to be linear or simply called linear regression model. If the values change in non-linear manner, then it is called curvilinear regression model. There can be several types of relationships. If increase in one variable causes increase in the other, it is called direct relationship. Or else, if increase in one causes decrease in the other, it is called inverse relationship. The following figures illustrate these types.

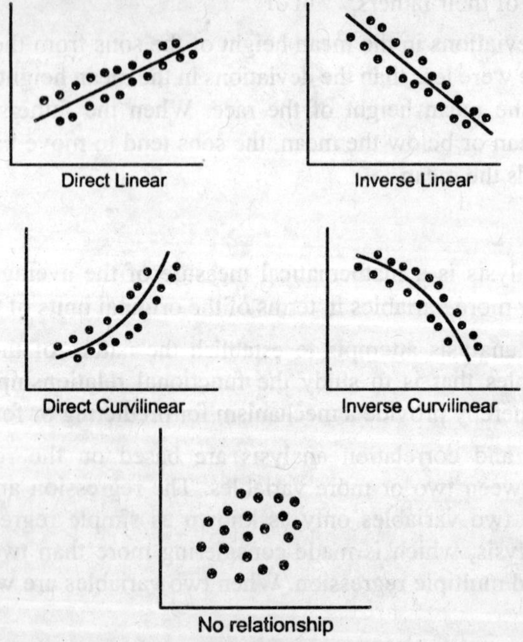

STRAIGHT LINE EQUATION

For any two sets of points (X_1, Y_1) and $X_2, Y_2)$, it is possible to draw a straight line by joining these two sets of points. The equation to this straight line will take the form of Y = a + b X. Here, 'a' is the intercepted portion of the straight line with Y-axis and 'b' is the slope of the straight line.

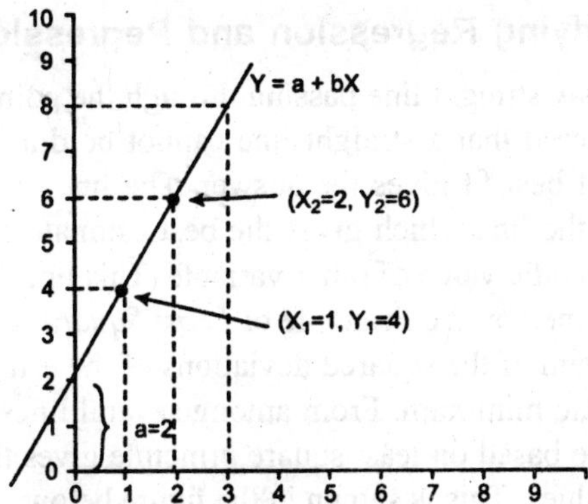

The two points are (1, 4) and (2, 6); through which the straight line has been drawn. Here the Y-intercept of the straight line is 2 units. So, a = 2. b is the slope of the straight line and is calculated from the formula

$$b = \frac{Y_2 - Y_1}{X_2 - X_1} = \frac{6-4}{2-1} = 2.$$

So the equation to the straight line becomes Y = 2 + 2 X.

Thus, for any value of X, the value of Y can be estimated. If suppose X = 3, then Y = 2 + 2 3 = 8. X is called the independent variable and Y is the dependent variable. This relationship is direct and the slope is positive. Similarly straight line with negative slope can be drawn as shown below.

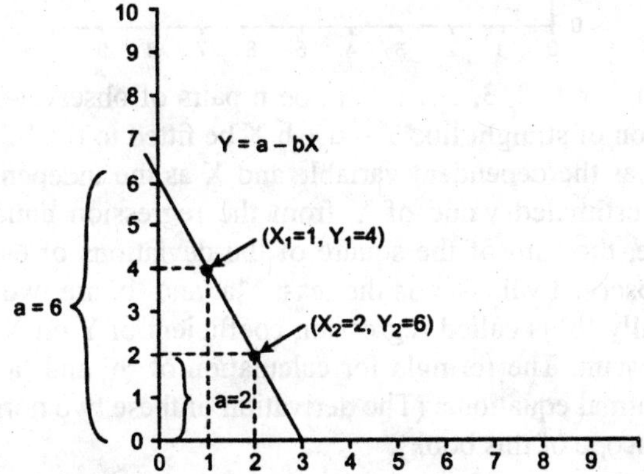

In the above figure, the relationship is inverse linear with a = 6 and b $= \frac{Y_2 - Y_1}{X_2 - X_1} = \frac{2-4}{2-1} = -2$. Hence the equation to the straight line becomes Y = 6 – 2 X. The slope is negative. We can see that, when X = 3, Y becomes equal to 0, so the straight line meets the X-axis at X= 3.

Method of Studying Regression and Regression Lines

We saw how to draw straight line passing through the points. But what if the points are so scattered that a straight line cannot be drawn to pass through them? Here line of best fit gives the answer. The line of best fit, called the regression line, is the line which gives the best estimate to the value of one variable for any specific value of other variable. This line of regression is the line of best fit obtained by the Principle of *Least Square*. This is based on the principle that the sum of the squared deviations of the estimated values from the true values is the minimum. From among several lines fitted through the data points, the line based on least square principle gives the minimum error to the estimated values. This is shown in the figure below.

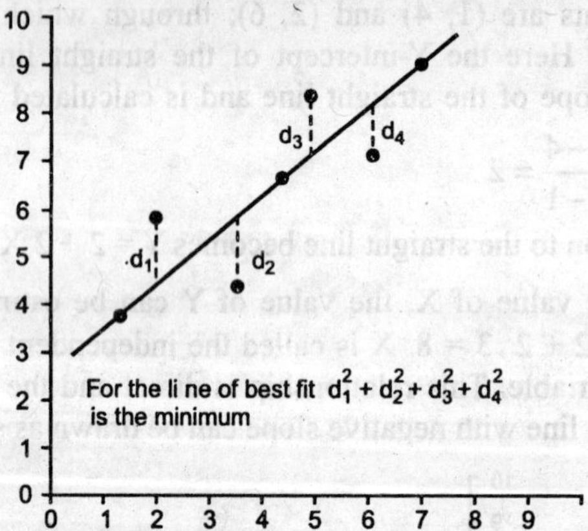

Let (X_i, Y_i), i = 1, 2, 3, ……….,n, be n pairs of observations present. A general equation of straight line Y = a + b X be fitted to the 'n' pairs of data, considering Y as the dependent variable and X as the independent variable. Let Y_e is the estimated value of Y from the regression equation. For the regression line, the sum of the square of the deviations of estimated value Y_e from the observed values Y is the least. 'a' and 'b' are two constants i.e. more specifically 'b' is called regression coefficient of Y on X and 'a' is the regression constant. The formula for calculation of 'b' and 'a' are given by solving two normal equations: (The derivation of these two normal equations is beyond the scope of this book).

$$\sum X = na + b\sum X \qquad \text{...(1)}$$

$$\sum XY = a\sum X + b\sum X^2 \qquad \text{...(2)}$$

On solution, we can get,

$$b = \frac{\sum XY - n\overline{X}\overline{Y}}{\sum X^2 - n\overline{X}^2} \text{ and } a = \overline{Y} - b\overline{X}$$

Many times, few direct formulae for b are also used, as given below:

$$b = \frac{\sum XY - \dfrac{\sum X \times \sum Y}{n}}{\sum X^2 - \dfrac{\left(\sum X\right)^2}{n}} = \frac{\dfrac{1}{n} \times \sum\left(X - \overline{X}\right)\left(Y - \overline{Y}\right)}{\sigma_X^2}$$

$$= \frac{Cov(X, Y)}{\sigma_X^2} = \frac{\sum\left(X - \overline{X}\right)\left(Y - \overline{Y}\right)}{\sum\left(X - \overline{X}\right)^2}$$

Once the values of a and b are known, then the regression equation of Y on X can be found by $Y_e = a + b\,X$

Example 13.1: Find the regression equation of Y on X for the following data. Plot these data. Plot the estimating equation on the scatter plot.

Grade Points	3	3	4	4	5	5	5	6	6	6	7	7	7	8	8	8	9	9	9	10
Starting salary in Thousands	8	7	9	6	10	9	8	12	10	9	11	10	10	13	14	10	12	15	11	12

Solution: Table below shows all calculations, which is self-explanatory.

S. No	X	Y	X^2	XY
1	3	8	9	24
2	3	7	9	21
3	4	9	16	36
4	4	6	16	24
5	5	10	25	50
6	5	9	25	45
7	5	8	25	40
8	6	12	36	72

S. No	X	Y	X^2	XY
9	6	10	36	60
10	6	9	36	54
11	7	11	49	77
12	7	10	49	70
13	7	10	49	70
14	8	13	64	104
15	8	14	64	112
16	8	10	64	80
17	9	12	81	108
18	9	15	81	135
19	9	11	81	99
20	10	12	100	120
Total =	**129**	**206**	**915**	**1401**

$$\overline{X} = \frac{129}{20} = 6.45 \text{ and } \overline{Y} = \frac{206}{20} = 10.3$$

The regression coefficient,

$$b = \frac{\sum XY - n\overline{X}\overline{Y}}{\sum X^2 - n\overline{X}^2} = \frac{1401 - 20 \times 6.45 \times 10.3}{915 - 20 \times 6.45^2} = 0.8716$$

The regression constant, a = $\overline{Y} - b\overline{X}$ = 10.3 – 0.8716 6.45 = 4.6781

Hence, the regression equation is Y_e = 4.6781 + 0.8716 X

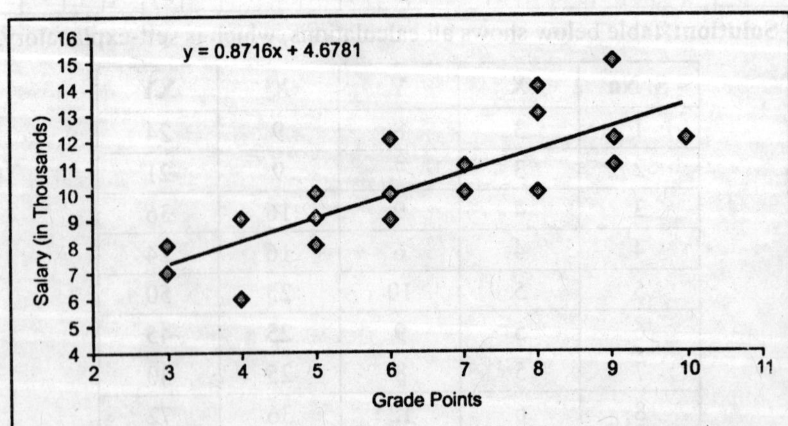

On the basis of the equation $Y_e = 4.6781 + 0.8716\,X$ the estimated values of Y i.e. Y_e are calculated and shown in the table below.

S. No	X	Y	Y_e	$Y - Y_e$	$(Y - Y_e)^2$
1	3	8	7.293	0.707	0.500
2	3	7	7.293	– 0.293	0.086
3	4	9	8.165	0.835	0.698
4	4	6	8.165	– 2.165	4.687
5	5	10	9.036	0.964	0.929
6	5	9	9.036	– 0.036	0.001
7	5	8	9.036	– 1.036	1.073
8	6	12	9.908	2.092	4.377
9	6	10	9.908	0.092	0.009
10	6	9	9.908	– 0.908	0.824
11	7	11	10.779	0.221	0.049
12	7	10	10.779	– 0.779	0.607
13	7	10	10.779	– 0.779	0.607
14	8	13	11.651	1.349	1.820
15	8	14	11.651	2.349	5.518
16	8	10	11.651	– 1.651	2.726
17	9	12	12.523	– 0.523	0.274
18	9	15	12.523	2.477	6.136
19	9	11	12.523	– 1.523	2.320
20	10	12	13.394	– 1.394	1.943
Total =	**129**	**206**	**206**	**0.00**	**35.185**

Note: Mark the difference between the estimated values and observed values and their sum is zero, $\sum(Y - Y_e) = 0$, which is a check that all our calculations are correct.

STANDARD ERROR OF ESTIMATE

After having developed a regression equation for two data sets X and Y for estimation purpose, it is obviously required to measure the reliability of the estimating equation. Because, in any given situation we possess only a sample data to determine the least square line, which may be taken only an estimate of the theoretical regression line calculated from the whole set of values

(X, Y) in the population. For this statisticians have developed the standard error of estimate symbolised as s_e. This standard error is very similar to the standard deviation studied earlier, because both are measures of dispersion only. Standard error of estimate is square root of the sum of the squared deviations of estimated values of dependent variable from the observed values divided by n-2 and the formula is:

$$S_e = \sqrt{\frac{\sum (Y - Y_e)^2}{n-2}}$$

It has also a direct formula given by,

$$s_e = \sqrt{\frac{\sum Y^2 - a\sum Y - b\sum XY}{n-2}}$$

Example 13.2: Find the standard error of estimate of the regression equation obtained in the above example and interpret the result.

Solution: Let us find the standard error of estimate by using both equations.

S. No	X	Y	X^2	Y^2	XY	Y_e	$Y - Y_e$	$(Y - Y_e)^2$
1	3	8	9	64	24	7.293	0.707	0.500
2	3	7	9	49	21	7.293	– 0.293	0.086
3	4	9	16	81	36	8.165	0.835	0.698
4	4	6	16	36	24	8.165	– 2.165	4.687
5	5	10	25	100	50	9.036	0.964	0.929
6	5	9	25	81	45	9.036	– 0.036	0.001
7	5	8	25	64	40	9.036	– 1.036	1.073
8	6	12	36	144	72	9.908	2.092	4.377
9	6	10	36	100	60	9.908	0.092	0.009
10	6	9	36	81	54	9.908	– 0.908	0.825
11	7	11	49	121	77	10.779	0.221	0.049
12	7	10	49	100	70	10.779	– 0.779	0.607
13	7	10	49	100	70	10.779	– 0.779	0.607
14	8	13	64	169	104	11.651	1.349	1.820
15	8	14	64	196	112	11.651	2.349	5.518
16	8	10	64	100	80	11.651	– 1.651	2.726
17	9	12	81	144	108	12.523	– 0.523	0.273
18	9	15	81	225	135	12.523	2.477	6.136

S. No	X	Y	X^2	Y^2	XY	Y_e	$Y - Y_e$	$(Y - Y_e)^2$
19	9	11	81	121	99	12.523	– 1.523	2.320
20	10	12	100	144	120	13.394	– 1.394	1.943
Total =	**129**	**206**	**915**	2220	1401	206	0.00	35.185

$$S_e = \sqrt{\frac{\sum(Y - Y_e)^2}{n-2}} = \sqrt{\frac{35.185}{20-2}} = 1.398$$

$$s_e = \sqrt{\frac{\sum Y^2 - a\sum Y - b\sum XY}{n-2}}$$

$$= \sqrt{\frac{2220 - 4.6781 \times 206 - 0.8716 \times 1401}{20-2}} = 1.398$$

Let us assume that the observed points are normally distributed around the regression line. Then we would expect to find 68.27% of the points within $\pm 1s_e$, 95.45% of the points within $\pm 2s_e$ and 99.73% of the points within $\pm 3s_e$. Here in the graph below, the boundary lines covering these three ranges have been shown. It is noted that within $\pm 1s_e$, there are 14 points, which is roughly 68.27% of the total number of points considered in the example. This assumption of normality will hold good, for large sample data only. In our example number of data is small, n=20. So accurate conclusions (68.27% of the points within $\pm 1s_e$, 95.45% of the points within $\pm 2s_e$ and 99.73% of the points within $\pm 3s_e$) are difficult to get. Nonetheless, the principle involved in prediction intervals is justified.

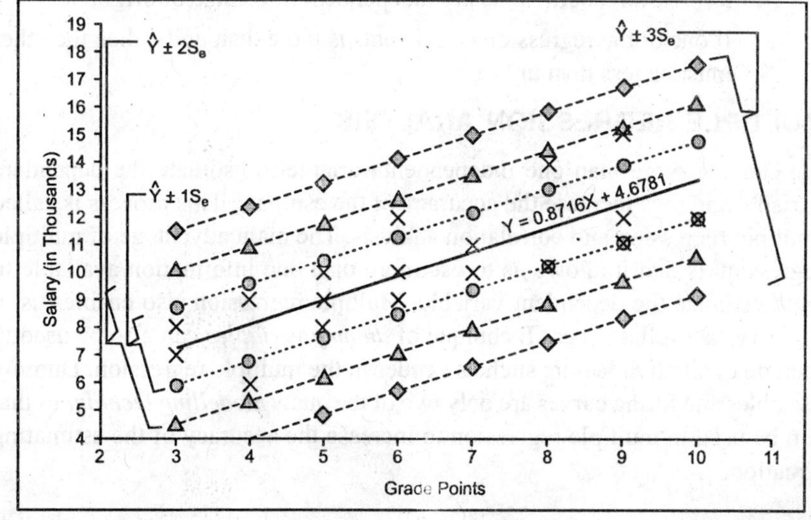

In order to get the required interval say 90% range of values, within which the salary of all students would lie, then instead of normal distribution, we would apply t-distribution. Suppose a boy got a grade point of 7.5, what would be his expected salary? For df = 20 – 2 = 18 and $\alpha = 0.05$; t = 1.734. So the range of values is: $Y_e \pm t \times S_e = 7.5 \pm 1.734 \times 1.398 = 9.924$ & 5.076. That means the boy would expect his salary with a minimum of 5.076 Thousand to a maximum of 9.924 Thousand.

TWO REGRESSION LINES

The regression line discussed so far Y = a + b X is the regression of Y on X, where b is the regression coefficient and is given by $b = \dfrac{\sum XY - n\overline{X}\overline{Y}}{\sum X^2 - n\,\overline{X}^2}$. What if it is required to find the regression of X on Y? The equation would be X = a' + b' Y. In the later case, $b' = \dfrac{\sum XY - n\overline{X}\,\overline{Y}}{\sum Y^2 - n\,\overline{Y}^2}$. So with the same set of data, two regression lines can be found. The following points describe some typical relationships:

1. $b \times b' = r^2$ = coefficient of determination
2. $b = r\dfrac{\sigma_y}{\sigma_x}$ and $b' = r\dfrac{\sigma_x}{\sigma_y}$
3. The correlation coefficient and regression coefficients have the same sign, i.e. r, b and b' have always the same sign.
4. Regression coefficients are independent of change of origin.
5. If one of the regression coefficients is more than unity, then the other must be less than unity.

MULTIPLE REGRESSION ANALYSIS

We can use more than one independent variable to estimate the dependent variable and thus increase the accuracy of the estimate. This process is called multiple regression and correlation analysis. The main advantage of multiple regression is that it allows us to use more than one information available to us to estimate the dependent variable. Multiple regression also enables us to fit curves as well as lines. Technique of *dummy variables* can also be used to include qualitative factors such as gender in the multiple regression. Dummy variables and fitting curves are only two of the many *modelling techniques* that can be used in multiple regression to increase the accuracy of the estimating equation.

In simple regression, the estimating equation Y_e describes the relationship between the two variables X and Y. In multiple regression, we extend that equation, adding one more term for each new variable. In symbolic form, for two independent variables, the equation becomes, Y_e . In simple regression, we use the line that describes the best fit as per least square method, i.e. the sum of squares of deviations of estimated values from actual values (or errors) should be minimum. The same principle is also adopted, but in place of a straight line, a plane is considered and the plane out of several possible planes that minimizes the sum of squares of the errors is the best fit. This can be well visualised from the figure below.

The three normal equations to determine the values of the numerical constants, a, b_1 and b_2 are:

$$=\sum Y = na + b_1 \sum X_1 + b_2 \sum X_2$$

$$=\sum X_1 Y = a \sum X_1 + b_1 \sum X_1^2 + b_2 \sum X_1 X_2$$

$$=\sum X_2 Y = a \sum X_2 + b_1 \sum X_1 X_2 + b_2 \sum X_2^2$$

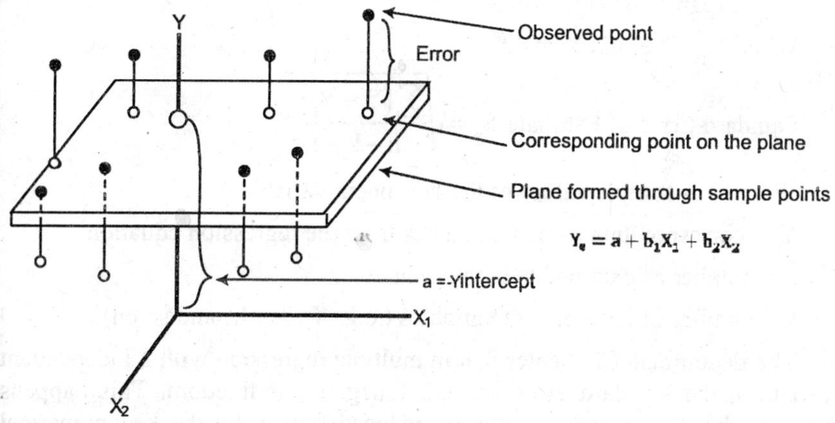

Example 13.3: Given the following set of data, (a) calculate the multiple regression plane (b) Predict Y when $X_1 = 2.8$ and $X_2 = 4.0$

Y	25	30	11	22	27	19
X_1	3.5	6.7	1.5	0.3	4.6	2.0
X_2	5.0	4.2	8.5	1.4	3.6	1.3

Solution: The following table is prepared, which is self-explanatory.

Y	X_1	X_2	X_1Y	X_2Y	X_1X_2	X_1^2	X_2^2	Y^2
25	3.5	5.0	87.5	125	17.5	12.25	25	625
30	6.7	4.2	201	126	28.14	44.89	17.64	900
11	1.5	8.5	16.5	93.5	12.75	2.25	72.25	121
22	0.3	1.4	6.6	30.8	0.42	0.09	1.96	484
27	4.6	3.6	124.2	97.2	16.56	21.16	12.96	729
19	2.0	1.3	38	24.7	2.6	4	1.69	361
134	**18.6**	**24**	**473.8**	**497.2**	**77.97**	**84.64**	**131.5**	**3220**

The normal equations will be as follows:

$134 = 6a + 18.6b_1 + 24.0b_2$

$473.8 = 18.6a + 84.64b_1 + 77.97b_2$

$497.2 = 24.0a + 77.97b_1 + 131.50b_2$

Solving these simultaneous equations, we get

$a = 20.392$, $b_1 = 2.340$ and $b_2 = -1.328$

So the regression equation is

When $X_1 = 2.8$ and $X_2 = 4.0$,

Standard Error of Estimate $S_e = \sqrt{\dfrac{\sum(Y - Y_e)^2}{n - k - 1}}$

Where Y = sample values of the dependent variable

Y_e = corresponding estimated values from the regression equation

n = number of data points in the sample

k = number of independent variables (e.g. X_1, X_2, X_3 and so on)

The denominator indicates that in multiple regression with k independent variables, the standard error has n-k-1 degrees of freedom. This happens because the degrees of freedom are reduced from n by the k+1 numerical constants, a_1, b_1, b_2, ... b_k that have all been estimated from the same sample. This Se is also called the *root-mean-square error* (or *root mse*).

Considering the above example, using the regression equation, we get the following table.

Y	X_1	X_2	Y_e	$Y - Y_e$	$(Y - Y_e)^2$
25	3.5	5	21.942	3.058	9.351
30	6.7	4.2	30.492	– 0.492	0.242

Y	X_1	X_2	Y_e	$Y-Y_e$	$(Y-Y_e)^2$
11	1.5	8.5	12.614	– 1.614	2.605
22	0.3	1.4	19.235	2.765	7.646
27	4.6	3.6	26.375	0.625	0.390
19	2	1.3	23.346	– 4.346	18.884
					39.118

$$\text{So, } S_e = \sqrt{\frac{\sum(Y-Y_e)^2}{n-k-1}} = \sqrt{\frac{39.118}{6-2-1}} = 3.611$$

Here also in case of multiple regression, we can construct an approximate confidence interval around the estimated value Y_e In the above example, when $X_1 = 2.8$ and $X_2 = 4.0$, Y_e. The appropriate t-value for $6 - 2 - 1 = 3$ degrees of freedom and 5% significance level is 3.182. So the limits of our confidence interval will be:

$$Y_e + t \times S_e = 21.632 + 3.182 \times 3.611 = 33.122$$
$$Y_e - t \times S_e = 21.632 - 3.182 \times 3.611 = 10.142$$

In multiple correlation, we can also measure the strength of the relationship among two variable using the *coefficient of multiple determination* r^2, or its square root r (the coefficient of multiple correlation). This coefficient of multiple determination is also the proportion of the total variation of Y that is explained by the regression plane. In MS-Excel the chart option gives the output showing the regression equation and R^2 value, after the trend line is fitted.

CURVE FITTING OR MODELLING TECHNIQUES

There can be several different regression equations we can fit with the group of explanatory variables with the variable that is to be explained and it depends on which explanatory variable we include and how we include them. Each one of such possible regression equations thus fitted is called a model. Modelling techniques are the various ways in which we can include the explanatory variables and check the appropriateness of our regression models. Although there are many different modelling techniques available, but we will concentrate only with two of the most commonly used devices.

USE OF DUMMY VARIABLES FOR QUALITATIVE DATA

The regression analysis done so far use only quantitative or numerical data. But what if a qualitative data is required to be incorporated into the regression equation? Here comes the use of dummy variables. Let us take a small example. Consider the starting salary of men versus that of women in a certain organization and the following is the data.

Male Employee (1)		Female Employee (2)	
Months Employed	Salary	Months employed	Salary
7	7.8	8	6.4
12	8.0	15	7.9
14	9.0	16	8.5
20	10.5	22	10.0
32	14.0	30	11.4
35	16.2		

Test whether men earn more than women at 1% level of significance.

H_0: $\mu_2 = \mu_2$ < Null hypothesis. There is no gender discrimination in the salary.

H_1: $\mu_1 > \mu_2$ < Alternative hypothesis. Women are discriminated against men.

Male Employee (1) **Female Employee (2)**

$n_1 = 6$ $n_2 = 5$

$\overline{x}_1 = 10.917$ $\overline{x}_2 = 8.84$

$s_1^2 = 11.898$ $s_2{}^2 = 3.713$

$$= s^2 = \frac{(n_1-1)s_1^2+(n_2-1)s_2^2}{n_1+n_2-2} = \frac{(6-1)\times 11.898+(5-1)\times 3.713}{6+5-2} = 8.260$$

$$\text{S.E.D.} = s\times\sqrt{\frac{1}{n_1}+\frac{1}{n_2}} = \sqrt{8.260}\times\sqrt{\frac{1}{6}+\frac{1}{5}} = 1.740$$

$$t_c = \frac{|\overline{x}_1-\overline{x}_2|}{\text{SED}} = \frac{|10.917-8.84|}{1.740} = 1.194$$

With 9 degrees of freedom, the critical value for an upper-tailed test with α=0.01 is 2.821. So null hypothesis is accepted and there appears to be no sex discrimination in starting salaries. But one thing to be kept in mind is we have altogether ignored the length of service of the employees. Let us now plot the scatter diagram for the same as shown below.

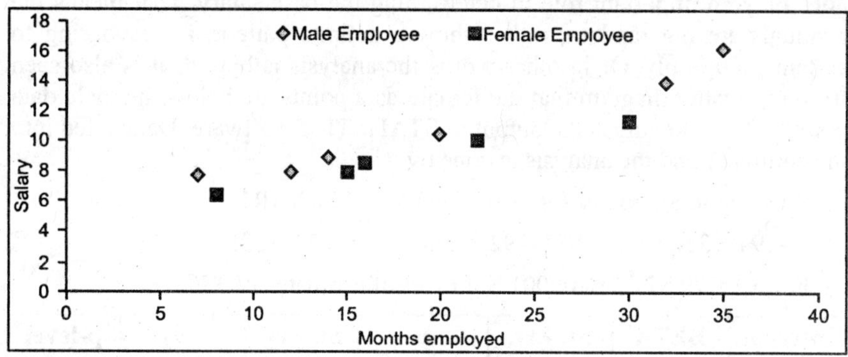

The scatter diagram shown above indicates that no doubt the starting salary increases with months of service, but salaries of male employees tend to be at a higher level than that of female employees. If we try to fit a linear regression between months employed(X) versus the base salary(Y) taking all the employees both male and female, we get $Y_e = 4.444 + 0.288\ X$ and the coefficient of determination $R^2 = 0.887$ (discussed in next chapter). That means about 88% variations in base salary is explained by the numbers of months of employment. Let us now find the fitted values(FIT) as per this regression equation and the difference of these fitted values (Y_e) from the actual observed values(Y) called the residuals(RES). These residuals are nothing but the error in the fit of the regression line at those points.

X	Y	Y_e = FIT	Y- Y_e = RES
7	7.8	6.46	1.34
8	**6.4**	**6.748**	**– 0.348**
12	8	7.9	0.1
14	9	8.476	0.524
15	**7.9**	**8.764**	**– 0.864**
16	**8.5**	**9.052**	**– 0.552**
20	10.5	10.204	0.296
22	**10**	**10.78**	**– 0.78**
30	**11.4**	**13.084**	**– 1.684**
32	14	13.66	0.34
35	16.2	14.524	1.676

A close look at the above table shows that, RES values pertaining to female employees (bold figures) are all negative, which indicates that gender does play an important role in determining the base salary. That means the residuals are not random, rather show a definite pattern, i.e. favouring to certain group only. Or in other words the analysis is biased. It is also seen from the scatter diagram that the female data points are below the male data points. The following is the output of STATISTICA software. Data is fed into the computer and the analysis is done by it.

Regression Summary for Dependent Variable: VAR1

$r = .94193520$ $r^2 = .88724192$ Adjusted $r^2 = .87471325$

$F(1, 9) = 70.817$ $p < .00001$ Std Error of estimate: 1.0386

Intercpt	BETA	St. Err. of BETA	B	St. Err. of B	t(9)	p-level
			4.444207	.727782	6.106507	.000178
VAR2	.941935	.111932	.288217	.034249	8.415278	.000015

But we have ignored the gender factor in our regression equation. Now the question is how to incorporate the gender variability into the regression equation? This is done by introducing a *dummy variable* (or *indicator variable*) into the method. For male employees, this value of dummy variable is given 0 and for female employees it is 1 as shown in the table below.

Months Employed (X_1)	Salary (Y)	Dummy Variable (X_2)	
7	7.8	0	Male Employees
12	8.0	0	
14	9.0	0	
20	10.5	0	
32	14.0	0	
35	16.2	0	
8	6.4	1	Female Employees
15	7.9	1	
16	8.5	1	
22	10.0	1	
30	11.4	1	

Now the regression equation fitted to the above data can be Y_e. Putting the values of the dummy variable the equation becomes:

Male employee: $Y_e = a + b_1X_1 + b_2X_2 = a + b_1X_1 + b_2\ 0 = Y_e$

Female employee: $Y_e = a + b_1X_1 + b_2X_2 = a + b_1X_1 + b_2\ 1 = Y_e$

Let us now fit a multiple regression equation to the above data and the following table is prepared, which is self-explanatory:

	Y	X_1	X_2	X_1Y	X_2Y	X_1X_2	X_1^2	X_2^2	Y^2
	7.8	7	0	54.6	0	0	49	0	60.84
	8	12	0	96	0	0	144	0	64
	9	14	0	126	0	0	196	0	81
	10.5	20	0	210	0	0	400	0	110.25
	14	32	0	448	0	0	1024	0	196
	16.2	35	0	567	0	0	1225	0	262.44
	6.4	8	1	51.2	6.4	8	64	1	40.96
	7.9	15	1	118.5	7.9	15	225	1	62.41
	8.5	16	1	136	8.5	16	256	1	72.25
	10	22	1	220	10	22	484	1	100
	11.4	30	1	342	11.4	30	900	1	129.96
Total:	109.7	211	5	2369.3	44.2	91	4967	5	1180.11

Using the normal equations, we get

$109.7 = 11a + 211\ b_1 + 5\ b_2$

$2369.3 = 211a + 4967\ b_1 + 91\ b_2$

$44.2 = 5a + 91\ b_1 + 5\ b_2$

Solving the equations, we can get, $a = 5.320$; $b_1 = 0.2798$ and $b_2 = -1.573$

So the regression equation is $Y_e\ 5.320 + 0.2798X_1 - 1.573X_2$

Again the STATISTICA software is used to find the regression equation and various other parameters like standard error of estimate and R^2 value. The output is given below.

Regression Summary for Dependent Variable: VAR1

$r = .98227402$ $r^2 = 0.96486225$ Adjusted $r^2 = 0.95607782$

$F(2,8) = 109.84$ $p < 0.00000$ Std Error of estimate: 0.61496

Intercept	BETA	St. Err. of BETA	B	St. Err. of B	t(9)	p-level
			5.32027	0.478660	11.11491	0.000004
VAR2	0.914493	0.066595	0.27982	0.020377	13.73226	0.000001
VAR3	– 0.279952	0.066595	–1.57299	0.374180	–4.20384	0.002981

Let us again find the residual values using this multiple regression equation. (STATISTICA can also be used to find the residuals as well)

Y	X_1	X_2	Y_e = FIT	$Y - Y_e$ = RES
7.8	7	0	7.279007	0.520993
8	12	0	8.678107	–0.67811
9	14	0	9.237747	–0.23775
10.5	20	0	10.91667	–0.41667
14	32	0	14.27451	–0.27451
16.2	35	0	15.11397	1.086034
6.4	8	1	5.985837	0.414163
7.9	15	1	7.944576	–0.04458
8.5	16	1	8.224396	0.275604
10	22	1	9.903316	0.096684
11.4	30	1	12.14188	–0.74188

Here we can see that the signs of residuals are random, indicating that this multiple regression equation where gender variability has been introduced as an additional factor is a better way of prediction. In case of multiple regression, the standard error is 0.615 and r^2-value is 0.96; that means about 96% variations are explained by introduction of gender variable into the regression model. But in case of simple regression, the standard error of estimate was 1.038 and r^2-value was only 0.88. Hence incorporating another independent variable has surely improved the regression model.

It may be noted that the choice of assigning 0 and 1 to the dummy variable is purely arbitrary. We can also assign 0 to female employees and 1 to male employees as well and there will not be any change in the values of the regression coefficients except the signs respectively. Dummy variable techniques can also be applied to problems of several qualitative variables.

MODELLING TECHNIQUE OR CURVE FITTING

So far we have studied only linear relationship between dependent variables with the independent variables. There can also be non-linear relationship among them. But which type of relationship is to be chosen or adopted? This is done by finding the residuals. Obviously, the regression equation, otherwise called the model, where more randomness in the residuals is found, is treated to be better than the rest. Added to this, the model which has less standard error of estimate(s_e) and higher coefficient of determination(R^2) is chosen to be the accepted model in curve fitting. As such calculations involve complicated calculation procedures, use of computer is made in finding these values. Considering an example, where the owner of a shop made an advertisement campaign in order to boost his sale. He found the following data regarding the weekly sale value versus the time since the advertisement campaign was over.

Time	Sale (Thousand Rs)	Time	Sale (Thousand Rs)	Time	Sale (Thousand Rs)	Time	Sale (Thousand Rs)
1	4618	4	4367	7	19746	10	65717
2	3741	5	5118	8	34215	11	86434
3	5836	6	8887	9	50306	12	105464

Let us now try to fit several types of curves to the above data and select the best one. Using the STATISTICA software, we can get the output for various types of curves.

The following is the output of linear regression model.

Regression Summary for Dependent Variable: VAR1

$r = .91147805$ $r^2 = .83079224$ Adjusted $r^2 = .81387147$

$F(1,10) = 49.099$ $p < .00004$ Std Error of estimate: 15518.

Intercpt	BETA	St. Err. of BETA	B	St. Err. of B	t(9)	p-level
			–26232.8	9550.585	–2.74672	.020594
VAR2	.911478	.130080	9092.8	1297.669	7.00706	.000037

So the equation is SALE = –26232.8 + 9092.8 TIME and $R^2 = 0.8308$

Although the t-value is high (7.00) and time accounts for about 83% of variations in sale values, the residuals (calculated in the table below) show a definite pattern and are far from random. They start with positive values and become negative in the middle part (Time 4 to 9) and then turn around again to become positive. Standard error of estimate is 15518.

TIME	SALE	FIT-1	RES-1
1	4618	-17140	21758
2	3741	-8047.2	11788.2
3	5836	1045.6	4790.4
4	4367	10138.4	–5771.4
5	5118	19231.2	–14113.2
6	8887	28324	–19437
7	19746	37416.8	–17670.8
8	34215	46509.6	–12294.6
9	50306	55602.4	–5296.4
10	65717	64695.2	1021.8
11	86434	73788	12646
12	105464	82880.8	22583.2

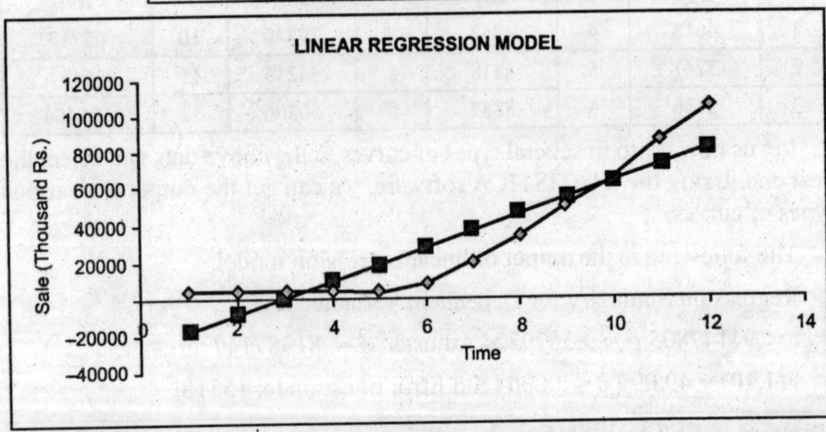

Now let us introduce another variable to the model of the form $X_2 = \text{TIME}^2$ and then run a multiple regression. The input data is shown below and the output of STATISTICA software.

SALE (Y)	TIME (X)	TIME^2 (X^2)
4618	1	1
3741	2	4
5836	3	9
4367	4	16

SALE (Y)	TIME (X)	TIME² (X²)
5118	5	25
8887	6	36
19746	7	49
34215	8	64
50306	9	81
65717	10	100
86434	11	121
105464	12	144

Regression Summary for Dependent Variable: VAR1

r = .99780813 r^2 = .99562106 Adjusted r^2 = .99464796

F(2, 9) = 1023.1 p < .00000 Std Error of estimate: 2631.4

Intercpt	BETA	St. Err. of BETA	B	St. Err. of B	t(9)	p-level
		13980.75	2719.616	5.14071	.000611	
VAR2	–.816115	.096419	–8141.52	961.867	–8.46429	.000014
VAR3	1.774657	.096619	1325.75	72.028	18.40575	.000000

The regression equation is SALE = 2719.616 – 8141.52 TIME + 1325.72 TIME² and it is noticed that the multiple regression of determination is r^2 = 0.9956, which indicates that the two variables together explain about 99.56% of the total variations. t-values are found to be -8.464 and 18.405, which justifies highly significant explanatory variables. Here the standard error of estimate is 2631.4 as against the previous simple linear model of 15518. So this curve model is no doubt superior to the straight-line model and let us now calculate the residuals and observe the pattern.

SALE	TIME	TIME²	FIT-2	RES-2
4618	1	1	–4096.18	8714.18
3741	2	4	–8260.54	12001.54
5836	3	9	–9773.46	15609.46
4367	4	16	–8634.94	13001.94
5118	5	25	–4844.98	9962.98
8887	6	36	1596.42	7290.58
19746	7	49	10689.26	9056.74

SALE	TIME	TIME²	FIT-2	RES-2
34215	8	64	22433.54	11781.46
50306	9	81	36829.26	13476.74
65717	10	100	53876.42	11840.58
86434	11	121	73575.02	12858.98
105464	12	144	95925.06	9538.94

Here we see all the residuals are positive. They don't exhibit any pattern. Because, we squared the X-variable and introduced it as a second variable into the model, the model is thus known as second-degree (or quadratic) regression model. There can be many other ways in which we can transform variables to get new variables, most computer statistical packages have these transformation built into them and hence need not be computed manually. Various models involving products, quotients, roots, power, logarithms, exponentials, trigonometric functions and many more can be formulated for regression analysis.

EXERCISES

1. The following is the data regarding the demand of some computer consumables for past 10 weeks as per a study made by one retail company, who was examining its inventory control policy. Fit a linear model with TIME as independent variable and DEMAND as dependent variable. Fit a quadratic model for the data. Is the later model better? Explain. Use available computer package.

Time:	1	2	3	4	5	6	7	8	9	10
Demand:	6.6	10.0	13.4	15.5	18.0	22.5	30.4	31.4	37.8	10.5

2. Below are some data on monthly expenditures(Rs.) and income (Rs.) and sex of head of family of 9 randomly chosen families. Develop a linear regression model between income and expenditure. Use dummy variable technique and explain if the model can be improved by introducing sex as additional independent variable.

Income:	45000	28070	26100	35500	18850	41000	26950	21500	36060
Expenditure:	37075	22800	24250	30400	17500	33550	25960	20360	35680
Sex:	M	M	F	M	M	F	F	M	F

3. The data given below is salinity (%) and dissolved oxygen (mg/l). Find out the regression line to estimate dissolved oxygen with respect to salinity. Test the regression coefficient. Estimate the value of dissolved oxygen when the salinity of water is 15.55%.

Salinity (%):	6	8	10	4	17	15
Dissolved Oxygen (mg l^{-1}):	8	6	6	10	4	3

4. In a rice field the leaf area and the chlorophyll content of successive growth stages are given below. Find out the regression equation considering leaf area as X and chlorophyll content as Y. Test the regression coefficient b. Find out the estimated chlorophyll content with respect to leaf area 2.75 (m2/m2).

Leaf Area (m2/m2)	2.00	3.35	4.35	3.26	2.65	2.25
Chlorophyll (mg/m2)	185.2	778.4	800.6	811.9	635.0	225.2

5. In a class the corresponding height and weight of 15 students are given below. Find out two lines of regression. Hence, find the correlation coefficient and show that $\sqrt{bb'} = r$.

Height (m)	1.25	1.40	1.22	1.23	1.27	1.37	1.20	1.15	1.32	1.11	1.13	1.16	1.33	1.25	1.17
Weight (kg)	56	60	50	40	46	50	40	35	40	30	35	30	36	41	32

6. Find the regression equation for the following data, plot the observed value and estimated value in a graph paper and hence draw the line of regression.

Length of Inflorescence(X):	20	18	17	19	21	15	16	14
No. of Flowers (Y):	14	13	14	15	17	11	12	10

7. Write brief accounts of Regression vs Correlation in about 200 words.
8. What is regression? Why are there two regression lines? Under what conditions there can be one regression line.
9. Define statistical correlation and regression. Deduce formulae for correlation coefficient and regression equations for two variables, X and Y. Write down the formulae for t-test in the above cases. Prove that the correlation coefficient lies between –1 and +1.
10. What do you understand by regression analysis? What are its advantages?
11. Describe the method of calculation of regression coefficients. Write down some properties of regression coefficients.
12. Differentiate between correlation and regression. Discuss the methods used to study correlation. Add a note on the coefficient of correlation.

CORRELATION

INTRODUCTION

Correlation analysis is the statistical tool we can use to describe the degree to which one variable is linearly related to another. There can be relationship between degree of safety measures taken and number of industrial accidents, between age and body mass index, between rainfall and runoff in a stream and so on. But the change in one variable causes a change in the other variable. This change implies that some sort of relation exists between the two sets of variables. In other words they are said to be correlated and the relationship between them is known as *Correlation.*

DEFINITION

According to Connor correlation is defined as *"If two or more quantities vary in sympathy so that movements in the one tend to be accompanied by corresponding movements in the other, then they are said to be correlated."*

Croxton and Cowden had said *"When the relationship is of a quantitative nature, the appropriate statistical tool for discovering and measuring the relationship and expressing it in a brief formula is known as correlation."*

UTILITY

1. Correlation study helps us to estimate the likely change in a variable with a particular amount of change in related variable.
2. The study of correlation reduces the range of uncertainty associated with decision-making.
3. Correlation study helps in identifying such factors, which can stabilize a turbulent situation.
4. It helps in prompting research and opening up of new vistas of knowledge.

ASSUMPTIONS

Two main assumptions are (i) Variables (x, y, ...) are normally distributed in population (ii) There is a cause-effect relationship in population; e.g. One cannot correlate between height of persons with their monthly incomes.

TYPES OF CORRELATION

1. **Positive Correlation:** If two sets of variables steadily vary in the same direction, correlation is direct and it is called positive correlation. Here one set of variable increases with the increase of the other set of variables. Consider the following series where the body weight of fishes increases with increase in their lengths.

Body Length in mm.	40	45	55	60	65	70
Body weight in gram.	7	7.5	8.3	9.5	10	12

2. **Negative Correlation:** If two sets of variables steadily vary in opposite direction, correlation is inverse and it is called negative correlation. Here one set of variables decreases with the increase of the other set of variables. Consider the following series where percentage germination of seeds decreases with an increase in the concentration of pesticide.

Conc. (ppm)	0	100	250	500	750	1000	1250	1500	2000
% Germination	84	81	65	52	39	32	28	17	6

3. **Linear Correlation:** When the amount of change in one set of variables bears a constant ratio to the amount of change in the other set of variables, correlation between them is linear.
4. **Non-Linear Correlation:** When the ratio of change is not constant, correlation is non-linear or curvilinear.

Mere existence of correlation between two sets of variables has no value in statistics. It has to be quantified by some means for any statistical analyses. So, correlation is generally measured by coefficient of determination and coefficient of correlation.

Strength or measure of any association that exists between two variables is determined by coefficient of determination. Because we have used a sample of points to develop the regression line, let us refer this measure as sample coefficient of determination. To find the coefficient of determination, two types of variation are used.

Variation of data points from their own mean = $\sum(Y - \bar{Y})^2$

Variation of the data points from the fitted regression line = $\sum(Y-Y_e)^2$

Coefficient of determination = $r^2 = 1 - \dfrac{\sum(Y-Y_e)^2}{\sum(Y-\bar{Y})^2}$

Let us consider the following sample data points.

X:	1	2	3	4	5	6	7	8	9	10
Y:	2	4	6	8	10	12	14	16	18	20

Below is the graph of the data set, which shows that the Y-intercept passes through origin and hence a = 0. Slope of the regression line is: b = 2.

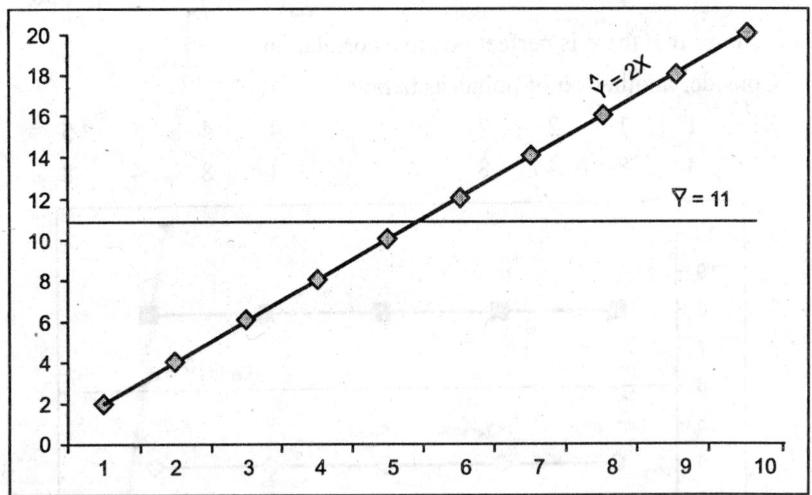

Regression equation is: $Y_e = 2X$. To find the coefficient of determination, let us construct the following table:

X	Y	$\sum(Y-\bar{Y})^2$	Y_e	$\sum(Y-Y_e)^2$
1	2	81	2	0
2	4	49	4	0
3	6	25	6	0
4	8	9	8	0
5	10	1	10	0
6	12	1	12	0

X	Y	$\sum(Y-\bar{Y})^2$	Y_e	$\sum(Y-Y_e)^2$
7	14	9	14	0
8	16	25	16	0
9	18	49	18	0
10	20	81	20	0
Total:	110	330		0

Mean = $110 \div 10 = 11 = \bar{Y}$

Coefficient of determination = $r^2 = 1 - \frac{\sum(Y-Y_e)^2}{\sum(Y-\bar{Y})^2} = 1 - \frac{0}{330} = 1.$

This shows that there is perfect positive correlation.

Consider another set of points as below:

X:	1	1	2	2	3	3	4	4	5	5
Y:	4	8	4	8	4	8	4	8	4	8

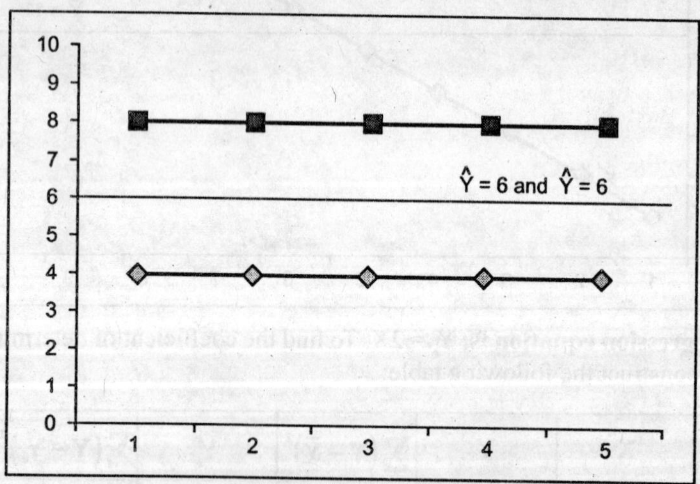

Here in this data set, slope of the regression line, b = 0. As can be seen from the graph, Y-intercept, a = 6. [*It is left to students as exercise to check values of a and b using simple linear regression formulae. Thus regression equation is* $Y_e = 6$]. To find the coefficient of determination, the following table is constructed.

X	Y	$\sum(Y-\overline{Y})^2$	Ye	$\sum(Y-Y_e)^2$
1	4	4	6	4
1	8	4	6	4
2	4	4	6	4
2	8	4	6	4
3	4	4	6	4
3	8	4	6	4
4	4	4	6	4
4	8	4	6	4
5	4	4	6	4
5	8	4	6	4
Total:	60	40		40

Mean = 60 ÷ 10 = 6 = $\overline{Y}$

Coefficient of determination = $r^2 = 1 - \dfrac{\sum(Y-Y_e)^2}{\sum(Y-\overline{Y})^2} = 1 - \dfrac{40}{40} = 0$.

This shows that there is correlation at all.

So, r^2, which determines the strength of linear relationship between two variables, can have values from 0 to 1 only. There is also a short cut method to find the r^2 value and is given by

$$r^2 = \frac{a\sum Y + b\sum XY - n\overline{Y}^2}{\sum Y^2 - n\overline{Y}^2}$$

INTERPRETATION OF r^2

We have already discussed about the regression line, which has the best fit linear equation based on method of least square. Obviously question arises about how to find the coefficient of determination by looking at the variation of Y that is explained by the regression line. To understand this concept, we have in total three variations:

Variation of Y from the mean: $Y - \overline{Y}$

Variation of Y from regression line: $Y - Y_e \rightarrow$ Unexplained variation

Variation of Y_e from mean value: $Y_e - \overline{Y} \rightarrow$ Explained variation

So, Total variation = Unexplained variation + Explained variation

As can be seen from the graph, considering a single point Y we have,

$$Y - \bar{Y} = (Y - Y_e) + (Y_e - \bar{Y})$$

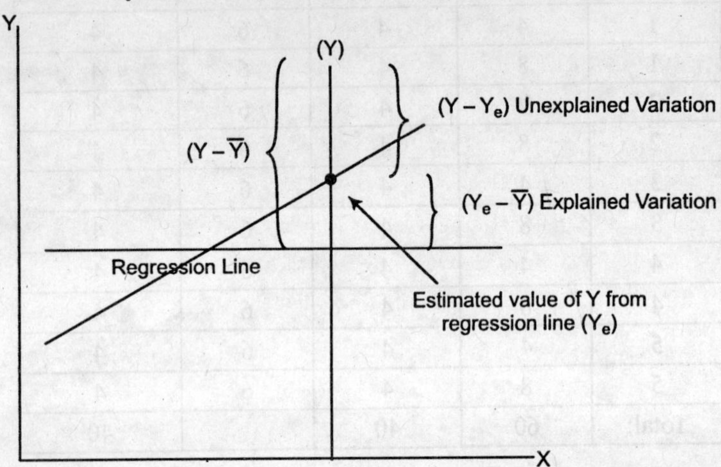

So the fraction of the total variation that remains unexplained is $\frac{\sum(Y - Y_e)^2}{\sum(Y - \bar{Y})^2}$. This value when subtracted from 1 gives the fraction of total variation that is explained, which is nothing but r^2. Hence, $r^2 = 1 - \frac{\sum(Y - Y_e)^2}{\sum(Y - \bar{Y})^2}$. It is this philosophy that elucidates how well Y is explained by X or associated with X. In other way r^2 is also equal to the fraction of the explained variation to the total variation, i.e. $r^2 = \frac{\sum(Y_e - \bar{Y})^2}{\sum(Y - \bar{Y})^2}$

COEFFICIENT OF CORRELATION

Coefficient of correlation is the square root of coefficient of determination. Sign of r indicates the direction of the relationship between two variables X and Y. That means, it takes up the sign of the regression coefficient b. In inverse relationship, r is negative and vice versa. So value of coefficient of correlation varies between -1 and +1. If in a relationship, r = 0.8, then r^2 = 0.64. So 64% of the variations in Y is explained by the regression line and the rest remains unexplained.

Coefficient of Correlation = $\pm\sqrt{\text{Coefficient of Determination}}$

i.e. = $\pm\sqrt{r^2}$

Note: Coefficient of correlation should be interpreted with caution. Take an example- let there is an association between qualification of couples and divorce rate, and the coefficient of correlation of a sample data is found to be 0.7, which can be taken as nearer to 1 than nearer to 0. But the coefficient of determination is 0.49 only. That means only 49% of divorce rate is explained by the couples' qualification, but 51% is due to some other reason.

It may be noted that the coefficient of correlation reveals how well, not how much, the two sets of data are associated with each other.

PROBABLE ERROR OF THE COEFFICIENT OF CORRELATION

After the calculation of coefficient of correlation, the next thing is to find out the extent to which it is dependable. For this purpose, the probable error of the coefficient of correlation is calculated. If the probable error is added to or subtracted from the coefficient of correlation, it would give two such limits within which we can reasonably expect the value of coefficient of correlation to vary for the entire population. It means that if from the same universe another set of samples is selected on the basis of random sampling, the coefficient of correlation between the two variables in this new sample would not fall outside the limits so established.

CONDITIONS

Probable Error is found under two conditions (i) Data are collected independently from normally distributed population (ii) Samples are only considered for analysis.

The formula for finding out the probable error of Karl Pearson's coefficient of correlation is:

$$\text{Probable error of coefficient of correlation} = 0.6745 \times \frac{1-r^2}{\sqrt{n}}$$

Thus if a coefficient of correlation 0.8 is observed by a study of 100 pairs of observations, the probable error would be

P.E. = $0.6745 \times [(1 - (0.8)^2) \div \sqrt{100}\,] = 0.6745 \times 0.36 \div 10 = 0.24282 \div 10 = 0.0243$

(i) It means that in the universe the limits of the coefficient of correlation should be

r ± P.E. or 0.8 ± 0.0243 or 0.7757 and 0.8243.

(ii) If the value of r is less than the P.E., there is no evidence of correlation.

(iii) If the value of r is more than six times the P.E., there is significant correlation.

i.e. if { r ÷ P.E. } > 6, then r is definitely significant.

Let us now take up some examples to fully understand the whole theory.

Example 14.1: The Chief accountant of a manufacturing company collected data on overhead expenses and units produced for 10 months and wanted to estimate a regression equation to predict future overhead.

Overhead	190	165	270	160	275	170	230	115	150	175
Units	40	42	55	35	55	38	47	30	36	40

(a) Develop the regression equation for the chief accountant

(b) Predict overhead when 50 units are produced.

(c) Calculate the standard error of estimate.

(d) Calculate sample coefficient of determination and sample coefficient of correlation.

(e) Calculate the probable error of r.

Solution: The table below is prepared, which is self-explanatory.

S. No	X	Y	XY	X^2	Y^2
1	40	190	7600	1600	36100
2	42	165	6930	1764	27225
3	55	270	14850	3025	72900
4	35	160	5600	1225	25600
5	55	275	15125	3025	75625
6	38	170	6460	1444	28900
7	47	230	10810	2209	52900
8	30	115	3450	900	13225
9	36	150	5400	1296	22500
10	40	175	7000	1600	30625
Total:	418	1900	83225	18088	385600

$$\text{Mean} = \overline{X} = \frac{418}{10} = 41.8$$

$$\text{Mean} = \overline{Y} = \frac{1900}{10} = 190$$

$$b = \frac{\sum XY - n\overline{X}\overline{Y}}{\sum X^2 - n\overline{X}^2} = \frac{83225 - 10 \times 41.8 \times 190}{18088 - 10 \times 41.8^2} = 6.181$$

$$a = \overline{Y} - b\overline{X} = 190 - 6.181 \times 41.8 = -68.366$$

(a) Thus the regression equation is, $Y_e = -68.366 + 6.181X$

(b) For production of 50 units(X = 50), overhead will be = -68.366 + 6.181 50 = 240.684

(c) $$s_e = \sqrt{\frac{\sum Y^2 - a\sum Y - b\sum XY}{n-2}}$$

$$= \sqrt{\frac{385600 - (-68.366) \times 1900 - 6.181 \times 83225}{10-2}} = 11.628$$

(d) $$r^2 = \frac{a\sum Y + b\sum XY - n\overline{Y}^2}{\sum Y^2 - n\overline{Y}^2}$$

$$= \frac{(-68.366) \times 1900 + (6.181) \times 83225 - 10 \times 190^2}{385600 - 10 \times 190^2} = 0.956$$

Coefficient of correlation, r = $\sqrt{0.956} = 0.978$

(e) Probable error of coefficient of correlation

$$= 0.6745 \times \frac{1-r^2}{\sqrt{n}} = 0.6745 \times \frac{1-0.956}{\sqrt{10}} = 0.009$$

Because,

$\frac{r}{PE} = \frac{0.978}{0.009} = 108.667 \gg 6$, There is highly significant correlation between the above set of data.

The regression equation developed so far is $Y_e = a + b X$, relates to sample data only. But what happens if the entire population data is considered? Let the population regression equation is represented by Y = A + B X. The sample regression equation may not exactly match with the population regression equation. But the sample regression coefficient b may be considered as population regression coefficient B and test the hypothesis about the value of B. The procedure is same as was done in the chapter of hypothesis testing. Let us take the same example of the regression equation (see Example 13.1) developed between grade points and starting salary, $Y_e = 4.6781 + 0.8716 X$, where sample size was 20. Let us suppose that past record has found the slope of population regression equation to be 1.00, i.e. B = 1.00. Then we would define:

Null Hypothesis: H_0: B = 1.00

Alternative Hypothesis: H_1: B ≠ 1.00

The test statistic or, the standard error of b,

$$s_b = \frac{s_e}{\sqrt{X^2 - n\bar{X}^2}} \text{ and } t_{\acute{a}(n-2)} = \frac{|b-B|}{s_b}$$

From the same example, we carry out the various calculations to find these values.

Grade Points	3	3	4	4	5	5	5	6	6	6	7	7	7	8	8	8	9	9	9	10
Starting salary in Thousands	8	7	9	6	10	9	8	12	10	9	11	10	10	13	14	10	12	15	11	12

S. No	X	Y	X²	Y²	XY
1	3	8	9	64	24
2	3	7	9	49	21
3	4	9	16	81	36
4	4	6	16	36	24
5	5	10	25	100	50
6	5	9	25	81	45
7	5	8	25	64	40
8	6	12	36	144	72
9	6	10	36	100	60
10	6	9	36	81	54
11	7	11	49	121	77
12	7	10	49	100	70
13	7	10	49	100	70
14	8	13	64	169	104
15	8	14	64	196	112
16	8	10	64	100	80

S. No	X	Y	X²	Y²	XY
17	9	12	81	144	108
18	9	15	81	225	135
19	9	11	81	121	99
20	10	12	100	144	120
Total =	**129**	**206**	**915**	**2220**	**1401**

$$s_e = \sqrt{\frac{\sum Y^2 - a\sum Y - b\sum XY}{n-2}} = \sqrt{\frac{2220-(4.6781)\times 209 - 0.8716\times 1401}{20-2}} = 1.398$$

$$s_b = \frac{s_e}{\sqrt{\sum X^2 - n\bar{X}^2}} = \frac{1.398}{\sqrt{915-20\times 6.45^2}} = 0.1535$$

$$t_{\acute{a}(n-2)} = \frac{|b-B|}{s_b} = \frac{|0.8716-1.00|}{0.1535} = 0.8365$$

For α = 0.005(or 99% confidence level) and 18 df table value of t = 2.878. Calculated value is well within the table value. So null hypothesis is accepted. That means, the slope of the population regression equation B = 1.00 still holds good. This procedure can also be used to find the confidence interval for the value of the population regression coefficient B, which is given by $b \times t \times s_b$.

Example 14.2: A real estate businessman has found the following information from the past data regarding appraised value of homes and their built-up area.

Area(Thousand sq ft):	1.1	1.5	1.6	1.6	1.4	1.3	1.1	1.7	1.9	1.5	1.3
Value(Rs Lakh):	75	95	110	102	95	87	82	115	122	98	90

(a) Estimate the least-square regression to predict appraised value from built-up area.

(b) If the businessman feels that a home's value goes up by Rs 50,000 for every additional 1,000 sq ft in area, test whether his conviction holds good or not. Use α = 0.10

Solution: The following table is prepared, which is self-explanatory:

S. No	X	Y	XY	X^2	Y^2
1	1.1	75	82.5	1.21	5625
2	1.5	95	142.5	2.25	9025
3	1.6	110	176	2.56	12100
4	1.6	102	163.2	2.56	10404
5	1.4	95	133	1.96	9025
6	1.3	87	113.1	1.69	7569
7	1.1	82	90.2	1.21	6724
8	1.7	115	195.5	2.89	13225
9	1.9	122	231.8	3.61	14884
10	1.5	98	147	2.25	9604
11	1.3	90	117	1.69	8100
Total:	16	1071	1591.8	23.88	106285

$$= \overline{X} = \frac{16}{11} = 1.4545 \text{ and } \overline{Y} = \frac{1071}{11} = 97.3636$$

The regression coefficient,

$$b = \frac{\sum XY - n\overline{XY}}{\sum X^2 - n\overline{X}^2} = \frac{1591.8 - 11 \times 1.4545 \times 97.3636}{23.88 - 11 \times 1.4545^2} = 55.958$$

The regression constant, a = $\overline{Y} - b\overline{X}$ = 97.3636 – 55.958 × 1.4545 = 15.97

(a) Thus the regression equation is, $Y_e = 15.97 + 55.958\,X$

(b) Sample regression slope, b = 55.958

Given that population regression slope, $B = \frac{50000}{1000} = 50$

$$s_e = \sqrt{\frac{\sum Y^2 - a\sum Y - b\sum XY}{n-2}} = s_e$$

$$= \sqrt{\frac{106285 - (15.97) \times 1071 - 55.958 \times 1591.8}{11-2}} = 3.448$$

$$s_b = \frac{s_e}{\sqrt{\sum X^2 - n\overline{X}^2}} = \frac{3.448}{\sqrt{23.88 - 11 \times 1.4545^2}} = 4.4244$$

$$t_{\acute{a}(n-2)} = \frac{|b - B|}{s_b} = \frac{|55.958 - 50|}{4.4244} = 1.347$$

For α = 0.10 (or 90% confidence level) and 9 df table value of t = 1.833. Calculated value is within the table value. So null hypothesis is accepted. That means, the slope of the population regression equation B = 50 holds good. That means the sample data does not provide any evidence to reject the businessman's conviction that a home's value goes up by Rs 50,000 for every additional 1,000 sq ft in area.

Example 14.3: The equations of two regressions obtained in a correlation analysis are as follows:

$3x + 12y = 19$ and $3y + 9x = 46$.

Find (i) two regression coefficients (ii) correlation coefficient (iii) mean values of x and y

Solution: Here two situations can arise:

1. Let us consider the first equation to be regression of x on y and the second be that of y on x.

   Considering the first equation to be regression of x on y we get,

   $$x = \frac{19}{3} - \frac{12}{3}y\text{; where } b = (-12) \div 3 = -4$$

   Considering the second equation to be regression of y on x we get,

   $$y = \frac{46}{3} - \frac{9}{3}x\text{; where } b' = (-9) \div 3 = -3$$

   Hence $r^2 = b \times b' = (-4) \times (-3) = 12$, which is not possible, because r^2 can take up values between 0 and 1.

2. So we have to consider the first equation to be the regression of y on x and the second to be that of x on y only. Hence,

   Considering the first equation to be regression of y on x we get,

   $$y = \frac{19}{12} - \frac{3}{12}x\text{; where } b = (-3) \div 12 = -1/4$$

   Considering the second equation to be regression of x on y we get,

   $$x = \frac{46}{9} - \frac{3}{9}y\text{; where } b' = (-3) \div 9 = -1/3$$

Hence $r^2 = b \times b' = (-1/3) \times (-1/4) = 1/12$, which is acceptable. So r

$$= \sqrt{\frac{1}{12}} = -0.2886$$

Here sign of b and b' are both negative. As b, b' and r bear same sign, so r = – 0.2886

As the mean values of x and y satisfy the two regression equations, we have,

$$\bar{y} = \frac{19}{12} - \frac{3}{12}\bar{x} \text{ and } \bar{x} = \frac{46}{9} - \frac{3}{9}\bar{y}\text{; solution of which gives:}$$

$$\bar{x} = 5.2121 \text{ and } \bar{y} = 0.2803$$

Comparison between Correlation and Regression

Correlation	Regression
Correlation means the relationship between two or more variables which vary such that change in one tends to be accompanied by corresponding change in other	Regression means stepping back or returning to the average value and is a mathematical measure expressing the average relationship between the two variables.
Correlation need not imply cause and effect relationship between the variables under study.	Regression analysis clearly indicates the cause and effect relationship between the variables. The variable corresponding to cause is taken as independent variable and the variable corresponding to effect is taken as dependent variable.
Correlation analysis is confined only to the study of linear relationship between the variables and so has limited applications.	Regression analysis has much wider application as it studies linear as well as non-linear relationship between the variables.
There may be correlation between two variables which is due to pure chance and has no practical relevance e.g. correlation between height of persons vs their salary.	There is no such thing like non-sense regression.
Correlation coefficient r(x,y) between two variables X and Y is a measure of the direction and degree of the linear relationship between two variables which is mutual. It is symmetric i.e. r(x,y) = r(y,x) and it is immaterial which of X and Y is dependent or independent.	Regression analysis aims at establishing the functional relationship between the two variables under study and then using this relationship to predict or estimate the value of the dependent variable for any given value of the independent variable. It also reflects upon the nature of the variable i.e. which is dependent variable and which is independent variable. Regression coefficients are not symmetric in X and Y. b(x,y) b(y,x)

Correlation	Regression
Correlation coefficient r(x,y) is a relative measure of the linear relationship between X and Y and is independent of the units of measurement. It is a pure number lying between –1 to +1.	Regression coefficients are absolute measures representing the change in the value of the variable for a unit change in the value of the other variable. Once the functional form of regression curve is known, by substituting the value of the dependent variable, we can obtain the value of independent variable and its value is in the units of the variable.

EXERCISES

1. When home sales are good, sale of home appliances like dishwashers, sewer connections, and refrigerators also rise. Following is a compilation of historical data on major appliance sales (thousand rupees) and sale of houses (lakh rupees).

House sale:	14.2	15.1	13.8	10.9	11.4	10.5	14.6	18.2	13.7
Appliance sale:	5.82	5.91	4.58	3.64	3.95	3.42	6.05	7.10	4.60

   (a) Develop an equation for the relationship between appliance sales (in thousands) and housing sales (in lakhs)

   (b) Interpret the slope of the regression line.

   (c) Compute and interpret the standard error of estimate.

   (d) Calculate the sample coefficient of determination and sample coefficient of correlation.

2. A firm administers a test to sales trainees before they go into the field. The management of the firm is interested in determining the relationship between the test scores and the sales made by the trainees at the end of one year in the field. The following data were collected for 10 sales personnel who have been in the field for one year.

Sales Person No.:	1	2	3	4	5	6	7	8	9	10
Test Score(T):	2.5	3.6	2.3	4.4	2.5	4.9	2.7	2.9	3.9	3.3
Units sold (S):	94	142	88	178	96	190	115	135	170	150

   (a) Find the least square regression line that could be used to predict sales from trainee test scores.

   (b) How much does the expected number of units sold increase for each 1-point increase in a trainee's test score.

(c) Use the least squares regression line to predict the number of units that would be sold by a trainee who received an average test score.

(d) Calculate the sample coefficient of determination and sample coefficient of correlation.

3. A leather industry is interested in removing pollutants from the exhaust at the factory, and cost is a concern. They collected data from other industries about the amount of money spent (in thousands) on environmental measures and the resulting amount of dangerous pollutants released (percentage of total emissions).

Money spent:	8.3	10.1	16.4	21.6	9.3	8.2	11.4	18.3	16.6
% of pollutants:	34.9	30.8	23.7	24.2	35.8	34.8	32.4	24.4	30.4
Money spent:	19.2	28.3	4.6	12.2					
% of pollutants:	26.4	14.8	30.5	27.9					

(a) Compute the regression equation.

(b) Predict the percentage of dangerous pollutants released when Rs. 20,500 is spent on control measures.

(c) Calculate the standard error of estimate.

(d) Calculate the sample coefficient of determination and sample coefficient of correlation.

4. A bank manager became interested in reducing the time spent by people in waiting at a counter. He observed and collected data on waiting time(Y) and number of bankers(X).

X:	2	3	5	4	2	6	1	3	4	3	3	4
Y:	12.8	10.3	3.2	3.5	10.6	3.2	7.8	10.5	8.9	10.3	9.6	12

(a) Develop a regression equation that best fits the data.

(b) Calculate the sample coefficient of determination and sample coefficient of correlation.

5. Write brief accounts of Regression vs Correlation in about 200 words.

6. Define statistical correlation and regression. Deduce formulae for correlation coefficient and regression equations for two variables X and Y. Write down the formulae for t-test in the above cases. Prove that the correlation coefficient lies between –1 and +1

7. What do you understand by regression analysis? What are its advantages?

8. Describe the method of calculation of regression coefficients. Write down some properties of regression coefficients.
9. Differentiate between correlation and regression. Discuss the methods used to study correlation. Add a note on the coefficient of correlation.
10. Explain correlation analysis. Describe different types and methods of studying correlation with suitable examples.
11. What is "Correlation"? Citing suitable examples, describe the type and degree of correlation, and mathematical estimation of 'r' between the bivariates.
12. What is correlation? Give the method of calculation of measuring correlation (i) Scatter diagram (ii) Karl Pearson's coefficient of correlation.

NON-PARAMETRIC TESTS

INTRODUCTION

The statistical tests discussed so far in earlier chapters were parametric tests. These tests specify certain conditions about the distribution of population from which the sample is drawn. In many situations these conditions are not ordinarily tested but they are assumed to hold good. On the validity of these assumptions, the meaningfulness of the result of a parametric test depends. Usual procedure of interpretation of parametric test is based on normal distribution and the data for analysis must be in interval scale. If the sample size is very small, the distribution of population is not clearly known or the data collected is not in interval scale, it is advisable to consider the non-parametric or distribution free tests for testing of data. A non-parametric test is based on a model that specifies very few general conditions and not about the distribution of the population from which the sample is drawn. The test is also carried out even if data are not in interval scale. Two major assumptions in non-parametric test is (i) The samples are drawn randomly (ii) The variable under condition is continuous.

In this chapter, some non-parametric tests are discussed based on single sample data, sample with pairwise data and two samples data drawn independently.

CHARACTERISTICS OF NON-PARAMETRIC TESTS

1. They do not suppose any particular distribution and the consequential assumptions.
2. They are rather quick and easy to use i.e., they do not require laborious computations since in many cases the observations are replaced by their rank order and in many others signs are used.
3. When the measurements are not accurate, as is necessary for standard tests of significance, then non-parametric methods come to the rescue, which can be used fairly satisfactorily.

4. Parametric tests cannot apply to ordinal or nominal scale data but non-parametric tests do not suffer from any such limitations.
5. The parametric tests of significance like 't' or 'F' make assumption about the homogeneity of the variances whereas this is not necessary for non-parametric tests of difference.
6. Non-parametric tests are often not as efficient or sharp as tests of significance for the parametric tests.
7. In non-parametric tests, ranks are used instead of the actual values of variables. So certain amounts of information are ignored.

SIGN TEST

The sign test is used to test whether the given population has a specific median M_d or not i.e. on the basis of sample, a researcher becomes able to decide whether the population from which the given sample is drawn has a specific median M_d or not. This test is similar to Student's t-test. In Student's t-test one decides inference on the basis of mean, but in the sign test one decides inference on the basis of median. For a symmetrical distribution mean is equal to median. The only assumption in this test is that the distribution of variable of interest is continuous.

METHOD

Let X_i (i = 1, 2, 3, ..., n) is the judgment or score of set of observations of sample size n.

Null hypothesis H_0: The population median is M_d. i.e. let M_d is the median of the population from which the sample X_i (i = 1, 2, 3......n) is drawn.

Statistic: Consider $d_i = X_i - M_d$ and note the sign of d_i, i.e. positive or negative or zero. If $d_i = 0$ i.e. tie and hence eliminate that observation thereby reducing the total number of observations by one. The number of positive signs and the number of negative signs are noted. The test is carried out at α level of significance. Usually $\alpha = 0.01$ or 0.05.

Decision Rule:

(i) Consider the alternative hypothesis H_1: P(+) > P(–). This is a one-sided alternative. Suppose it is observed that there are K_1 negative signs in the sample of n observations. From the table it is observed that $P(K \leq K_1 |n, p) = T_1$. If $T_1 < \alpha$, then reject the null hypothesis and accept the alternative H_1.

(ii) Consider the alternative hypothesis H_1: P(–) > P(+). This is a one-sided alternative. Suppose it is observed that there are K_2 positive signs in the sample of n observations. From the table it is seen that $P(K \le K_2 | n, p) = T_2$. If $T_2 < \alpha$, reject the null hypothesis H_0 and accept the alternative H_1.

(ii) Consider the alternative hypothesis H_1: $P(-) \neq P(+)$. This is a two-sided alternative. Suppose it is observed that there are K_1 negative and K_2 positive signs in the sample of n observations and $K_1 < K_2$. From the table it is seen that $P(K \le K_1 | n, p) = T_1$. If T_1 is less than $\alpha/2$ the null hypothesis H_0 is rejected and the alternative hypothesis is accepted.

Example 15.1: A random sample of 15 medical students were given a test to measure their level of authoritarianism with the following results.

Student Number	Authoritarianism Score	Student Number	Authoritarianism Score	Student Number	Authoritarianism Score
1	75	6	95	11	88
2	90	7	132	12	124
3	85	8	74	13	110
4	110	9	82	14	76
5	115	10	104	15	98

Test at the 0.05 level of significance, the null hypothesis that the median score for the population sampled is 100. Determine the p value.

Solution:

Assumption. Assume that the scores are given on a continuous variable.

Hypothesis. H_0 : The population median is 100 i.e. $P(+) = P(-)$

H_1 : The population median is not 100 i.e. $P(+) \neq P(-)$

Let $\alpha = 0.05$ (Level of significance)

Test statistic: Now let us construct the following table wherein the scores above (+) and below (–) the hypothesized median (= 100) based on the given data are shown.

Student No.	1	2	3	4	5	6	7	8	9	10	11	12	13	14	15
Auth. Score	75	90	85	110	115	95	132	74	82	104	88	124	110	76	98
Difference from the median	75 –100 = -25	90 - 100= -10	85 - 100= -15	110- 100= +10	115- 100= +15	95 - 100= -5	132- 100= +32	74 - 100= -26	82 - 100= -18	104- 100= +4	88 - 100= -12	124- 100= +24	110- 100= +10	76 - 100= -24	98 - 100= -2
Sign	-	-	-	+	+	-	+	-	-	+	-	+	+	-	-

It seems reasonable to conclude that the observations in the above table constitute a set of n (=15) independent random variables from the population with parameter p. If we let K = the test statistic, the sampling distribution of K is the binomial probability distribution with parameter p = 0.5 if null hypothesis is true.

Decision rule: The decision rule depends on the alternative hypothesis. H_1 : P(+) ≠ P(-), reject H_0 if (given that H_0 is true) the probability of obtaining a value of K as extreme as or more extreme than was actually computed is equal to or less than $\alpha/2$. (= 0.025)

So reject H_0 if the p value for the computed test statistic is less than or equal to 0.05 / 2 = 0.025.

Calculation of test statistic: We may determine the probability of observing K (=6) or fewer number of + signs (The less frequently occurring sign in the above constructed table) when given a sample size of 15 and parameter p = 0.5 by referring to table for Binomial distribution that

$P(K \leq 6 \mid 15, 0.5) = 0.3036$ (from the Binomial Table), which is more than 0.025.

Conclusion: Null hypothesis that the median score is 100 is accepted.

SIGN TEST FOR PAIRED DATA

Here in case of paired data, the sample observations are taken both before and after any intervention in the system. The difference in the paired sample observations will show – or + signs. If there is no difference before and after intervention in the system, then there will be as many – signs as there are + signs in the entire population. That means the probability of getting two signs will be same and equal to 50%, i.e. p = 0.5 and so q = 0.5. Though this is a case of binomial distribution, normal approximation may be considered. So the standard error would be $\sqrt{\frac{pq}{n}}$. The test statistic (z) is compared with the standard normal variate at α=0.01 or 0.05. An example below clarifies.

Example 15.2: The following data show the satisfaction levels (as a percentage) of employees of a public sector company before and after their company was bought by a private firm. Did the private firm rise to the employees' cause ?

Employee No:	**1**	**2**	**3**	**4**	**5**	**6**	**7**	**8**	**9**	**10**
Before:	98.4	96.6	82.4	96.3	75.4	82.5	81.5	91.3	90.3	92.3
After:	82.3	95.3	94.1	97.2	77.4	82.4	81.5	84.4	89.3	90.5
Employee No:	11	12	13	14	15	16	17	18	19	20
Before:	98.1	95.6	81.4	95.3	74.4	81.5	80.5	90.3	89.3	91.3
After:	98.1	95.8	81.2	93.5	73.1	81.5	81.5	84.4	90.1	90.5
Employee No:	21	22	23	24	25	26	27	28	29	30
Before:	88.4	86.6	72.4	86.3	65.4	72.5	71.5	81.3	89.3	82.3
After:	89.0	86.6	71.6	86.3	64.5	72.5	81.5	80.9	92.1	82.3
Employee No:	31	32	33	34	35	36	37	38	39	40
Before:	88.4	86.6	72.4	86.3	65.4	72.5	71.5	81.3	89.3	82.3
After:	88.4	85.9	72.4	97.2	62.7	70.4	70.9	81.3	87.3	81.5

Solution:

Employee No:	**1**	**2**	**3**	**4**	**5**	**6**	**7**	**8**	**9**	**10**
Before:	98.4	96.6	82.4	96.3	75.4	82.5	81.5	91.3	90.3	92.3
After:	82.3	95.3	94.1	97.2	77.4	82.4	81.5	84.4	89.3	90.5
Difference:	16.1	1.30	-11.7	-0.90	-2.0	0.10	0.00	6.9	1.00	1.80
Employee No:	11	12	13	14	15	16	17	18	19	20
Before:	98.1	95.6	81.4	95.3	74.4	81.5	80.5	90.3	89.3	91.3
After:	98.1	95.8	81.2	93.5	73.1	81.5	81.5	84.4	90.1	90.5
Difference:	0.00	-0.20	0.20	1.80	1.30	0.00	-1.00	5.90	-0.80	0.80
Employee No:	21	22	23	24	25	26	27	28	29	30
Before:	88.4	86.6	72.4	86.3	65.4	72.5	71.5	81.3	89.3	82.3
After:	89.0	86.6	71.6	86.3	64.5	72.5	81.5	80.9	92.1	82.3
Difference:	-0.60	0.00	0.80	0.00	0.90	0.00	-10.0	0.40	-2.80	0.00
Employee No:	31	32	33	34	35	36	37	38	39	40
Before:	88.4	86.6	72.4	86.3	65.4	72.5	71.5	81.3	89.3	82.3
After:	88.4	85.9	72.4	97.2	62.7	70.4	70.9	81.3	87.3	81.5
Difference:	0.00	0.70	0.00	-10.90	-2.70	2.10	0.60	0.00	2.00	0.80

As there are ten 0-values, those may be neglected. So there are 19 +'s (success) and 11 –'s (failure). Consider n = 30

p = 19 ÷ 30 = 0.633 and q = 11 ÷ 30 = 0.367. We can see that, np = 19 and nq=11 and both are greater than 5. So normal approximation can be applied.

Null Hypothesis H_0: P = 0.5 (Hypothesized proportion having satisfaction of privatization)

Alternative Hypothesis H_1: Q = 0.5

$\alpha = 0.05$

Standard error of proportion $= \acute{o}_p = \sqrt{\frac{PQ}{n}} = \sqrt{\frac{0.5 \times 0.5}{30}} = 0.091$

$$Z = \frac{|p - P|}{\acute{o}_p} = \frac{|0.633 - 0.5|}{0.091} = 1.79$$

But table value of Z is $Z_t = 1.96$, which shows that the calculated value of Z is within the table value i.e. in the acceptance region. So, null hypothesis is accepted. The employees are satisfied with the privatization of the company. The private firm has satisfied the employees.

It may be noted that we are able to use normal approximation to the binomial as our sampling distribution because np and nq are both greater than 5. When these conditions are not met, binomial distribution is to be applied.

MANN-WHITNEY U TEST

The Mann-Whitney test is based on ranking of the data. It is also called Mann-Whitney-Wilcoxon test. This is a two samples test and an alternative test of parametric Fisher's t-test. When sample size is small and the assumption about normality does not remain valid, the Mann-Whitney U test becomes more appropriate.

Assumption

(i) Two samples are drawn independently and observations may not be pair wise.

(ii) The variable is continuous.

(iii) The measurement scale is at least ordinal.

(iv) Comparison is made with respect to population median.

Method

Let X_i (i =1, 2, 3, ..., n) be the set of observations in the first sample drawn from a population with mean μ_x. Let Y_i (i = 1, 2, 3, ..., m) be the set of observations in the second sample, independently drawn from a population with mean μ_y.

For computation of test statistic we combine the two samples and rank all observations in ascending order i.e. smallest to largest. Then the two samples are separated with their ranks. If two or more observations are same then they are allotted with mean rank.

The test statistic(U) is:

$$U = n_1 n_2 + \frac{n_1(n_1+1)}{2} - R_1 \text{ ; where}$$

n_1 = number of items in sample 1

n_2 = number of items in sample 2

R_1 = sum of the ranks of the items in sample 1

R_2 = sum of the ranks of the items in sample 2

Mean of the sampling distribution of U: $\grave{\imath}_U = \frac{n_1 n_2}{2}$

Standard Error of the U statistic: $\acute{o}_U = \sqrt{\frac{n_1 n_2 (n_1 + n_2 + 1)}{12}}$

Null Hypothesis H_0: There is no difference between two population means i.e. $\mu_1 = \mu_2$

Alternative Hypothesis H_1: $\mu_1 \neq \mu_2$

α is taken as 0.01 or 0.05. Normal approximation can be made for the analysis.

Example 15.3: A researcher conducted an experiment to assess the effects of prolonged inhalation of cadmium oxide. Twelve animals considered as experimental units, while 10 similar animals considered as controls. The variable of interest was hemoglobin level following the experiment. The results are shown in the table below. Can it be concluded that prolonged inhalation of cadmium oxide reduces hemoglobin level.

Hemoglobin Determinations (grams) for 22 laboratory animals.

Exposed animals (X)	14.3	14.2	13.8	16.5	14.1	16.6	15.8	15.6	14.1	15.3	15.7	16.7
Unexposed animals (Y)	17.4	16.2	17.1	17.5	15.0	16.0	16.9	15.0	16.3	16.8		

Solution:

***Assumptions*:** We presume that the assumptions of the Mann-Whitney U test are met.

Null Hypothesis H_0: There is no difference between two population means i.e.

Alternative Hypothesis H_1: $\mu_1 \neq \mu_2$

Where μ_x is the mean for a population of animals exposed to cadmium oxide and μ_y is the mean of a population for animals not exposed to the substance. Suppose we let a = 0.05.

Test statistic: To compute the test statistic we combine the two samples and rank all observations from smallest to largest while keeping track of the sample to which each observation belongs. Tied observations are assigned a rank equal to the mean of the rank positions for which they are tied. The results of this step are shown below:

X	Rank	Y	Rank
13.8	1		
14.1	2.5		
14.1	2.5		
14.2	4		
14.3	5		
		15	6.5
		15	6.5
15.3	8		
15.6	9		
15.7	10		
15.8	11		
		16	12
		16.2	13
		16.3	14
16.5	15		
16.6	16		
16.7	17		
		16.8	18
		16.9	19
		17.1	20
		17.4	21
		17.5	22
Total:	101		

$$U = n_1 n_2 + \frac{n_1(n_1+1)}{2} - R_1 = 12 \times 10 + \frac{12(12+1)}{2} - 101 = 97$$

Where n_1 is the number of sample X observations and R_1 is the sum of the ranks assigned to the sample observations from the population of X values. The choice of which sample's values we label X is arbitrary.

Mean of the sampling distribution of U: $ì_U = \frac{n_1 n_2}{2} = \frac{12 \times 10}{2} = 60$

Standard Error of the U statistic: $ó_U = \sqrt{\frac{n_1 n_2 (n_1 + n_2 + 1)}{12}}$

$$= \sqrt{\frac{12 \times 10(12+10+1)}{12}} = 15.166$$

$$Z = \frac{|U - _U|}{ó_U} = \frac{|97 - 60|}{15.166} = 2.440$$

Decision: Table value of Z for 5% significance level is 1.96. So calculated value of Z is more than the table value and falls in the rejection region. So null hypothesis is rejected. Mean of samples before and after exposure to cadmium oxide are not same. That means prolonged exposure to cadmium oxide has certainly affected the hemoglobin level.

***Note*:** Same result will also be obtained, had we taken Y-values as n_1. In that case the test statistic would be $U = n_1 n_2 + \frac{n_2(n_2+1)}{2} - R_2$

KRUSKAL-WALLIS TEST

Kruskal-Wallis test is an extension of the Mann-Whitney test, where more than two samples data can be dealt. The following example illustrates the procedure:

Example 15.4: The following table shows sample retail prices for three brands of T-shirts. Use the Kruskal-Wallis test to determine whether there is any difference among the retail prices of the brands throughout the country. Use 0.01 level of significance.

Brand-A:	99	64	101	85	79	88	97	95	90	100
Brand-B:	83	102	125	61	91	96	94	89	93	75
Brand-C:	89	98	56	105	87	90	87	101	76	89

Sample: Let us club all data and arrange in ascending order and rank them.

Value	Rank	Value	Rank	Value	Rank
56	1	88	11	96	21
61	2	89	13	97	22
64	3	89	13	98	23
75	4	89	13	99	24
76	5	90	15.5	100	25
79	6	90	15.5	101	26.5
83	7	91	17	101	26.5
85	8	93	18	102	28
87	9.5	94	19	105	29
87	9.5	95	20	125	30

Separating these ranks according to groups, we get

Brand-A		Brand-B		Brand-C	
Value	Rank	Value	Rank	Value	Rank
64	3	61	2	56	1
79	6	75	4	76	5
85	8	83	7	87	9.5
88	11	89	13	87	9.5
90	15.5	91	17	89	13
95	20	93	18	89	13
97	22	94	19	90	15.5
99	24	96	21	98	23
100	25	102	28	101	26.5
101	26.5	125	30	105	29
Total:	161		159		145

Using a test statistic

$$= K = \frac{12}{n(n+1)}\sum \frac{R_j^2}{n_j} - 3(n+1)$$

$$= \frac{12}{30(30+1)}\left[\frac{161^2}{10}+\frac{159^2}{10}+\frac{145^2}{10}\right]-3(30+1)=0.196$$

Null Hypothesis H_0: $\mu_1 = \mu_2 = \mu_3$ (Population means are all equal)

Alternative Hypothesis H_1: $\mu_1 \neq \mu_2 \neq \mu_3$ There are differences among the three populations)

Use $\alpha = 0.01$

Degrees of freedom = 3 – 1 = 2. Sampling distribution of K-statistic can be assumed to follow chi-square distribution, when all sample sizes are at least 5. χ^2 value for 2 df at 1% significance level is 9.210, which is greater than the calculated value of K. So null hypothesis is accepted. That means there is no significant difference among the three population means.

KOLMOGOROV-SMIRNOV TEST(K-S TEST)

K-S test, named after statisticians A.N.Kolmogorov and N.V.Smirnov, is a simple non-parametric test for establishing if there is any significant difference between an observed and expected frequency. This is, hence, another measure of goodness of fit as was done in case of chi-square test studied earlier. But K-S test is more powerful and is easier to use, because it does away with grouping of data. The statistic used is D_n, which is a distribution free statistic and depends on sample size n, but independent of the expected frequency distribution. The following example illustrates the method.

Example 15.5: A firm manager has collected the following salary statistics on his field sales persons. Both expected frequencies based on normal probability distribution and observed frequencies are given below. At the 0.10 level of significance, can the manager conclude that the distribution of salary earnings is normal?

	Earnings (in thousands)						
	25-30	**31-36**	**37-42**	**43-48**	**49-54**	**55-60**	**61-66**
Observed frequency:	9	22	25	30	21	12	6
Expected frequency:	6	17	32	35	18	13	4

Solution: K-S statistic is: $D_n = \max |F_e - F_o|$

K-S test is always one-tailed.

Null Hypothesis H_0: Salary earnings are normally distributed (i.e. Population data are normal)

Alternative Hypothesis H_1: Not a good description of normal distribution pattern.

Use $\alpha = 0.10$

The following table explains the required calculations:

Observed Frequency	Observed Cumulative Frequency	Observed Relative Cumulative Frequency (F_o)	Expected frequency	Expected Cumulative Frequency	Expected Relative Cumulative Frequency (F_e)	$\lvert F_o - F_e \rvert$ Absolute Deviation
(1)	*(2)*	*(3)*	*(4)*	*(5)*	*(6)*	*(7)*
9	9	0.072	6	6	0.048	0.024
22	31	0.248	17	23	0.184	0.064
25	56	0.448	32	55	0.440	0.008
30	86	0.688	35	90	0.720	0.032
21	107	0.856	18	108	0.864	0.008
12	119	0.952	13	121	0.968	0.016
6	125	1.000	4	125	1.000	0.000

From the above table the maximum value in Col-(7) is: $D_n = 0.064$.

The critical value of D is found from the K-S table. For sample size n = 7, $D_t = 0.436$. Here the calculated value of D is within the table value ($0.064 < 0.436$). So the Null hypothesis is accepted. That means the manager can safely conclude from the sample data that the salary earnings of the entire staff are normally distributed. There is no reason to reject this.

RANK CORRELATION

Rank Correlation is the amount or degree of association between two sets of qualitative observations which can be ranked or graded through scores. For example, it may be required to find if there is any association between the efficiency of employees before say 10 years and now for considering promotional benefits to them. Or we may like to find correlation between size of agricultural labor force and volume of production of crop in the districts of a state, between marks secured by a group of students in two different subjects etc. As the rank correlation coefficient depends on the ranks only, it does not depend on the specific distribution of the variables. It is quick and non-laborious method of estimating association between two variables. This measure is called the Spearman Rank Correlation Coefficient, in honor of the statistician Charles Edward Spearman, who developed it in 1904.

Coefficient of Rank Correlation is given by:

$$r_s = 1 - \frac{6\sum d^2}{n(n^2-1)};$$

where, r_s = coefficient of rank correlation

n = number of paired observations

d = difference between the ranks for each pair of observations.

Now the question arises as to how to test the value of rank correlation coefficient? We apply the same method of hypothesis testing. The null hypothesis is, no such correlation exists in the population from which samples are drawn, that is, the population rank correlation coefficient (ρ_s) is zero.

Null hypothesis H_0: $\rho_s = 0$ [There is no relation in the ranked data of the population]

Alternative hypothesis H_1: $\rho_s \neq 0$ [There is a correlation in the ranked data of the population]

For small samples, the distribution of r_s is not normal and also t-distribution is not applicable. Instead we use Spearman's rank Correlation Values to determine the acceptance and rejection for such hypotheses. However, for large samples(n>30), the sampling distribution of r_s is approximately normal, with a mean 0 and standard deviation of $\frac{1}{\sqrt{n-1}}$, otherwise known as Standard Error of the Coefficient of Rank Correlation. An example is now taken to understand.

Example 15.6: The following marks were awarded by two examiners, evaluating the same paper to 16 students. Examine if the examiners have the same standard of judgment:

S. No of Student:	**(1)**	**(2)**	**(3)**	**(4)**	**(5)**	**(6)**	**(7)**	**(8)**
1st Examiner:	60	50	43	58	60	46	70	80
2nd Examiner:	52	57	47	63	58	50	68	75
Sl No of Student:	(9)	(10)	(11)	(12)	(13)	(14)	(15)	(16)
1st Examiner:	53	60	43	58	83	70	79	62
2nd Examiner:	50	57	46	53	82	68	73	57

Solution: The marks of the two examiners are sorted in ascending order and is given below:

Marks by 1st Examiner	S. No	Rank	Marks by 2nd Examiner	S. No	Rank
43	*3*	1.5	46	*11*	1
43	*11*	1.5	47	*3*	2
46	*6*	3	50	*6*	3.5
50	*2*	4	50	*9*	3.5
53	*9*	5	52	*1*	5
58	*4*	6.5	53	*12*	6
58	*12*	6.5	57	*2*	8
60	*1*	9	57	*10*	8
60	*5*	9	57	*16*	8
60	*10*	9	58	*5*	10
62	*16*	11	63	*4*	11
70	*7*	12.5	68	*7*	12.5
70	*14*	12.5	68	*14*	12.5
79	*15*	14	73	*15*	14
80	*8*	15	75	*8*	15
83	*13*	16	82	*13*	16

Then these ranks are presented in a single table below. The difference in ranks are then found out.

S. No	Marks by 1st Examiner	Rank	Marks by 2nd Examiner	Rank	Difference of ranks (d)	d^2
(1)	(2)	(3)	(4)	(5)	(6) = (3) – (5)	$(7) = (6)^2$
1	60	9	52	5	4	16
2	50	4	57	8	–4	16
3	43	1.5	47	2	–0.5	0.25
4	58	6.5	63	11	–4.5	20.25
5	60	9	58	10	–1	1
6	46	3	50	3.5	–0.5	0.25
7	70	12.5	68	12.5	0	0
8	80	15	75	15	0	0
9	53	5	50	3.5	1.5	2.25

S. No	Marks by 1st Examiner	Rank	Marks by 2nd Examiner	Rank	Difference of ranks (d)	d^2
10	60	9	57	8	1	1
11	43	1.5	46	1	0.5	0.25
12	58	6.5	53	6	0.5	0.25
13	83	16	82	16	0	0
14	70	12.5	68	12.5	0	0
15	79	14	73	14	0	0
16	62	11	57	8	3	9
						$\sum d^2 = 66.5$

$$\text{Rank correlation coefficient} = r_s = 1 - \frac{6\sum d^2}{n(n^2-1)} = 1 - \frac{6 \times 66.5}{16(16^2-1)} = 0.902$$

Null hypothesis H_0: $\rho_s = 0$ [Two examiners have same level of judgment, i.e. there is no difference]

Alternative hypothesis H_1: $\rho_s \neq 0$ [Two examiners have different levels of judgment]

This is a small sample two-tailed test. Hence, Spearman's Rank Correlation table will be followed. Table value of the coefficient for 5% significance level and n=16 is 0.500. So r_s is more than the table value and hence is in the rejection region. Null hypothesis is rejected. The judgment levels of two examiners are not same.

Example 15.7: The following are the salary and age data for forty Ph.D. candidates graduating this year from SOA University. At the 0.05 significance level, does the rank correlation of age and salary suggest that older candidates get higher starting salaries?

S. No	(1)	(2)	(3)	(4)	(5)	(6)	(7)	(8)	(9)	(10)
Salary in Thousands	67	60	57	59	50	55	59	63	69	72
Age	29	25	30	35	27	31	32	38	28	34

Sl No	(11)	(12)	(13)	(14)	(15)	(16)	(17)	(18)	(19)	(20)
Salary in Thousands	78	69	52	73	67	64	59	56	58	60
Age	32	28	30	36	29	22	26	27	30	35
Sl No	**(21)**	**(22)**	**(23)**	**(24)**	**(25)**	**(26)**	**(27)**	**(28)**	**(29)**	**(30)**
Salary in Thousands	69	78	74	71	72	78	69	52	73	67
Age	34	29	30	31	29	25	30	35	27	31
Sl No	**(31)**	**(32)**	**(33)**	**(34)**	**(35)**	**(36)**	**(37)**	**(38)**	**(39)**	**(40)**
Salary in Thousands	64	60	57	59	50	55	67	64	60	57
Age	32	34	29	30	31	29	25	32	37	29

Solution: The following table is prepared showing the data in ascending order and the corresponding ranks.

S. No	Salary in Thousands	Rank	S. No	Age	Rank
5	50	1.5	16	22	1
35	50	1.5	2	25	3
13	52	3.5	26	25	3
28	52	3.5	37	25	3
6	55	5.5	17	26	5
36	55	5.5	5	27	7
18	56	7	18	27	7
3	57	9	29	27	7
33	57	9	9	28	9.5
40	57	9	12	28	9.5
19	58	11	1	29	14
4	59	13.5	15	29	14
7	59	13.5	22	29	14
17	59	13.5	25	29	14
34	59	13.5	33	29	14

S. No	Salary in Thousands	Rank	S. No	Age	Rank
2	60	17.5	36	29	14
20	60	17.5	40	29	14
32	60	17.5	3	30	20.5
39	60	17.5	13	30	20.5
8	63	20	19	30	20.5
16	64	22	23	30	20.5
31	64	22	27	30	20.5
38	64	22	34	30	20.5
1	67	25.5	6	31	25.5
15	67	25.5	24	31	25.5
30	67	25.5	30	31	25.5
37	67	25.5	35	31	25.5
9	69	29.5	7	32	29.5
12	69	29.5	11	32	29.5
21	69	29.5	31	32	29.5
27	69	29.5	38	32	29.5
24	71	32	10	34	33
10	72	33.5	21	34	33
25	72	33.5	32	34	33
14	73	35.5	4	35	36
29	73	35.5	20	35	36
23	74	37	28	35	36
11	78	39	14	36	38
22	78	39	39	37	39
26	78	39	8	38	40

The table below shows the difference of ranks.

S. No	Salary in Thousands	Rank	Age	Rank	Difference in rank (d)	d^2
(1)	(2)	(3)	(4)	(5)	(6)=(3)-(5)	$(7)=(6)^2$
1	67	25.5	29	14	11.5	132.25
2	60	17.5	25	3	14.5	210.25
3	57	9	30	20.5	-11.5	132.25
4	59	13.5	35	36	-22.5	506.25
5	50	1.5	27	7	-5.5	30.25
6	55	5.5	31	25.5	-20	400
7	59	13.5	32	29.5	-16	256
8	63	20	38	40	-20	400
9	69	29.5	28	9.5	20	400
10	72	33.5	34	33	0.5	0.25
11	78	39	32	29.5	9.5	90.25
12	69	29.5	28	9.5	20	400
13	52	3.5	30	20.5	-17	289
14	73	35.5	36	38	-2.5	6.25
15	67	25.5	29	14	11.5	132.25
16	64	22	22	1	21	441
17	59	13.5	26	5	8.5	72.25
18	56	7	27	7	0	0
19	58	11	30	20.5	-9.5	90.25
20	60	17.5	35	36	-18.5	342.25
21	69	29.5	34	33	-3.5	12.25
22	78	39	29	14	25	625
23	74	37	30	20.5	16.5	272.25
24	71	32	31	25.5	6.5	42.25
25	72	33.5	29	14	19.5	380.25
26	78	39	25	3	36	1296
27	69	29.5	30	20.5	9	81
28	52	3.5	35	36	-32.5	1056.25
29	73	35.5	27	7	28.5	812.25

S. No	Salary in Thousands	Rank	Age	Rank	Difference in rank (d)	d^2
30	67	25.5	31	25.5	0	0
31	64	22	32	29.5	-7.5	56.25
32	60	17.5	34	33	-15.5	240.25
33	57	9	29	14	-5	25
34	59	13.5	30	20.5	-7	49
35	50	1.5	31	25.5	-24	576
36	55	5.5	29	14	-8.5	72.25
37	67	25.5	25	3	22.5	506.25
38	64	22	32	29.5	-7.5	56.25
39	60	17.5	37	39	-21.5	462.25
40	57	9	29	14	-5	25
						$\sum d^2 = 10977$

Rank correlation coefficient $r_s = 1 - \frac{6\sum d^2}{n(n^2-1)} = 1 - \frac{6 \times 10977}{40(40^2-1)} = -0.0297$

Null hypothesis H_0: $\rho_s = 0$ [There is no relation between salary and age]

Alternative hypothesis H_1: $\rho_s \neq 0$ [Age affects the salary. Older scholars get higher salaries]

$\alpha = 0.01$

This is a large sample one-tailed test. Hence, normal approximation can be made.

Standard Error of the Coefficient of Rank Correlation

$= Z_c = \frac{r_s - \rho_s}{SE} = \frac{-0.0297 - 0}{0.16} = -0.1856$

Table value of Z for 1% significance level is 2.33(one tailed). So calculated value is more than the table value and hence is in the rejection region. Null hypothesis is rejected. That means the age affects the starting salary, i.e. older students get higher salaries.

EXERCISES

1. The following data show employees' rates of defective work before and after a change in the wage incentive plan. Compare the following two sets of data to see whether the change lowered the defective units produced. Use 1% level of significance.

Before:	8	7	6	9	7	10	8	6	5	8	10	7
After:	6	5	4	7	6	9	8	9	6	5	6	8

2. Use sign test to see whether there is a difference between the number of days required to collect an account receivable before and after a new collection policy. Use 5% significance level.

Before:	33	36	41	30	37	45	32	29	30	32	38	40
After:	33	29	36	32	37	45	35	30	32	34	40	38

3. An owner has departmental stores at three locations. Based on the sample data collected from three shops regarding the number of customers who visit the shop for purchasing, test at 5% significance level that the owner gets same number of customers using Kruskal-Wallis test.

Location 1:	99	64	101	85	79	88	97	95	90	100
Location 2:	89	98	56	105	87	90	87	101	76	89
Location 3:	75	93	89	94	96	91	61	125	102	83

4. The following shows the annual hours missed due to illness for the 24 men and women at a company. At 0.01 significance level, find if there is any difference attributed to gender using Mann-Whitney U Test.

Men:	38	34	33	47	58	83	18	36	41	37	24	48
Women:	31	44	25	30	70	63	54	42	36	22	25	50

5. A university that accepts students from rural and urban high schools is interested in whether the different backgrounds lead to a difference in first-year CGPA. Data are given below for 10 randomly selected first-year students of rural and 11 students from urban background. Use Mann-Whitney U Test at 5% significance level.

Urban:	3.19	2.05	2.82	2.16	3.84	4.0	2.91	2.75	3.01	1.98	2.58
Rural:	3.45	3.16	2.84	2.09	2.11	3.08	3.97	3.85	3.72	2.73	

6. The following table shows sample retail prices for three brands of shoes. Use Kruskal-Wallis test to determine whether there is any difference among the retail prices of the brands throughout the city. Use 0.01 level of significance.

Brand-A: 689 690 692 681 676 688 685 695 697 686 700

Brand-B: 678 693 681 687 689 671 690 696 682 685

Brand-C: 680 688 686 685 679 680 684 685 690 692

7. Describe sign test, stating clearly the assumptions made.
8. Explain briefly the distinction between parametric test and non-parametric test.
9. Describe Mann-Whitney U test. Give the assumptions used in this test.
10. Describe the procedure of Wilcoxon Signed Rank test. Give its null hypothesis and assumptions.
11. The following pulmonary function data were collected on children with muscular dystrophy before and after a period of respiratory therapy. Scores are expressed as percent of predicted normal values for height, weight and body surface measurement.

Before	74	65	84	89	84	65	78	86	83	82
After	79	78	100	92	104	70	81	84	85	90

Using sign rank test, determine whether the therapy is effective at 5% level of significance. What is the p value?

12. The following are the ranks of the ages of 20 surgical patients and the dose (Y) of an analgesic agent required to block one spinal segment.

Rank of age in years (X)	Rank of dose requirement (Y)	Rank of age in years (X)	Rank of dose requirement (Y)	Rank of age in years (X)	Rank of dose requirement (Y)	Rank of age in years (X)	Rank of dose requirement (Y)	Rank of age in years (X)	Rank of dose requirement (Y)
1	1	5	7	9	10	13	10	17	18
2	6	6	9	10	11	14	17	18	16
3	3	7	4	11	14	15	21	19	11
4	5	8	16	12	7	16	16	20	13

Test at the 0.05 level of significance, the null hypothesis the the median dose is 15. Determine the p value.

13. In one psychometric analysis a random sample of 25 handicapped students were measured for their manual dexterity and the following scores were obtained. Does it provide sufficient evidence to conclude that the mean score for the population is not 45? Test at 5% significance level. Determine the p value.

32 52 21 39 23 55 35 27 37 41 34 51
51 35 46 40 31 19 41 33 52 36 34 46 41

14. Ten patients with bronchitis participated in an experiment to evaluate the effectiveness of a drug. The following table shows the change in FEV_1 (forced expired volume in 1 second) values (expressed as liters) two hours after drug administration. Determine whether the drug results in mean change in FEV_1 of less than 0.19. Test it at 5% level of significance and find the p value.

Subject	1	2	3	4	5	6	7	8	9	10
Drug	0.12	0.16	0.21	0.28	0.10	0.17	0.20	0.24	0.25	0.09

15. A plant survivor ranked a sample of eight workers on the number of hours of overtime worked and length of employment. Is the rank correlation between the two measures significant at the 0.01 level?

Hours of overtime: 6.0 9.0 3.0 5.0 4.0 8.0 2.0 7.0

Years employed: 2.0 7.0 5.5 3.0 8.0 9.0 5.5 4.0

16. The following are the years of experience(X) and average customer satisfaction(Y) for 10 service providers. Is there a significant rank correlation between the two measures? Use the 0.05 level of significance.

X: 6.4 5.9 6.2 7.0 3.5 1.9 9.5 4.8 7.3 2.5

Y: 5.4 8.7 4.8 4.3 5.0 6.2 5.2 6.4 6.9 5.3

17. Mr. Panda has an automobile repair factory. He is interested in improving his estimates of repair time required and believes that the best predictor is the number of operating hours on the engine since its last major repair. Below are the data on ten engines Mr. Panda worked on recently. At the 0.10 level of significance, does the rank correlation suggest a strong relationship?

Engine	Hours since Last major repair	Hours required to repair
1.	1000	40
2.	1200	55
3.	950	42
4.	1455	65
5.	2050	70
6.	1300	50

Engine	Hours since Last major repair	Hours required to repair
7.	1650	45
8.	1750	68
9.	500	38
10.	2200	68

18. The following table shows the observed frequencies, along with the frequencies to be expected under a normal distribution.

	Test Score				
	31-40	41-50	51-60	61-70	71-80
Observed Frequency:	30	100	440	500	130
Expected Frequency:	40	170	500	390	100

(a) Calculate the K-S statistic.

(b) Can we conclude that these data do in fact come from a normal distribution? Use the 0.10 level of significance.

TIME SERIES ANALYSIS

Time series analysis is a quantitative method used to determine patterns in data collected over time and for future forecast. It is also used to detect patterns of change in statistical information over regular time intervals. For example- there occurs sudden increase in price of essential commodities during natural calamities. The sale of refrigerators falls during winter season whereas the sale of woollen clothes increases during the same period. The five-year plans in India have brought about a steady increase in per capita income of the populace. Such patterns of change are used to arrive at an estimate for future. Thus, time-series analysis helps us to cope with uncertainty about the future. This study is more important for business processes and stock markets. Periodic fluctuations in sensex affect the stock price. The quality of forecast that the management can make is strongly related to the information that can be extracted and used from the past. Consumers prefer to buy commodities during off season, when prices are usually at their low. Examples are many in daily life. This chapter deals with such phenomenon known as time series analysis.

Few examples of periodic change are illustrated below through graphical representations for better appreciation of the concept. Bivariate data are plotted, where time is the independent variable.

Power Consumption in Odisha (in million units)

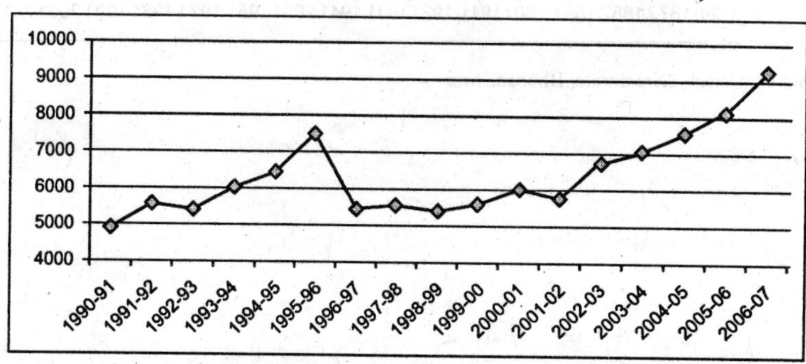

Source: Orissa Agricultural Statistics, 2006-07

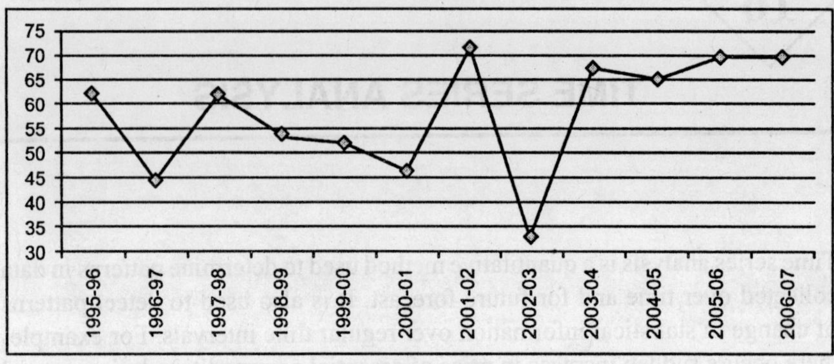

Source: Orissa Agricultural Statistics, 2006-07

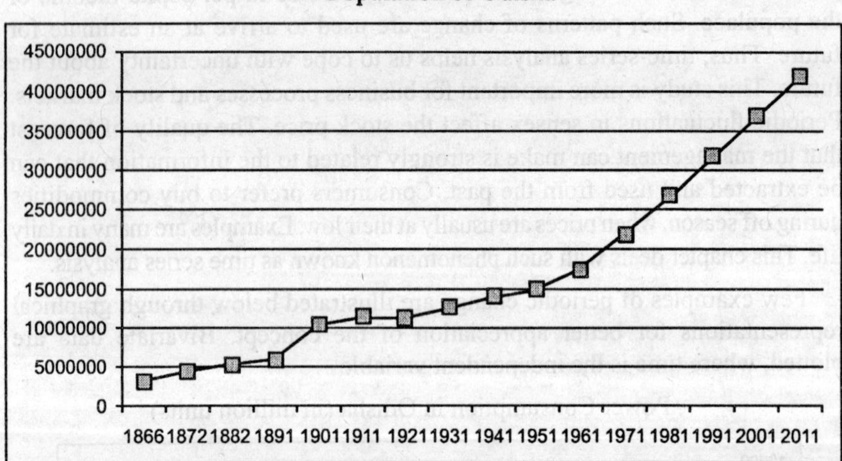

Source: Census Directorate, Bhubaneswar.

Monsoon Rainfall in Baliguda Block of Odisha State (in mm)

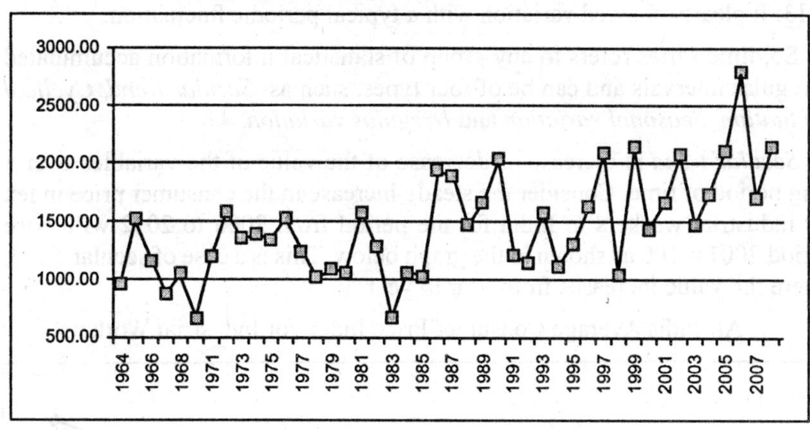

Source: Revenue & Disaster Management Department, Govt of Odisha

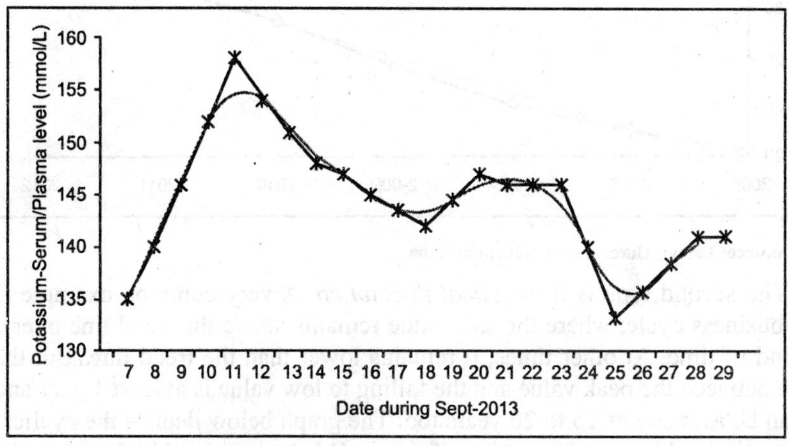

Source: Apollo Hospital, Bhubaneswar

From the graphs above, it can be seen that there is periodic variation of values. These variations are due to the effect of a number of movements, which are its characteristics. For example- production of rice declined to about 32 lakh tones during 2002-03 from 72 lakh tones during its previous year. Several reasons can be attributed to such decrease in production, which needs detailed analysis. Similarly, there is a steady growth in population of Odisha from 1961 to 2011, whereas the same is not steady from 1866 to 1961. Another example is taken for one patient named Mrs. Lily Ray, who was diagnosed with pancreatic disorder by the Apollo Hospital authorities, Bhubaneswar and her potassium-Serum / Plasma level was monitored daily during September 2013. It clearly showed variation with a typical periodic fluctuation.

So, time series refers to any group of statistical information accumulated at regular intervals and can be of four types, such as- *Secular trend, Cyclical fluctuation, Seasonal variation* and *Irregular variation.*

Secular trend is increase or decrease of the value of the variables over a long period of time. Consider the steady increase in the consumer price index for industrial workers in India for the period from 2006 to 2012 with base period 2001 = 100 as shown in the graph below. This is a case of secular trend, where the value increases from year to year.

All India Average Consumer Price Index for Industrial Workers

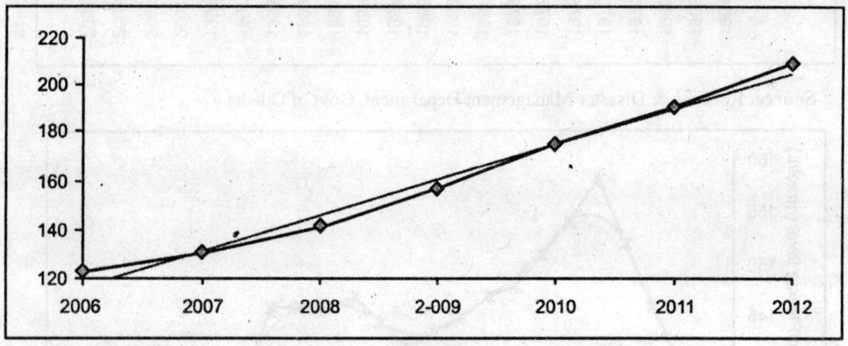

Source: Labour Bureau, Government of India

The second type is the *cyclical fluctuation.* A very common example is the business cycle, where the sale value remains above the trend line over a period of time. At other times, it remains lower than the trend line. But the time between the peak value and the falling to low value is at least 1 year and it can be as many as 15 to 20 years too. The graph below depicts the cyclical fluctuation, where the production of rice in Odisha remained higher than the trend line at certain years and remained below the trend line at the remaining years. It may be noted that the cyclical movements do not follow any regular pattern but move in a somewhat unpredictable manner.

Production of Rice in Odisha (in Lakh tones)

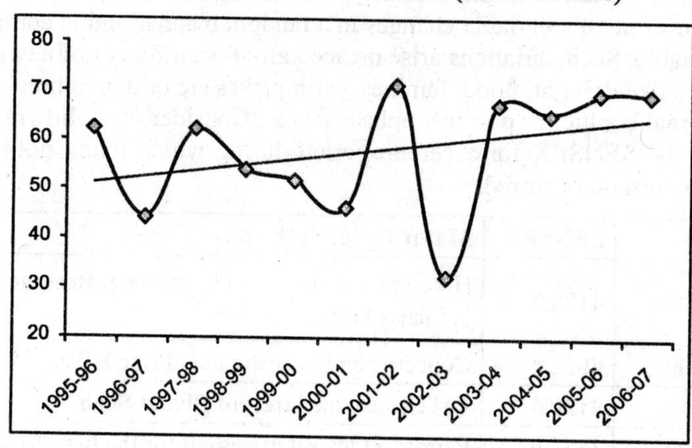

Source: ***Orissa Agricultural Statistics, 2006-07***

Seasonal variation is another type of variation, where the pattern of change occurs within a year that tends to be repeated from year to year. For example- a hospital may expect a substantial increase in the number of patients suffering from water-borne diseases during rainy season and increase in sun-stroke patients during summer. Because, there occurs a regular pattern, it is possible to forecast for the future. Monthly average rainfall and temperature data show seasonal variations. The graph below shows that rainfall pattern of Tumudibandha block of Kandhamal district of Odisha. It is noted that the peak occurs at every third quarter of the year, indicating that there is a seasonal variation in the rainfall pattern.

Rainfall of Tumudibandha block during 1991-93

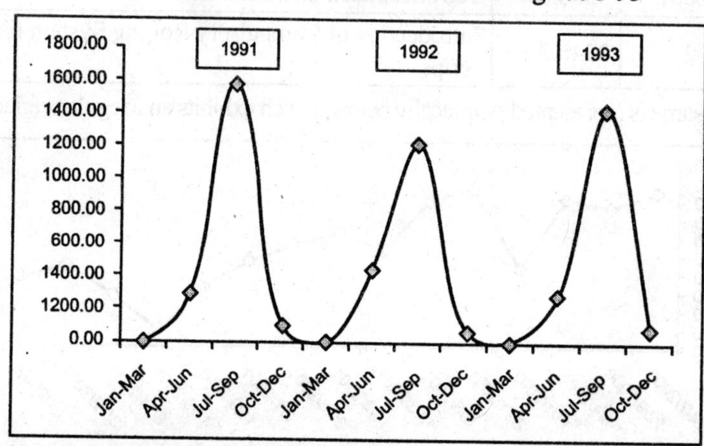

Source: Revenue & Disaster Management Department, Govt of Odisha

Irregular variation is the fourth type of change in time-series analysis, where the value of a variable changes in a random manner and is completely unpredictable. Such variations arise on account of occurrences of any natural calamities like draught, flood, famines, earthquakes etc or due to break out of war, internal conflict or political upheavals etc. Consider the following table showing the SENSEX for a certain period during which many political or economic turmoil occurred.

Date	SENSEX	Major Event of the day
24/7/2000	4188.3	Hike in bank Rate by 1% and rash Reserve ration 0.5 basis Points
22/9/2000	4032.4	Concerns on Domestic Fuel Price Hike
8/1/2001	4120.4	TMT Crash on Arrest of Bharat Shah
11/1/2002	3362.9	Petrol and Diesel prices cut for the first time
29/1/2001	4234.6	Gujarat Quake Pulls down Sensex
2/3/2001	4095.2	Short selling By Bear Cartel
13/3/2001	3540.7	Tehelka Revelation
16/3/2001	3745.7	Tehelka Revelations About Defense Bribery
30/3/2001	3604.4	Arrest Of Ketan Parekh
2/7/2001	3426	Ban on Badla
11/9/2001	3150	Terrorist attack on US
17/9/2001	2681	Fears of War
8/11/2001	3060	US Attack on Afghanistan
13/12/2001	3388.6	Terrorist attack on Parliament
1/1/2002	3246.2	Introduction of Compulsory Rolling System for all scrips

The same is represented graphically below, which exhibits an irregular variation.

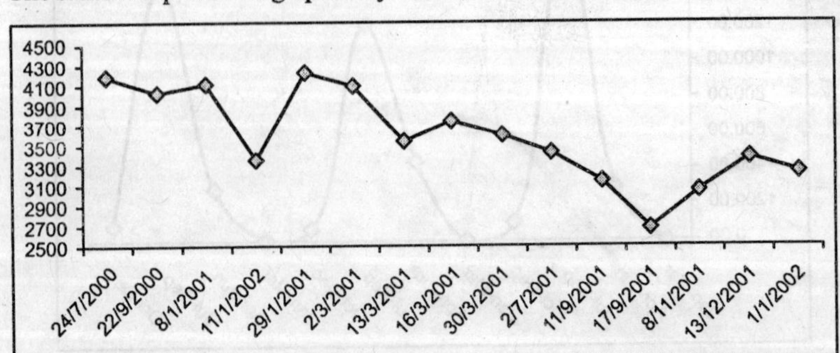

It may be noted that the SENSEX had its low at 2681 immediately following the WTC terrorist attack in New York.

REASONS FOR STUDYING TIME SERIES ANALYSIS

The time series analysis plays an important part in studying movements of economic and business data, while its major role is in the theory of forecasting. Because future forecast depends on the analysis of the past data and behaviour.

- Time series analysis is useful in planning and allocating budgets for different sectors of economy.
- The determination of trend line or curve helps in forecasting and the future value is estimated through extrapolation.
- Study of cyclical or seasonal fluctuations helps in devising policies for the manufacturers to decrease the production and produce alternative products to suit the seasons.
- Comparison of different time series helps us to establish and compare the intensity of growth or decline in the long run.
- Time series data of the same phenomenon at different places may be compared to know the peculiarities embedded in their nature.
- In many situations, studying the secular trend of a time series helps in eliminating the trend component from the series.
- Composite forces of movements affect the values of a time series and hence those should be separated out to enable the user to study, understand and measure each of them separately, to assess their relative impact and if necessary, to combine them suitably for future projections.

Now, let us discuss the method of combining the factor forces which are responsible in giving rise to a resultant force (Y_t), the value in a time series at time t. Generally two types of models are used in time series:

(1) Additive Model: $Y_t = T + C + S + I$,

(2) Multiplicative Model: $Y_t = TCSI$

where T, C, S and I are trend, cyclical, seasonal and irregular components respectively. In addition, the multiplicative model can be transformed into additive model by taking logarithm of both sides. However, all time series need not have all four components always.

Moreover, trends can be linear or curvilinear. Not all relationships in time series follow linear trends. For example, the net state domestic product from 1999-00 to 2007-08 clearly follows a curvilinear trend as shown here.

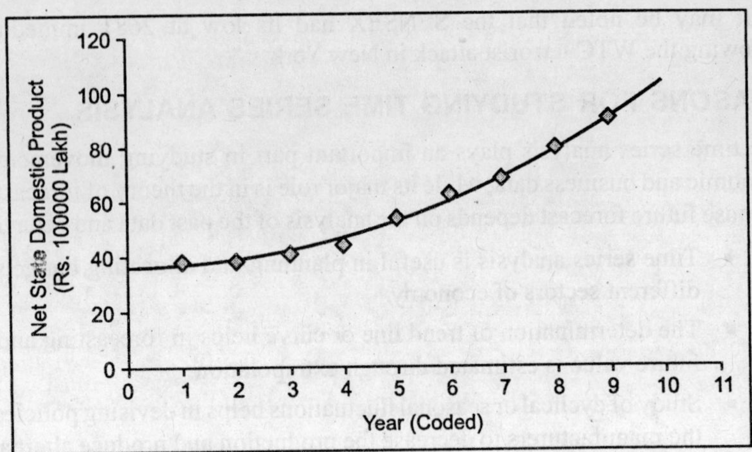

Source: Directorate of Economics and Statistics, Odisha

Another vivid example of curvilinear trend is the life cycle of a new product launched in the market. Initially, the sale becomes slow. As the product gains popularity, the sale grows rapidly. Then the product gets firmly established in the market and its sale grows at a somewhat constant rate. Finally, as the product reaches the end of its cycle due to launching of its competitive brands, the sale diminishes.

FITTING LINEAR TREND BY METHOD OF LEAST SQUARES

The general equation for a linear trend fitting is given by:

$Y_e = a + bX$

Where, Y_e = estimated value of the dependent variable

X = independent variable (time)

a = Y-intercept (value of Y when X=0)

b = slope of the trend line.

As has already been discussed in the chapter "Regression", the best-fitting line is obtained by using the method of least squares, which are again stated below;

Slope of Best-Fitting Regression Line $= b = \dfrac{\sum XY - n\bar{X}\bar{Y}}{\sum X^2 - n\bar{X}^2}$

Y-intercept of the Best-Fitting Regression Line $= a = \bar{Y} - b\bar{X}$

Where, Y = values of dependent variable

X = values of independent variable

$\overline{Y}$ = mean of the values of dependent variable

$\overline{X}$ = mean of the values of independent variable

n = number of data points in time series

a = Y-intercept

b = slope of the regression line.

CODING OF THE INDEPENDENT VARIABLE (TIME)

Because of the regularity of time-series data, it becomes possible to simplify the calculation procedure by coding the independent variable (time). Many times, the X-axis values take the form of years, months, weeks etc. But this coding of data eliminates the tedious process of calculating such big numbers encountered in X-axis. The mid-value of independent variable is taken as zero and accordingly, the values of the X-variable above its mid value becomes negative and values of X-variable below the mid value becomes positive. But in case of even number of variables, the mid-point occurs at middle of two central two values of X-variable. This is illustrated below.

Odd number of X-variables

Even number of X-variables

Odd number of X-variables		
X (1)	$X-\overline{X}$ (2)	Coded data (3)
2001	2001-2004	–3
2002	2002-2004	–2
2003	2003-2004	–1
2004	2004-2004	0
2005	2005-2004	+1

Even number of X-variables			
X (1)	$X-\overline{X}$	$(X-\overline{X})\times 2$	Coded data (3)
2001	2001-2003 $\frac{1}{2}$ =	$-2\frac{1}{2}\times 2=$	–5
2002	2002-2003 $\frac{1}{2}$ =	$-1\frac{1}{2}\times 2=$	–3
2003	2003-2003 $\frac{1}{2}$ =	$-\frac{1}{2}\times 2=$	–1
2004	2004-2003 $\frac{1}{2}$ =	$\frac{1}{2}\times 2=$	1
2005	2005-2003 $\frac{1}{2}$ =	$1\frac{1}{2}\times 2=$	3

Odd number of X-variables			Even number of X-variables			
2006	2006-2004	+2	2006	2006-2003 $\frac{1}{2}$ =	$2\frac{1}{2}\times 2=$	5
	2007-2004	+3				
Sum of X-values $=\sum X = 14{,}028$ 2007 $=\bar{x}(\text{coded}) = 0$ $\bar{X} = \frac{\sum X}{n} = \frac{14028}{7} = 2004$			Sum of X-values = $\sum X = 12{,}021$ $\bar{x}(\text{coded}) = 0$ $\bar{X} = \frac{\sum X}{n} = \frac{12021}{6} = 2003\frac{1}{2}$ To simplify the procedure, the coded values are again multiplied by 2.			

By the process shown above, the tedious method of squaring the large X-variables are avoided. Let us denote new X-variables in the coded data as x(small letter). The equations take the form of:

Slope of trend line,

$$b = \frac{\sum XY - n\bar{X}\bar{Y}}{\sum X^2 - n\bar{X}^2} = \frac{\sum xY - n\bar{x}\bar{Y}}{\sum x^2 - n\bar{x}^2} = \frac{\sum xY - n \times o \times \bar{Y}}{\sum x^2 - n \times 0^2} = \frac{\sum xY}{\sum x^2}$$

Y-intercept, $a = \bar{Y} - b\bar{X} = \bar{Y} - b\bar{x} = \bar{Y} - b \times 0 = \bar{Y}$

Let us now take up an example for prediction a future value using this coding procedure.

EXAMPLE ON LINEAR TREND

The yield of the crop Arhar during kharif season in Odisha are shown below (data taken from Orissa Agricultural Statistics).

Year:	1999	2000	2001	2002	2003	2004	2005	2006
Yield in kg/ hectare:	778	807	828	848	885	886	890	920

Predict the yield for the year 2007 using linear regression model.

The following table is prepared, which is self-explanatory:

X (1)	Y (2)	$x - \bar{X}$ (3)	x (4) = (3) × 2	xY (5) = (4) × (2)	x^2 (6) = $(4)^2$
1999	778	$1999-2002\frac{1}{2}=-3\frac{1}{2}$	$-3\frac{1}{2}\times 2=-7$	–5446	49
2000	807	$2000-2002\frac{1}{2}=-2\frac{1}{2}$	$-2\frac{1}{2}\times 2=-5$	–4035	25
2001	828	$2001-2002\frac{1}{2}=-1\frac{1}{2}$	$-1\frac{1}{2}\times 2=-3$	–2484	9
2002	848	$2002-2002\frac{1}{2}=-\frac{1}{2}$	$-\frac{1}{2}\times 2=-1$	–848	1
2003	885	$2003-2002\frac{1}{2}=+\frac{1}{2}$	$\frac{1}{2}\times 2=1$	885	1
2004	886	$2004-2002\frac{1}{2}=+1\frac{1}{2}$	$+1\frac{1}{2}\times 2=3$	2658	9
2005	890	$2005-2002\frac{1}{2}=+2\frac{1}{2}$	$+2\frac{1}{2}\times 2=5$	4450	25
2006	920	$2006-2002\frac{1}{2}=+3\frac{1}{2}$	$+3\frac{1}{2}\times 2=7$	6440	49
				$\sum X = 16020$	$\Sigma X^2 =$ 168

Sum of X-variables = $\sum X = 16020$

Sum of Y-variables = $\sum Y = 6842$

$$\bar{X} = \frac{\sum X}{n} = \frac{16020}{8} = 2002\frac{1}{2} \text{ and } \bar{Y} = \frac{\sum Y}{n} = \frac{6842}{8} = 855.25$$

So, Slope of trend line, $b = \frac{\sum xY}{\sum x^2} = \frac{1620}{168} = 9.643$ and Y-intercept, $a = \bar{Y}$ $= 855.25$

Thus, the general equation of the trend is $Y_e = a + bx = 855.25 + 9.643x$

Projecting the trend line to the next level, i.e. the yield value for the year 2007 (code will be

$$= 2\times\left(2007 - 2002\frac{1}{2}\right) = 9.0;\ Y_e = 855.25 + 9.643\times 9.0 = 942.037 \text{ kg/hectare.}$$

SECOND DEGREE TREND IN TIME SERIES (NON-LINEAR)

In many instances, trend may not occur linearly and hence, may not describe the actual trend line. Such time series are best described not by straight lines, but by curves, like parabola. The parabolic curve is described mathematically by a second degree equation given by:

$Y_e = a + bx + cx^2$

where, Y_e = estimate of the independent variable

a, b and c = numerical constants

x = coded values of X-variables

By use of least square method, the second degree equation constants can be found from the following three equations:

$$\sum Y = an + c\sum x^2 \text{ ; as } \sum x = 0$$

$$\sum x^2 Y = a\sum x^2 + c\sum x^4 \text{ ;as } \sum x^3 = 0$$

$$\sum xY = b\sum x^2 \text{ as } a\sum x = 0 \ \&\ c\sum x^3 = 0$$

EXAMPLE ON PARABOLIC TREND

Consider the following table showing the net state domestic product (in 10^5 lakh rupees) from 1990-00 to 2007-08 in Odisha.

S. No	Year	Net State Domestic Product (in Rs.10^5 Lakh)
1	1999-00	38.2
2	2000-01	38.2789
3	2001-02	41.06525
4	2002-03	44.20777
5	2003-04	54.04738
6	2004-05	62.49696
7	2005-06	68.2952
8	2006-07	79.42995
9	2007-08	89.86739

Source: Directorate of Economics & Statistics, Odisha, BBSR

Let us now fit a parabolic curve to the data. First the time period (year) is coded from 1 to 9. As there is an odd number of data set, there is no need to multiply the variable by 2. The following table is prepared showing necessary calculations, which is self-explanatory.

Y	X	$X-\bar{X}=x$	x^2	x^4	xY	x^2Y
(1)	(2)	(3)	(4)	(5)	(6)	(7)
38.2	1999	1999 - 2003 = -4	16	256	-152.8	611.2
38.2789	2000	2000 - 2003 = -3	9	81	-114.837	344.5101
41.06525	2001	2001 - 2003 = -2	4	16	-82.1305	164.261
44.20777	2002	2002 - 2003 = -1	1	1	-44.2078	44.20777
54.04738	2003	2003 - 2003 = 0	0	0	0	0
62.49696	2004	2004 - 2003 = 1	1	1	62.49696	62.49696
68.2952	2005	2005 - 2003 = 2	4	16	136.5904	273.1808
79.42995	2006	2006 - 2003 = 3	9	81	238.2899	714.8696
89.86739	2007	2007 - 2003 = 4	16	256	359.4696	1437.878
$\sum Y$ = 515.8888	$\sum X$ = 18207		$\sum x^2$ = 60	$\sum x^4$ = 708	$\sum xY$ = 402.8718	$\sum x^2Y$ = 3652.604

$$= \bar{X} = \frac{\sum X}{n} = \frac{18207}{9} = 2003$$

By substituting the values in the equations we get,

$515.8888 = 9a + 60c$

$3652.604 = 60a + 708c$

$402.8718 = 60\ b$

By solving the three equations for three unknowns we get, a = 52.7031; b = 6.7145 and c = 0.6927

Hence the parabolic equation fitted to the data is given by:

$Y_e = a + bx + cx^2 = 52.7031 + 6.7145\ x + 0.6927\ x^2$

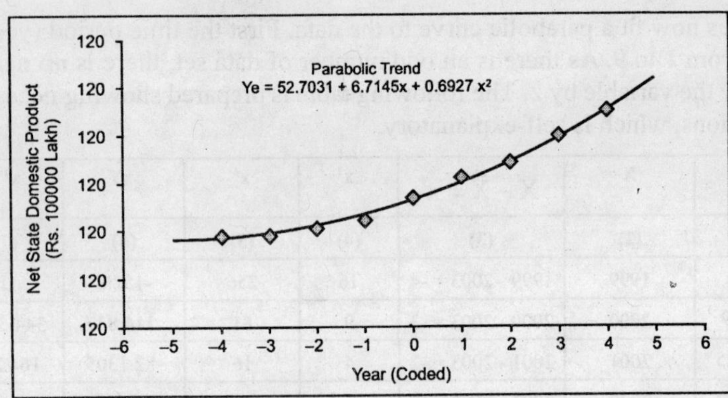

Forecast based on this parabolic trend can be made for the year 2008. For this, first the year 2008 must be coded, i.e. $X - \overline{X} = x = 2008 - 2003 = 5$. Putting this value of x, we get $Y = 52.7031 + 6.7145 \times 5 + 0.6927 \times 5^2 = 103.5931$. So the predicted value of the net state domestic product is 103.5931×10^5 lakh rupees.

A point of caution must be maintained here while predicting by parabolic trend line, i.e. as seen from the curve the trend line shows an increasing trend and any future prediction will show values in increasing manner, which may be poor in estimating. Because there may be other factors which might be responsible in slowing or reversing the growth rate of the variable and not considered in the parabolic trend fitting line. Moreover, the growth may stabilize after certain years and any future predictions beyond that will yield erroneous result. It is therefore necessary to exercise particular care when using a second degree equation while forecasting. There is a Wall Street saying: "no tree grows to the sky", meaning no stock price rises forever. At the initial period, growth may be non-linear like parabolic, exponential etc, but this early growth seldom continues.

Merits of Method of Least Squares

1. This method is mathematically sound and attracts attention of the analyst.
2. It can be conveniently applied for forecasting or predicting values for future time periods.
3. This method is free from human bias.

Demerits of Method of Least Square

1. The most important limitations of this method is that not all mathematical equations can be fitted to the data by this method, as this method presupposes that all trend lines can be represented by mathematical functions.

2. This method does not take into account of the periodic fluctuations.
3. Any new addition of data will require fresh calculations.
4. It involves tedious calculations and is time-consuming, though many computer packages are available now-a-days to ease the procedure.

A. SECULAR TREND

This component represents basic tendency of the series. It is a long term movement in a time series. Factors like population change, change in habits and costumes of people, major changes arising out of change in government policies or technological development give rise to such trend. Secular trend can be either linear or non-linear. In order to determine secular trend, the short term fluctuations are to be eliminated or smoothened out by employing methods such as:

A1. Free hand curve method

A2. Semi-average method

A3. Method of moving averages

A4. Method of least squares.

A1. Free hand curve method

This is the simplest of all methods and easy to understand, but subject to its reliability, because the free hand curve varies from person to person. It is otherwise known as eye-inspection method, where a smooth curve is drawn through the plotted points in such a way that it represents general tendency of the series. Observe the graph below showing the total power consumption in Odisha. The graph is smoothened by free hand, where the fluctuations in one direction are roughly compensated by the fluctuations in the other thereby determining an average path.

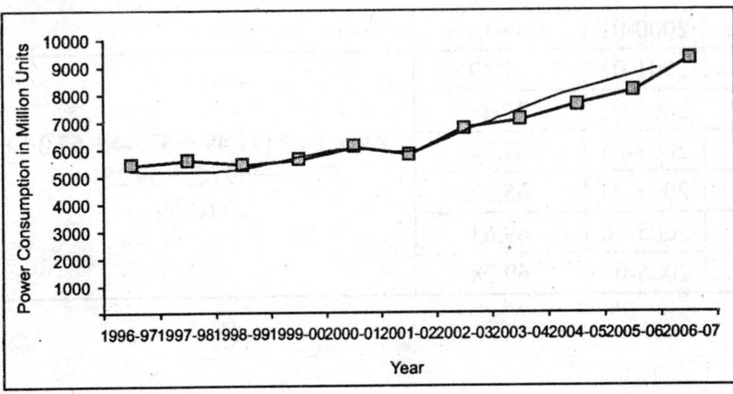

There is yet another method called the **method of selected points**, where the free-hand curve is passed through a few pre-selected points. These points are so selected that the typical growth factor is observed in the series. This is a refinement over the previous method. Observe in the above graph that the points corresponding to the years 1996-97, 1998-99, 2000-01, 2003-04 and 2006-07 are points selected at the outset prior to joining with free hand.

A2. Semi-Average Method

In this method, the time series is divided into two parts and the arithmetic mean of the values in each part is calculated. These two mean values are plotted against the middle time period for each half and joined by a straight line, which show the trend. In case of an odd number of time periods, the value of the middle time period is considered for both the halves, or else it may be totally omitted. Some statisticians consider half of the middle time period value on each half. However, this method is not free from criticisms. If there exists any of extreme value in any half, it greatly affects the points and consequently the trend. Hence this method is not helpful if the trend is not linear. Consider the following example on rice production (lakh tones) in Odisha from 1995-96 to 2006-07 (even number of data):

S. No	Year	Rice Production (lakh tonnes)	
1	1995-96	62.26	Average = (62.26 + 44.38 + 62.05 + 53.90 + 51.87 + 46.14) ÷ 6 = 53.43
2	1996-97	44.38	
3	1997-98	62.05	
4	1998-99	53.9	
5	1999-00	51.87	
6	2000-01	46.14	
7	2001-02	71.49	Average = (71.49 + 32.44 + 67.34 + 65.37 + 69.63 + 69.28) ÷ 6 = 62.59
8	2002-03	32.44	
9	2003-04	67.34	
10	2004-05	65.37	
11	2005-06	69.63	
12	2006-07	69.28	

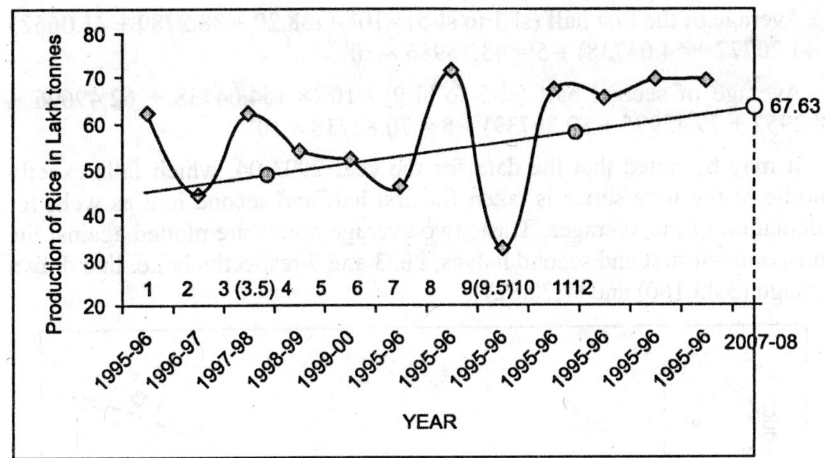

Source: Orissa Agricultural Statistics, 2006-07

Average of the rice production values from 1995-96 to 2000-01 is 53.43 lakh tonnes and that of 2001-02 to 2006-07 is 62.59 lakh tonnes. These two points are plotted corresponding to the middle of two time period halves, i.e. 3.5 and 9.5 respectively and then joined by a straight line.

Straight line equation to the trend line can be obtained by considering the two points (3.5, 53.43) and (9.5,62.59), which gives rise to $\frac{Y-53.43}{X-3.5}=\frac{62.59-53.43}{9.5-3.5}$ => Y = 1.527X + 48.09. Extending the straight line, thus obtained, beyond 2006-07 will give the forecasted production value for the year 2007-08 (for X = 13) which is equal to 1.527 X 13 + 48.09 = 67.94 lakh tonnes.

Take another example having an odd number of data as in the following:

S. No	Year	Net State Domestic Product in Rs.Lakh (at current price)
1	1999-00	38.20 10^5
2	2000-01	38.2789 10^5
3	2001-02	41.06525 10^5
4	2002-03	44.20777 10^5
5	2003-04	54.04738 10^5
6	2004-05	62.49696 10^5
7	2005-06	68.2952 10^5
8	2006-07	79.42995 10^5
9	2007-08	89.86739 10^5

Source: Directorate of Economics & Statistics, Odisha, BBSR

Average of the first half (sl-1 to sl-5) $=10^5 \times (38.20 + 38.2789 + 41.06525 + 44.20777 + 54.04738) \div 5 = 43.15986 \times 10^5$

Average of second half (sl-5 to sl-9) $=10^5 \times (54.04738 + 62.49696 + 68.2952 + 79.42995 + 89.86739) \div 5 = 70.82738 \times 10^5$

It may be noted that the data for the year 2003-04, which falls exactly middle of the time series is taken for first half and second half as well, for calculation of the averages. These two average points are plotted against the mid-points of first and second halves, i.e. 3 and 7 respectively i.e. line drawn through (3,43.160) and (7,70.827).

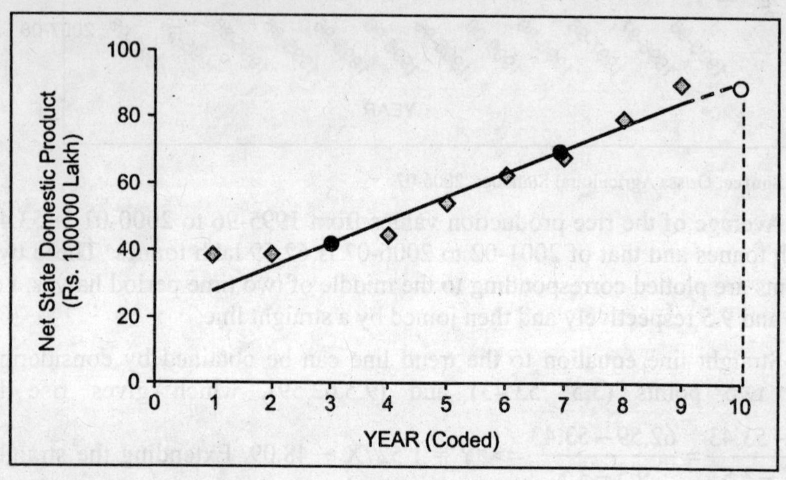

A3. Method of Moving Averages

The fluctuations in the data may be smoothened out by means of a moving average in a series of successive simple average of a specified number of consecutive values in a time-ordered data. Moving average is usually carried out by taking 3-year or 4-year or 5-year values at a time. These moving average values are trend values corresponding to the middle of the time periods. The following table shows moving average methods employed in the original data set by taking 3-years and 5-years averages.

Year	Rice Production (lakh tones)	3-yearly moving average		5-yearly moving average	
1995-96	62.26				
1996-97	44.38	(62.26 + 44.38 + 62.05) ÷ 3 =	56.23		
1997-98	62.05	(44.38 + 62.05 + 53.9) ÷ 3 =	53.44	(62.26 + 44.38 + 62.05 + 53.9 + 51.87) ÷ 5 =	54.89
1998-99	53.9	(62.05 + 53.9 + 51.87) ÷ 3 =	55.94	(44.38 + 62.05 + 53.9 + 51.87 + 46.14) ÷ 5 =	51.67
1999-00	51.87	and	50.64	(62.05 + 53.9 + 51.87 + 46.14 + 71.49) ÷ 5 =	57.09
2000-01	46.14	so	56.50	and	51.17
2001-02	71.49	on	50.02	so	53.86
2002-03	32.44		57.09	on	56.56
2003-04	67.34		55.05		61.25
2004-05	65.37		67.45	(32.44 + 67.34 + 65.37 + 69.63 + 69.28) ÷ 5 =	60.81
2005-06	69.63	(65.37 + 69.63 + 69.28) ÷ 3 =	68.09		
2006-07	69.28				

But when it is required to carry out the moving average procedure considering an even number of years at a time, the average values do not fall into any time period given in the data. So the moving average values thus obtained are imposed upon by another even number moving average values calculated from it. For example- let us find a 4-year moving average values. Then again a 2-year moving values are calculated from this 4-year moving average values and placed at proper time periods. The following example will clarify.

Year	Rice Production (lakh tones)	4-yearly moving average (Not centered)		Total of two successive moving average (Centered)		4-yearly moving average (Centered)	
(1)	(2)	(3)		(4)		(5) = (4) ÷ 2	
1995-96	62.26						
1996-97	44.38	(62.26 + 44.38 + 62.05 + 53.9) ÷ 4 =	55.65				

Year	Rice Production (lakh tones)	4-yearly moving average (Not centered)		Total of two successive moving average (Centered)		4-yearly moving average (Centered)	
1997-98	62.05	(44.38 + 62.05 + 53.9 + 51.87) ÷ 4 =	53.05	(55.65 + 53.05) =	108.70	108.72 =	54.35
1998-99	53.9	(62.05 + 53.9 + 51.87 + 46.14) ÷ 4 =	53.49	(53.05 + 53.49) =	106.54	106.542 =	53.27
1999-00	51.87	and	55.85	(53.49 + 55.85) =	109.34	109.342 =	54.67
2000-01	46.14	so	50.49	and	106.34	and	53.17
2001-02	71.49	on	54.35	so	104.84	so	52.42
2002-03	32.44		59.16	on	113.51	on	56.76
2003-04	67.34		58.70		117.86		58.93
2004-05	65.37	(67.34 + 65.37 + 69.63 + 69.28) ÷ 4 =	67.91	(58.70 + 67.91) =	126.60	126.602 =	63.30
2005-06	69.63						
2006-07	69.28						

Data Source: Orissa Agricultural Statistics, 2006-07

As can be seen, the simple moving average of say m-period gives trend values at any point as the arithmetic mean of the m-values centred at that point. So, the simple moving average is the result of fitting straight lines to m consecutive values. This gives equal weight to all values, whose average is to be taken. But unequal weights can also be used to eliminate fluctuations in a time series, which help for fitting a polynomial of higher degree (to first m-values and so on.

The period of moving average is determined on the basis of the length of the cycle. If the extent of the moving average equals the period of cycle(or oscillation), then the periodic fluctuations (if regular) are completely eliminated from the series. However, in case of irregular periodic fluctuations, the average period of cycles is computed to determine an approximate period for the application of moving average technique. This eliminates the cyclical component considerably, though not entirely.

Most important is the fixing up of the period of moving average, m and the degree of the polynomial, p (in case a polynomial provides a good fit). Situations where the data contain only trend and irregular components, moving averages of different periods by trial and error is carried out until the plotted data are reasonably smoothened.

Merits of Moving Average

1. The most advantage of this technique is that if few more data is added to the original data, the previous calculations are not affected and rather few more trend values are obtained for analysis.
2. If the period of moving average coincides with the period of oscillation (cycle), then the periodic fluctuation is eliminated.
3. This method gives a correct picture of the long-term trend if the trend is linear.
4. In this method, no hypothetical mathematical function determines the path of the trend. But the trend path follows a general movement of data.
5. This method is simple and easy to calculate.

Demerits of Moving Average

1. A strong demerit of this method is that for a moving average of period of 2m + 1, the trend values for the first and last m periods cannot be obtained.
2. As the trend path does not follow any mathematical function, any further mathematical analysis cannot be done by this method, e.g. forecasting or predicting values for future periods.
3. If the trend is not linear, the trend values calculated through this method may not show the true tendency of the data.
4. It is subjected to human bias, as the choice of the period is left to the person carrying out the analysis.

Example 16.1: The rainfall in the month of July in Tikabali block of Kandhamal district for the period from 1993 to 2001 is furnished below. Estimate the trend values by 3-yearly moving average method.

Year	1993	1994	1995	1996	1997	1998	1999	2000	2001
July Rainfall (mm)	245.00	339.00	199.00	252.00	300.00	270.00	415.00	350.00	745.00

Source: Revenue & Disaster Management Department, Govt of Odisha

Solution: The following table is prepared, which is self-explanatory.

Year	Rainfall	3-yearly moving average
1993	245.00	
1994	339.00	261.00
1995	199.00	263.33
1996	252.00	250.33
1997	300.00	274.00
1998	270.00	328.33
1999	415.00	345.00
2000	350.00	503.33
2001	745.00	

The values are plotted as below.

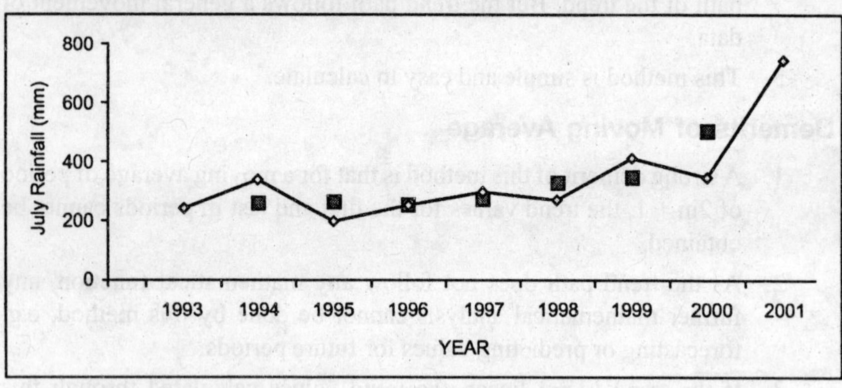

Example 16.2: The following table shows the total number of employments in registered factories in Odisha for the period from 1985 to 1997. Estimate the trend values by 5-yearly moving average method.

Year	Total employment in registered factories (number)
1985	144671
1986	139810
1987	156500
1988	156265
1989	156404
1990	155018

Year	Total employment in registered factories (number)
1991	166462
1992	176092
1993	180343
1994	189480
1995	186538
1996	177860
1997	178925

Source: Economic Survey, Orissa 2005-06

Solution: The following table is prepared which is self-explanatory.

Year	Total employment in registered factories (number)	5-yearly moving average values
1985	144671	
1986	139810	
1987	156500	150730
1988	156265	152799.4
1989	156404	158129.8
1990	155018	162048.2
1991	166462	166863.8
1992	176092	173479
1993	180343	179783
1994	189480	182062.6
1995	186538	182629.2
1996	177860	
1997	178925	

The values are plotted in a graph as shown below.

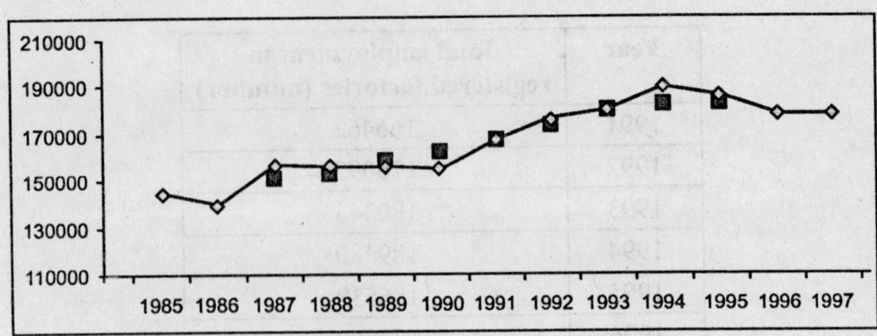

A4. Method of Least Square

(a) Fitting Linear Trend by Method of Least Squares

The general equation for a linear trend fitting is given by:

$$Y_e = a + bX$$

where, Y_e = estimated value of the dependent variable

X = independent variable (time)

a = Y-intercept (value of Y when X=0)

b = slope of the trend line

As it has already been discussed in the chapter "Regression", the best-fitting line is obtained by using the method of least squares, which are again stated below;

Slope of Best-Fitting Regression Line = $b = \dfrac{\sum XY - n\bar{X}\bar{Y}}{\sum X^2 - n\bar{X}^2}$

Y-intercept of the Best-Fitting Regression Line = $a = \bar{Y} - b\bar{X}$

where Y = values of dependent variable

X = values of independent variable

$\bar{Y}$ = mean of the values of dependent variable

$\bar{X}$ = mean of the values of independent variable

n = number of data points in time series

a = Y-intercept

b = slope of the regression line.

CODING OF THE INDEPENDENT VARIABLE (TIME)

Because of the regularity of time-series data, it becomes possible to simplify the calculation procedure by coding the independent variable (time). Many times, the X-axis values take the form of years, months, weeks etc. But this coding of data eliminates the tedious process of calculating such big numbers encountered in X-axis. The mid-value of independent variable is taken as zero and accordingly, the values of the X-variable above its mid value becomes negative and values of X-variable below the mid value becomes positive. But in case of even number of variables, the mid-point occurs at middle of two central values of X-variable. This is illustrated below.

Odd number of X-variables		
X (1)	$x = X - \bar{X}$ (2)	Coded data (3)
2001	2001-2004=	−3
2002	2002-2004=	−2
2003	2003-2004=	−1
2004	2004-2004=	0
2005	2005-2004=	+1
2006	2006-2004=	+2
2007	2007-2004=	+3

Sum of x-values = $\sum X = 14{,}028$

$= \bar{x}(\text{coded}) = 0$

$= \bar{X} = \frac{\sum X}{n} = \frac{14028}{7} = 2004$

Even number of X-variables			
X (1)	$X - \bar{X}$	$(X - \bar{X}) \times 2 = x$	Coded data (3)
2001	$2001\text{-}2003\frac{1}{2} =$	$-2\frac{1}{2} \times 2 =$	-5
2002	$2002\text{-}2003\frac{1}{2} =$	$-1\frac{1}{2} \times 2 =$	-3
2003	$2003\text{-}2003\frac{1}{2} =$	$-\frac{1}{2} \times 2 =$	-1
2004	$2004\text{-}2003\frac{1}{2} =$	$= \frac{1}{2} \times 2 =$	1
2005	$2005\text{-}2003\frac{1}{2} =$	$= 1\frac{1}{2} \times 2 =$	3
2006	$2006\text{-}2003\frac{1}{2} =$	$2\frac{1}{2} \times 2 =$	5

Sum of X-values = $\sum X = 12{,}021$

$= \bar{x}(\text{coded}) = 0$

$= \bar{X} = \frac{\sum X}{n} = \frac{12021}{6} = 2003\frac{1}{2}$

To simplify the procedure, the coded values are again multiplied by 2.

By the process shown above, the tedious method of squaring the large X-variables are avoided. Let us denote new X-variables in the coded data as x(small letter). The equations take the form of:

Slope of trend line,

$$b = \frac{\sum XY - n\bar{X}\bar{Y}}{\sum X^2 - n\bar{X}^2} = \frac{\sum xY - n\bar{x}\bar{Y}}{\sum x^2 - n\bar{x}^2} = \frac{\sum xY - n \times o \times \bar{Y}}{\sum x^2 - n \times 0^2} = \frac{\sum xY}{\sum x^2}$$

Y-intercept, $a = \bar{Y} - b\bar{X} = \bar{Y} - b\bar{x} = \bar{Y} - b \times 0 = \bar{Y}$

Let us now take up an example to prediction a future value using this coding procedure.

Example 16.3: The yield of the crop Arhar during kharif season in Odisha are shown below (data taken from Odisha Agricultural Statistics).

Year:	1999	2000	2001	2002	2003	2004	2005	2006
Yield in kg/hectare:	778	807	828	848	885	886	890	920

Predict the yield for the year 2007 using linear regression model.

Solution: The following table is prepared, which is self-explanatory.

X (1)	Y (2)	$x - \bar{X}$ (3)	x (4)=(3)2	xY (5) = (4) × (2)	x^2 (6) = $(4)^2$
1999	778	$1999 - 2002\frac{1}{2} = -3\frac{1}{2}$	$-3\frac{1}{2} \times 2 = -7$	–5446	49
2000	807	$2000 - 2002\frac{1}{2} = -2\frac{1}{2}$	$-2\frac{1}{2} \times 2 = -5$	–4035	25
2001	828	$2001 - 2002\frac{1}{2} = -1\frac{1}{2}$	$-1\frac{1}{2} \times 2 = -3$	–2484	9
2002	848	$2002 - 2002\frac{1}{2} = -\frac{1}{2}$	$-\frac{1}{2} \times 2 = -1$	–848	1
2003	885	$2003 - 2002\frac{1}{2} = +\frac{1}{2}$	$\frac{1}{2} \times 2 = 1$	885	1
2004	886	$2004 - 2002\frac{1}{2} = +1\frac{1}{2}$	$+1\frac{1}{2} \times 2 = 3$	2658	9
2005	890	$2005 - 2002\frac{1}{2} = +2\frac{1}{2}$	$+2\frac{1}{2} \times 2 = 5$	4450	25
2006	920	$2006 - 2002\frac{1}{2} = +3\frac{1}{2}$	$+3\frac{1}{2} \times 2 = 7$	6440	49
				$\sum xY =$ 1620	$\sum x^2 =$ 168

Sum of X-variables = $\sum X = 16020$

Sum of Y-variables = $\sum Y = 6842$

$$\bar{X} = \frac{\sum X}{n} = \frac{16020}{8} = 2002\frac{1}{2} \text{ and } \bar{Y} = \frac{\sum Y}{n} = \frac{6842}{8} = 855.25$$

So, Slope of trend line, $b = \frac{\sum xY}{\sum x^2} = \frac{1620}{168} = 9.643$ and Y-intercept, $a = \bar{Y}$ = 855.25

Thus, the general equation of the trend is $Y_e = a + bx = 855.25 + 9.643x$

Projecting the trend line to the next level, i.e. the yield value for the year 2007 (code will be = $2007\text{-}2002\frac{1}{2} = 4.5$); $Y_e = 855.25 + 9.643 \times 4.5$ = 898.6435 kg/hectare.

(b) Second Degree Trend in Time Series (Non-Linear)

In many instances, trend may not occur linearly and hence, may not describe the actual trend line. Such time series are best described not by straight lines, but by curves, like parabola. The parabolic curve is described mathematically by a second degree equation given by:

$$Y_e = a + bx + cx^2$$

where, Y_c = estimate of the dependent variable

a, b and c = numerical constants

x = coded values of X-variables

By use of least square method, the second degree equation constants can be found from the following three equations:

$$\sum Y = an + c\sum x^2 \text{ as } b\sum x = 0$$

$$\sum x^2Y = a\sum x^2 + c\sum x^4$$

$$b = \frac{\sum xY}{\sum x^2} \text{ as } a\sum x = 0 = c\sum x^3$$

Example 16.4: The following table shows the net state domestic product (in 10^5 lakh rupees) from 1990-00 to 2007-08 in Odisha:

S. No	Year	Net State Domestic Product (in Rs.10^5 Lakh)
1	1999-00	38.2
2	2000-01	38.2789
3	2001-02	41.06525
4	2002-03	44.20777
5	2003-04	54.04738
6	2004-05	62.49696
7	2005-06	68.2952
8	2006-07	79.42995
9	2007-08	89.86739

Source: Directorate of Economics & Statistics, Odisha, BBSR

Fit a parabolic curve to the data and predict the value for the year 2008-09.

Solution: First the time period (year) is coded from 1 to 9. As there is an odd number of data set, there is no need to multiply the variable by 2. The following table is prepared showing necessary calculations, which is self-explanatory.

Y	X	$X-\bar{X}=x$	x^2	x^4	xY	x^2Y
(1)	(2)	(3)	(4)	(5)	(6)	(7)
38.2	1999	1999-2003 = –4	16	256	–152.8	611.2
38.2789	2000	2000-2003 = –3	9	81	–114.837	344.5101
41.06525	2001	2001-2003 = –2	4	16	–82.1305	164.261
44.20777	2002	2002-2003 = –1	1	1	–44.2078	44.20777
54.04738	2003	2003-2003 = 0	0	0	0	0
62.49696	2004	2004-2003 = 1	1	1	62.49696	62.49696
68.2952	2005	2005-2003 = 2	4	16	136.5904	273.1808
79.42995	2006	2006-2003 = 3	9	81	238.2899	714.8696
89.86739	2007	2007-2003 = 4	16	256	359.4696	1437.878
$\sum Y$ = 515.8888	$\sum X$ = 18207		$\sum x^2 = 0$	$\sum x^4$ = 708	$\sum xY$ = 402.8718	$\sum x^2 Y$ = 3652.604

$$= \bar{X} = \frac{\sum X}{n} = \frac{18207}{9} = 2003$$

By substituting the values in the equations we get,

$515.8888 = 9a + 60c$

$3652.604 = 60a + 708c$

$$b = \frac{402.8718}{60}$$

By solving the three equations for three unknowns we get, a = 52.7031; b = 6.7145 and c = 0.6927

Hence the parabolic equation fitted to the data is given by:

$Y_e = a + bx + cx^2 = 52.7031 + 6.7145\,x + 0.6927\,x^2$

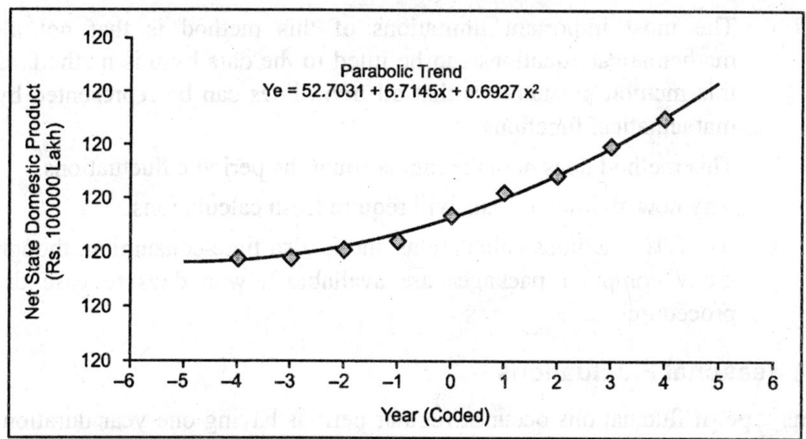

Forecast based on this parabolic trend can be made for the year 2008-09.

For this first the year 2008 must be coded, i.e. $X - \overline{X} = x = 2008 - 2003 = 5$. Putting this value of x, we get $Y_e = 52.7031 + 6.7145 \times 5 + 0.6927 \times 5^2 = 103.5931$. So the predicted value of the net state domestic product is 103.5931×10^5 lakh rupees.

A point of caution must be maintained here while predicting by parabolic trend line, i.e. as seen from the curve, the trend line shows an increasing trend and any future prediction will show values in increasing manner, which may be poor in estimating. Because there may be other factors which might be responsible in slowing or reversing the growth rate of the variable and not considered in the parabolic trend fitting line. Moreover, the growth may stabilize after certain years and any future predictions beyond that will yield erroneous result. It is therefore necessary to exercise particular care when using a second degree equation while forecasting. There is a Wall Street saying: "no tree grows to the sky", meaning no stock price rises forever. At the initial period, growth may be non-linear like parabolic, exponential etc, but this early growth seldom continues.

Merits of Least Square Method

1. This method is mathematically sound and attracts attention of the analyst.
2. It can be conveniently applied for forecasting or predicting values for future time periods.
3. This method is free from human bias.

Demerits of Least Square Method

1. The most important limitations of this method is that not all mathematical equations can be fitted to the data by this method, as this method presupposes that all trend lines can be represented by mathematical functions.
2. This method does not take into account the periodic fluctuations.
3. Any new addition of data will require fresh calculations.
4. It involves tedious calculations and is also time-consuming, though many computer packages are available now a days to ease the procedure.

(B) Seasonal Fluctuations

This type of fluctuations occur at regular periods having one year duration. More common example is sale of winter clothes, which increases during winter months only. Similarly, sale of sarees increases during festive occasions, sale of cold drinks increases during summer months etc. So isolation of seasonal fluctuation from the time series is essential in order to study its effect on the size of the variable. Study of seasonal variation helps business people to plan the production schedule in more efficient manner so as to cater to the public demands according to season. But determination of seasonal variation is based on the assumption that the irregular or random disturbances of different years are independent of each other. Here also four methods are employed for its study, e.g.

1. Method of simple average
2. Method of moving average
3. Method of least square
4. Link relative method.

B1. Method of Simple Average

In this method, the observed value of the series is expressed by

$Y_t = T + S + C + I$

Under the situations, where trend and cyclical component are both absent, then monthly average through years will eliminate irregular component, because of averaging. If monthly averages are taken as 100, then the percentages of various monthly averages can be computed, which are otherwise called as **seasonal indices**. On the contrary, if only cyclical component is present, then a sufficient number of years should be taken to cover the period of the cycle and then the method can be applied to eliminate both the cyclical and irregular components. If only the trend is present, then it is estimated by any suitable method and then subtracted from the original data, which will thus enable eliminating this component. After that method of average can be applied to compute the seasonal indices.

Although the method of simple average is easy and simple to calculate, this method is not usually adopted for the simple reason that, the assumption of absence of trend may not always be justified. Moreover, cyclical effect and irregular disturbances may not always be eliminated in the process. With determination of trend from annual data free from seasonal fluctuations, it is not always easy to determine its effect on successive months.

Example 16.5: Given below the rainfall figures (in mm) of Anugul district in Odisha for the period from 2000 to 2003. Compute the seasonal indices by simple average method.

	2000	**2001**	**2002**	**2003**
Jan	0.00	0.00	8.60	0.00
Feb	42.40	0.00	0.00	4.40
Mar	2.60	21.90	1.80	6.70
Apr	24.20	14.40	46.80	23.80
May	95.40	49.80	93.60	37.80
Jun	188.30	291.90	88.80	221.60
Jul	337.20	588.00	70.60	209.80
Aug	318.20	354.40	210.20	328.40
Sep	162.20	63.60	230.90	180.80
Oct	19.60	147.00	13.00	250.00
Nov	2.60	8.20	5.60	23.80
Dec	0.00	0.00	0.00	32.40

Source: Special Relief Commissioner, Odisha

Solution: The calculations are shown in the table below, which is self-explanatory.

	2000	2001	2002	2003	Monthly Average	Seasonal Indices
	(1)	(2)	(3)	(4)	(5)	(6) = (5) ÷ 100.44375 × 100
Jan	0.00	0.00	8.60	0.00	2.15	2.14
Feb	42.40	0.00	0.00	4.40	11.70	11.65
Mar	2.60	21.90	1.80	6.70	8.25	8.21
Apr	24.20	14.40	46.80	23.80	27.30	27.18
May	95.40	49.80	93.60	37.80	69.15	68.84
Jun	188.30	291.90	88.80	221.60	197.65	196.78
Jul	337.20	588.00	70.60	209.80	301.40	300.07
Aug	318.20	354.40	210.20	328.40	302.80	301.46
Sep	162.20	63.60	230.90	180.80	159.38	158.67
Oct	19.60	147.00	13.00	250.00	107.40	106.93
Nov	2.60	8.20	5.60	23.80	10.05	10.01
Dec	0.00	0.00	0.00	32.40	8.10	8.06
				Total:	1205.33	1200.00

$$\text{Average of monthly average} = \frac{1205.33}{12} = 100.44375$$

2. Method of Moving Average

Two methods are employed here:

(a) Multiplicative Model: This method is otherwise known as *Ratio to Moving Average Method*. The model is $Y_t = T \times S \times C \times I$. OR, $Y_t = T \; S \times I$, where cyclical component is absent. It is assumed that the period of seasonal variation is 12 months and the variation has the same pattern every year i.e. S is constant. A 12 monthly moving average eliminates the S & I components leaving T & C components and the S × I component is given by:

$$\frac{\text{Original value}}{\text{Moving average}} = \frac{T \times S \times C \times I}{T \times C} = S \times I$$

Then the monthly average of S × I over the years will eliminate the I component. Calculation of seasonal indices is done as before. In this method the fluctuation of indices is less and the indices seem to be more representative of data. But the limitation of this method is that complete elimination of S × I component may not always hold good. Elimination of S × I component may be accompanied by some part of the cyclical component too.

(b) Additive Model: This is given by $Y_t = T + C + S + I$. If C component is absent or negligible, then the trend T is calculated by moving average method and then this moving average values are subtracted from Y_t to obtain S + I component. The monthly average of S + I will then eliminate I and determine the seasonal variations (S)

Example 16.6: Quarterly rainfall data for Bhubaneswar in Odisha are given below. Calculate the seasonal (quarterly) indices.

(a) Using multiplicative model or ratio to moving average method:

	Qr I	Qr II	Qr III	Qr IV
1961	117.80	83.40	85.10	240.60
1962	97.20	154.30	159.00	300.90
1963	19.90	67.30	207.40	319.80
1964	31.80	79.60	140.80	252.60

Source: Special Relief Commissioner, Odisha

Solution: The following table is prepared, which is self-explanatory:

Year	Quarterly Rainfall	4-Quarter moving average	4-Quarter moving average centered	(5) = (2) ÷ (4)
(1)	(2)	(3)	(4)	(5)
1961	117.80			
	83.40	131.73		
	85.10	126.58	129.15	0.66
	240.60	144.30	135.44	1.78
1962	97.20	162.78	153.54	0.63
	154.30	177.85	170.31	0.91
	1590	158.53	168.19	0.95
	300.90	136.78	147.65	2.04

Year	Quarterly Rainfall	4-Quarter moving average	4-Quarter moving average centered	(5) = (2) ÷ (4)
1963	19.90	148.88	142.83	0.14
	67.30	153.60	151.24	0.45
	207.40	156.58	155.09	1.34
	319.80	159.65 ,	158.11	2.02
1964	31.80	143.00	151.33	0.21
	79.60	126.20	134.60	0.59
	140.80			
	252.60			

Computation of seasonal indices:

Quarter	1961	1962	1963	1964	Quarterly Average	Adjusting factor	Quarterly Index
I		0.63	0.14	0.21	0.33	1.023018	33.75
II		0.91	0.45	0.59	0.65	1.023018	66.49
III	0.66	0.95	1.34		0.98	1.023018	100.25
IV	1.78	2.04	2.02		1.95	1.023018	199.48
				Total:	3.90		400.00

Average of quarterly average = $\frac{3.90}{4} = 0.975$

Adjusting factor = 4 ÷ 3.9 = 1.023018

(b) Using additive model

Year	Quarterly Rainfall	4-Quarter moving average	4-Quarter moving average centered	(5)= (2) – (4)
(1)	(2)	(3)	(4)	(5)
1961	117.80			
	83.40	131.73		
	85.10	126.58	129.15	–44.05
	240.60	144.30	135.44	105.16
1962	97.20	162.78	153.54	–56.34

Year	Quarterly Rainfall	4-Quarter moving average	4-Quarter moving average centered	(5)= (2) – (4)
	154.30	177.85	170.31	–16.01
	1590	158.53	168.19	–9.19
	300.90	136.78	147.65	153.25
1963	19.90	148.88	142.83	–122.93
	67.30	153.60	151.24	–83.94
	207.40	156.58	155.09	52.31
	319.80	159.65	158.11	161.69
1964	31.80	143.00	151.33	–119.53
	79.60	126.20	134.60	–55.00
	140.80			
	252.60			

Computation of seasonal variations:

Year / Quarter	1961	1962	1963	1964	Quarterly seasonal variation = Quarterly Average
I		–56.34	-122.93	–119.53	–99.60
II		–16.01	-83.94	–55.00	–51.65
III	–44.05	–9.19	52.31		–0.31
IV	105.16	153.25	161.69		140.03

3. Method of Least Square (Ratio to Trend Method)

This method assumes a multiplicative model and constancy of S component for a month. Initially, the trend component t is estimated by least square method and then it is eliminated by its division from the value. Then the ratio (S) averaged out to eliminate I component. This also eliminates C component by judicious selection of period of years in question. Thus seasonal indices are left with, from which seasonal indices are computed. This method is advantageous because the trend values are obtained for each time period. Shortcoming of this method is that if the data contain apparent cyclical fluctuations, then the accurate picture of trend, whether linear or non-linear cannot be depicted. This method presumes complete elimination of C and I components by process of averaging. But this can only be achieved if cyclical variations are known to be absent or do not have pronounced swings.

Example 16.7: Calculate the seasonal indices by ratio to trend method for the data given in example 16.6.

	Qr-I	Qr-II	Qr-III	Qr-IV
1961	117.80	83.40	85.10	240.60
1962	97.20	154.30	159.00	300.90
1963	19.90	67.30	207.40	319.80
1964	31.80	79.60	140.80	252.60

Solution: The following table is prepared, which is self-explanatory:

Quarter / Year	Qr-I	Qr-II	Qr-III	Qr-IV	Yearly Total	Monthly Average (Y)	Deviation from mid-year (X)	XY	X^2	Trend values
1961	117.8	83.4	85.1	240.6	526.9	43.91	–3	-131.73	9	51.16
1962	97.2	154.3	159	300.9	711.4	59.28	–1	-59.28	1	49.80
1963	19.9	67.3	207.4	319.8	614.4	51.20	1	51.20	1	48.43
1964	31.8	79.6	140.8	252.6	504.8	42.07	3	126.20	9	47.07
					Total:	196.46	0.00	-13.61	20.00	

The straight line equation fitted: Y = a + bX

Normal equations are: $\sum Y = na + b\sum X$

$$\sum XY = a\sum X + b\sum X^2$$

i.e., 196.46 = 4a + b 0

–13.61 = a 0 + b 20

By solving we get, a = 49.11 and b = –0.680

Thus the estimated trend line is: Y = 49.11 – 0.68X

The trend values are obtained by substituting –3, –1, +1 and +3 successively in the above equation and entered in the Col-11 of the above table.

Annual change is: b = –0.68

So, quarterly change = $\frac{-0.68}{4} = -0.17$

Calculation of Quarterly Trend value

In the table above, the trend value for the year 1961 is 51.16, which corresponds to mid-period between 2nd and 3rd quarter. So the quarterly change of -0.17

should be divided equally between 2nd and 3rd quarter to calculate the trend values for the 2nd and 3rd quarter.

So the trend value

for 2nd quarter is $51.16 - \frac{-0.17}{2} = 51.2408$

for 3rd quarter is $51.16 + \frac{-0.17}{2} = 51.0707$

for 1st quarter is 51.2408 – (–0.17)=51.41

for 4th quarter is 51.0707 + (–0.17) = 50.9006

In the same manner, the trend values for different quarters of the year 1962, 1963 and 1964 are calculated, which are given below.

Trend Values

Quarter / Year	Qr-I	Qr-II	Qr-III	Qr-IV
1961	51.411	51.241	51.071	50.901
1962	50.050	49.880	49.710	49.540
1963	48.689	48.519	48.349	48.179
1964	47.328	47.158	46.988	46.818

Seasonal Indices

Quarter ÷ Year	Qr-I	Qr-II	Qr-III	Qr-IV	
1961	117.851.411 =2.291	83.451.241 =1.628	85.151.071 =1.666	240.6 ÷ 50.901 = 4.727	
1962	97.2÷50.05 =1.942	154.3÷49.55 =3.093	159.0÷49.71 =3.199	300.9 ÷ 49.54 = 6.074	Average of averages = (1.629 + 1.949 + 3.038 + 5.708) ÷ 4 =3.006
1963	19.9÷48.689 =0.409	67.3÷48.519 =1.387	207.4÷48.349 =4.290	319.8 ÷ 48.179 = 6.638	
1964	31.8÷47.328 =0.672	79.6÷47.158 =1.688	140.8÷46.988 =2.996	252.6 ÷ 46.818 = 5.395	
Total:	5.314	7.796	12.151	22.834	
Average:	5.314÷4=1.329	7.796÷4=1.949	12.151÷4=3.038	22.834 ÷ 4 = 5.708	
Seasonal Index	$\frac{1.329}{3.006} \times 100 =$ = 44.196	$\frac{1.949}{3.006} \times 100 =$ = 64.839	$\frac{3.038}{3.006} \times 100 =$ = 101.059	$\frac{5.708}{3.006} \times 100 =$ = 189.907	

4. Link Relative Method

The link relatives are expressed as percentages by dividing the figures for each period with the figure for the previous period. Then the average of the link averages are calculated for each period by any method of averaging like arithmetic mean, median etc. Next is to chain the averages of link relatives on the basis of the first period. The chain relative is given by: (Link relative for that period Chain relative for the preceding period) / 100. Initially the chain relative for the first period is taken as 100. Then the chain relative for the first period is calculated on the basis of the last period. Its difference is taken from the assumed chain relative of the first period i.e. 100. This difference is then distributed among different periods because the difference is the result of long term effect of changes. If the year has k periods and D is the difference, then the subtraction will be $t\frac{D}{k}$ from the $(t+1)^{th}$ period and thus the corrected chain relatives are obtained. Finally, the average of the corrected chain relatives are calculated and the corrected chain relatives are expressed as percentages of this average. These percentages thus obtained are seasonal indices.

The link average method uses data more completely than the moving average method. But the calculations are very tedious.

Example 16.8: Calculate the seasonal indices by the method of link relatives using the data given in the Example 16.7.

Quarter / Year	Qr-I	Qr-II	Qr-III	Qr-IV
1961	117.8	83.4	85.1	240.6
1962	97.2	154.3	159	300.9
1963	19.9	67.3	207.4	319.8
1964	31.8	79.6	140.8	252.6

Solution: The following table is prepared, which is self-explanatory.

Quarter / Year	Qr-I	Qr-II	Qr-III	Qr-IV
1961	—	83.4 ÷ 117.8100 = 70.79	85.1 ÷ 83.4100 = 102.03	240.6 ÷ 85.1100 = 282.72
1962	97.2 ÷ 240.6100 = 40.39	154.3 ÷ 97.2100 = 158.74	159 ÷ 154.3100 = 103.04	300.9 ÷ 159100 = 189.24
1963	19.9 ÷ 300.9100 = 6.61	67.3 ÷ 19.9100 = 338.19	207.4 ÷ 67.3100 = 308.17	319.8 ÷ 207.4100 = 154.19

Quarter / Year	Qr-I	Qr-II	Qr-III	Qr-IV
1964	31.8 ÷ 319.8100 = 9.94	79.6 ÷ 31.8100 = 250.31	140.8 ÷ 79.6100 = 176.88	252.6 ÷ 140.8100 = 179.40
Arithmetic Average:	(40.399 + 6.613 + 9.944) ÷ 3 = 18.98	(70.798 + 158.745 + 338.191 + 250.314) ÷ 4 =204.51	(102.038 + 103.046 + 308.172 + 176.884) ÷ 4 = 172.54	(282.726 + 189.245 + 154.195 + 179.403) ÷ 4 = 201.39
Chain Relatives:	100 (Assumed)	204.51100÷100 =204.51	172.54204.51÷100 =352.86	201.39352.86÷100 =710.62
Corrected Chain Relatives:	100	204.51 – 1 8.72^ =195.79	352.86 – 2 8.72^ =335.42	710.62 – 3 8.72^ =684.46
Average of corrected chain relatives: $\frac{100+195.79+335.42+684.46}{4} = 328.92$				
Seasonal Index:	100÷328.92100 =30.403	195.79÷328.92100 =59.525	335.42÷328.92100 =101.976	684.46÷328.92100 =208.093

Chain relative of the first quarter on the basis of the last quarter is given by: $\frac{18.98 \times 710.62}{100} = 134.88$

So the total difference works out to be: 134.88 – 100 = 34.88

Difference per quarter is: $\frac{34.88}{4} = 8.72$

$$\text{Seasonal Index} = \frac{\text{Corrected chain relative}}{\text{Average of corrected chain relatives}} \times 100$$

$$= \frac{\text{Corrected chain relative}}{328.92} \times 100$$

Note: The sum of all the seasonal indices becomes: 30.403 + 59.525 + 101.976 + 208.093 = 400, which is a check against the calculations done above. In case of monthly indices, this sum should be equal to 1200. In case, this does not happen, then the calculated seasonal indices are normalised by multiplying with $\frac{400}{\text{Sum of indices}}$ [for quarterly indices];

or $\frac{1200}{\text{Sum of indices}}$ [for monthly indices] ·

While determining seasonal index, a point of caution needs to be mentioned here. Unless there is sufficient reason to believe that the data possess a trend which can be explained by a mathematical equation, the moving average method should be preferred.

Applications of seasonal indices have certain limitations. The assumptions made in its computation sometimes may not hold good. There may be influences of random fluctuations on the seasonal indices. Moreover, the computed seasonal indices represent an average pattern for the years, which may not be exactly expected to get repeated in a particular year.

Notwithstanding its limitations, seasonal indices have many advantages. It is useful to control the business operations during current year and it provides an aid to understand the current monthly variations by comparing with corresponding seasonal indices. It is helpful in business planning, economic forecasting and managerial control. It is also helpful in scheduling purchases, personnel requirement, seasonal financing, labour intake and advertisement programs etc.

DATA DE-SEASONALISING

In case of multiplicative model: Deseasonalised data $= \dfrac{\text{Original data}}{\text{Seasonal data}} \times 100$

In case of additive model: De-seasonalised data

$$= \text{Original data} - \frac{\text{Seasonal index}}{100}$$

Example 16.9: Deseasonalise the data given in example 16.6 using seasonal indices computed by ratio to moving average method.

Solution: First solve the problem given in example 16.6 by ratio to moving average method and then the following is the result obtained.

Quarter	1961	1962	1963	1964	Quarterly Average	Quarterly Index
I		0.63	0.14	0.21	0.33	0.33÷0.975100 =33.58
II		0.91	0.44	0.59	0.65	0.65÷0.975100 =66.39
III	0.66	0.95	1.34		0.98	0.98÷0.975100 =100.54
IV	1.78	2.04	2.02		1.95	1.95÷0.975100 =199.50

Now the values of each quarter are divided by the corresponding seasonal indices and multiplied by 100 to get the deseasonalised data as given in the following table.

Quarter /Year	Qr-I	Qr-II	Qr-III	Qr-IV
1961	117.80 ÷ 33.58100 = 350.80	83.40 ÷ 33.58100 = 248.36	85.10 ÷ 33.58100 = 253.42	240.60 ÷ 33.58100 = 716.50
1962	97.20 ÷ 66.39100 = 146.41	154.30 ÷ 66.39100 = 66.39	159.00 ÷ 66.39100 = 146.41	300.90 ÷ 66.39100 = 66.39
1963	19.90 ÷ 100.54100 = 19.79	67.30 ÷ 100.54100 = 66.94	207.40 ÷ 100.54100 = 206.29	319.80 ÷ 100.54100 = 318.08
1964	31.80 ÷ 199.50100 = 15.94	79.60 ÷ 199.50100 = 199.50	140.80 ÷ 199.50100 = 15.94	252.60 ÷ 199.50100 = 199.50

Example 16.10: Sale of a certain text book and the seasonal indices at a book shop are given below. During the first quarter the sale was found to be Rs 20,000. Determine how much worth of books should the shop-owner keep ready to meet the demand of book lovers in succeeding quarters.

Quarter	Seasonal Index (%)
Jan-Mar	20
Apr-Jun	35
Jul-Sep	270
Oct-Dec	75

Solution: The normal sale during Jan-Mar is: 20000 ÷ 20 × 100 = Rs. 100000

Hence the estimated worth of the book to be kept in stock during each succeeding quarter are:

Quarter	Seasonal Index (%)	Estimated sales of book (Rs)
Apr-Jun	35	10000035 ÷ 100 = 35000
Jul-Sep	270	100000270 ÷ 100 = 270000
Oct-Dec	75	10000075 ÷ 100 = 75000

Example 16.11: The seasonal indices of sale of ice-cream in a shop are as follows. The average yearly sale is estimated to be Rs. 4,50,550. Estimate monthly sales of the shop if there is no trend component,

Month	Seasonal Index	Month	Seasonal Index
Jan	62	Jul	108
Feb	87	Aug	103
Mar	102	Sep	93
Apr	127	Oct	88
May	157	Nov	73
Jun	137	Dec	63

Solution: Average monthly sale = 450550 ÷ 12 = 37545.83

Estimated sale for a particular month = Average monthly sale Seasonal Index / 100

Month	Seasonal Index	Estimated sales = Average monthly sale Seasonal Index ÷ 100
Jan	62	37545.83 × 62 ÷ 100 = 23278.42
Feb	87	37545.83 × 87 ÷ 100 = 32664.87
Mar	102	37545.83 × 102 ÷ 100 = 38296.75
Apr	127	37545.83 × 127 ÷ 100 = 47683.20
May	157	37545.83 × 157 ÷ 100 = 58946.95
Jun	137	37545.83 × 137 ÷ 100 = 51437.79
Jul	108	37545.83 × 108 ÷ 100 = 40549.50
Aug	103	37545.83 × 103 ÷ 100 = 38672.20
Sep	93	37545.83 × 93 ÷ 100 = 34917.62
Oct	88	37545.83 × 88 ÷ 100 = 33040.33
Nov	73	37545.83 × 73 ÷ 100 = 27408.46
Dec	63	37545.83 × 63 ÷ 100 = 23653.87

Example 16.12: The quarterly trend equation of the sales of certain commodity in a shop during 2001-05 is given by: $y = 25 + 1.4t + 0.4t^2$; where y=quarterly sale in '000 rupees, t=quarterly time periods(origin taken at 2nd quarter of 2003). Estimate the quarterly sales for the quarter of 2006.

Quarter:	I	II	III	IV
Seasonal Index:	130	70	40	160

Solution: Assuming a multiplicative model, the problem is solved and the following table is prepared, which is self-explanatory.

Quarter of 2006	Seasonal Index	t	Trend value = y = 25+1.4t+0.4t²	Estimated Sales ('000 Rs)
I	130	10	79	79 × 130÷100 = 102.70
II	70	11	88.8	88.8 × 70 ÷ 100 = 62.16
III	40	12	99.4	99.4 × 40 ÷ 100 = 39.76
IV	160	13	110.8	110.8 × 160 ÷ 100 = 177.28

Seasonal variation makes a complete cycle within each year and does not affect one year any more than another. But there is another type of variation in the time series that oscillates above and below the trend line for periods longer than a year, which is called *Cyclical Variation*. Cyclical variation is measured by two methods, i.e. (a) The first method expresses the variation as a percentage of the trend (b) The second method (*relative cyclical residual*) calculates the variation as a percent deviation from the trend. The methods are discussed below.

C. Cyclical Variation

Only secular trend, cyclical and irregular components are considered in a time series depicting the annual data, because seasonal variation makes a complete, regular cycle within each year and thus does not affect one year any more than another. Secular trend can be described using a trend line and hence it is possible to isolate the remaining cyclical and irregular components from the trend. It is assumed that the cyclical component explains most of the variation left unexplained by the trend component, though in real life it is not true. However, methods are available for analysis of cyclical component for such time series like Fourier analysis and spectral analysis, which are beyond the scope of this book. Moreover, predicting the cyclical variation is also not discussed in the book. Procedure for describing the past cyclical variations only is discussed here.

The two ways by which the cyclical variation can be measured are (a) *Percent of Trend* method and (b) *Relative Cyclical Residual* method described below.

(a) In a time series comprising annual data, the fraction of the trend is obtained by dividing the actual value(Y) by the corresponding trend value for each value, which when multiplied by 100 to gives the percent of trend.

$$\text{Percent of Trend} = \frac{Y}{Y_e} \times 100$$

Y = actual time series value

Y_e = estimated trend value for the same point in the time series.

(b) In the relative cyclical residual method, the percentage deviation from the trend is found for each value by using the formula:

$$\text{Relative Cyclical Residual} = \frac{Y - Y_e}{Y_e} \times 100$$

The problem taken up below explains the methods.

Example 16.13: A gas company has supplied 0.612, 0.680, 0.714, 0.850 and 0.884 billion cubic metre of gas respectively during the years 2001 to 2005.

(a) Find the linear estimating equation that best describes these data.

(b) Calculate the percent to trend for these data.

(c) Calculate the relative cyclical residual for these data.

(d) In which years does the largest fluctuation from trend occurred? Is it the same for both methods?

Solution: Students are advised to develop the linear regression by themselves.

Year (X)	x	Y	xY	x^2	Ye	$\frac{Y}{Y_e} \times 100$	$\frac{Y - Y_e}{Y_e} \times 100$
(1)	(2)	(3)	(4)	(5)	(6)	(7)	(8)
2001	–2	0.612	–1.224	4	0.0652	938.65	838.65
2002	–1	0.680	–0.680	1	0.6766	100.50	0.502
2003	0	0.714	0.000	0	0.748	95.45	–4.545
2004	1	0.850	0.850	1	0.8194	103.73	3.734
2005	2	0.884	1.768	4	0.8908	99.24	–0.763
Sum:	0	3.740	0.714	10			

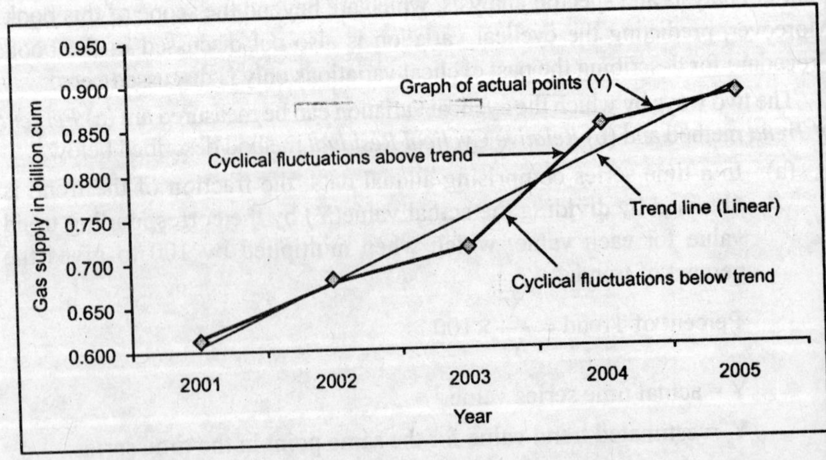

(a) $a = \bar{Y} = \frac{3.740}{5} = 0.748$

$$b = \frac{\sum xY}{\sum x^2} = \frac{0.714}{10} = 0.0714$$

$$Y_e = 0.748 + 0.0714x$$

(b) Percent to trend are calculated in Col. 7 in above table.

(c) Relative cyclical residual are calculated in Col-8 in above table.

(d) Largest fluctuation (by both methods) was during 2003.

D. Irregular Variation

The final component of a time series is the irregular variation. This is the left over unpredictable factor in the series after successive removal of trend, seasonal and cyclical components. Irregular variation occurs over short intervals and follows a random pattern. Because of its unpredictability, no mathematical explanation is attached to it. In most cases, irregular variation is difficult, if not impossible to predict, but no attempt is made to fit a line to account for this. The variate difference method provides a method to estimate the random component of a time series.

Let us now take up a problem, which involve all four components of a time series.

Example 16.14: Given below are the sales figures of a garment shop in ('000 Rupees) for the period from 2001 to 2005. De-seasonalise the time series. Develop the trend line. Find the cyclical variation around the trend line. Plot the data and predict the sale for the year 2006.

Year	Quarter	Quarterly Sales	Year	Quarter	Quarterly Sales
2001	I	15	2004	I	16
	II	20		II	24
	III	8		III	10
	IV	17		IV	20
2002	I	14	2005	I	17
	II	19		II	25
	III	9		III	13
	IV	17		IV	24

Year	Quarter	Quarterly Sales	Year	Quarter	Quarterly Sales
2003	I	16			
	II	23			
	III	12			
	IV	21			

Solution: The following tables are prepared, which are self-explanatory. First the data are deseasonalised, then the trend line is found and then the cyclical variation around the trend line is found. The trend line is extended to find the sales figures for the year 2006.

Year	Quarter	Actual Sales	4-Quarter Moving Total	4-Quarter Moving Average	4-Quarter Centred Moving Average	Percentage of Actual to Moving Average
(1)	(2)	(3)	(4)	(5)	(6)	(7) = (3) ÷ (6) 100
2001	I	15	—			
	II	20	—			
	III	8	60	15.00	14.875	53.78
	IV	17	59	14.75	14.625	116.24
2002	I	14	58	14.50	14.625	95.73
	II	19	59	14.75	14.750	128.81
	III	9	59	14.75	15.000	60.00
	IV	17	61	15.25	15.750	107.94
2003	I	16	65	16.25	16.625	96.24
	II	23	68	17.00	17.500	131.43
	III	12	72	18.00	18.000	66.67
	IV	21	72	18.00	18.125	115.86
2004	I	16	73	18.25	18.000	88.89
	II	24	71	17.75	17.625	136.17
	III	10	70	17.50	17.625	56.74
	IV	20	71	17.75	17.875	111.89

Year	Quarter	Actual Sales	4-Quarter Moving Total	4-Quarter Moving Average	4-Quarter Centred Moving Average	Percentage of Actual to Moving Average
2005	I	17	72	18.00	18.375	92.52
	II	25	75	18.75	19.250	129.87
	III	13	79	19.75	—	
	IV	24	—	—	—	

Year/ Quarter	2001	2002	2003	2004	2005	Quarterly Average	Seasonal Index
(1)	(2)	(3)	(4)	(5)	(6)	(7) = Average (2):(6)	(8) = (7) ÷ 0.993 100
I		0.96	0.96	0.89	0.93	0.933432	94.0032
II		1.29	1.31	1.36	1.30	1.315706	132.5008
III	0.54	0.60	0.67	0.57		0.592964	59.71567
IV	1.16	1.08	1.16	1.12		1.129815	113.7803
					Sum:	3.972	400

$$\text{Average of Quarterly Average} = \frac{3.972}{4} = 0.993$$

A linear trend is assumed.

Year	Quarter	Actual Sales	Seasonal Index	Deseasonalised Sales (Y)	Coding the Time Variable	x	xY	x²
(1)	(2)	(3)	(4)	(5)=(3) ÷ (4)100	(6)	(7) = (6)2	(8) = (7) (5)	(9) = (7)^2
2001	I	15	94.00	15.96	-9.5	-19	-303.18	361
	II	20	132.50	15.09	-8.5	-17	-256.60	289
	III	8	59.72	13.40	-7.5	-15	-200.95	225
	IV	17	113.78	14.94	-6.5	-13	-194.23	169
2002	I	14	94.00	14.89	-5.5	-11	-163.82	121
	II	19	132.50	14.34	-4.5	-9	-129.06	81
	III	9	59.72	15.07	-3.5	-7	-105.50	49
	IV	17	113.78	14.94	-2.5	-5	-74.71	25

Year	Quarter	Actual Sales	Seasonal Index	Deseasonalised Sales (Y)	Coding the Time Variable	x	xY	x^2
2003	I	16	94.00	17.02	-1.5	-3	-51.06	9
	II	23	132.50	17.36	-0.5	-1	-17.36	1
	III	12	59.72	20.10	0.5	1	20.10	1
	IV	21	113.78	18.46	1.5	3	55.37	9
2004	I	16	94.00	17.02	2.5	5	85.10	25
	II	24	132.50	18.11	3.5	7	126.79	49
	III	10	59.72	16.75	4.5	9	150.71	81
	IV	20	113.78	17.58	5.5	11	193.36	121
2005	I	17	94.00	18.08	6.5	13	235.10	169
	II	25	132.50	18.87	7.5	15	283.02	225
	III	13	59.72	21.77	8.5	17	370.09	289
	IV	24	113.78	21.09	9.5	19	400.77	361
			Sum:-	340.84			423.93	2660.00

$$a = Y_e = \frac{\sum Y}{n} = \frac{340.84}{20} = 17.04$$

$$b = \frac{\sum xY}{\sum x^2} = \frac{423.93}{2660.00} = 0.16$$

Equation of Trend Line is: 17.04 + 0.16X

The trend values are calculated below:

Year	Quarter	Actual Sales	x	Deseasonalised Sales (Y)	$Y_e = 17.04 + 0.16 \times$ Col(4)	Percent of Trend $\frac{Y}{Y_e} \times 100$
(1)	(2)	(3)	(4)	(5)	(6)	(7)
2001	I	15	-19	15.96	14.01	113.87
	II	20	-17	15.09	14.33	105.31
	III	8	-15	13.40	14.65	91.44
	IV	17	-13	14.94	14.97	99.81

Year	Quarter	Actual Sales	x	Deseasonalised Sales (Y)	$Y_e = 17.04 + 0.16 \times$ Col(4)	Percent of Trend $\frac{Y}{Y_e} \times 100$
2002	I	14	-11	14.89	15.29	97.41
	II	19	-9	14.34	15.61	91.88
	III	9	-7	15.07	15.93	94.63
	IV	17	-5	14.94	16.25	91.97
2003	I	16	-3	17.02	16.56	102.76
	II	23	-1	17.36	16.88	102.82
	III	12	1	20.10	17.20	116.82
	IV	21	3	18.46	17.52	105.35
2004	I	16	5	17.02	17.84	95.41
	II	24	7	18.11	18.16	99.76
	III	10	9	16.75	18.48	90.64
	IV	20	11	17.58	18.79	93.52
2005	I	17	13	18.08	19.11	94.62
	II	25	15	18.87	19.43	97.09
	III	13	17	21.77	19.75	110.22
	IV	24	19	21.09	20.07	105.10

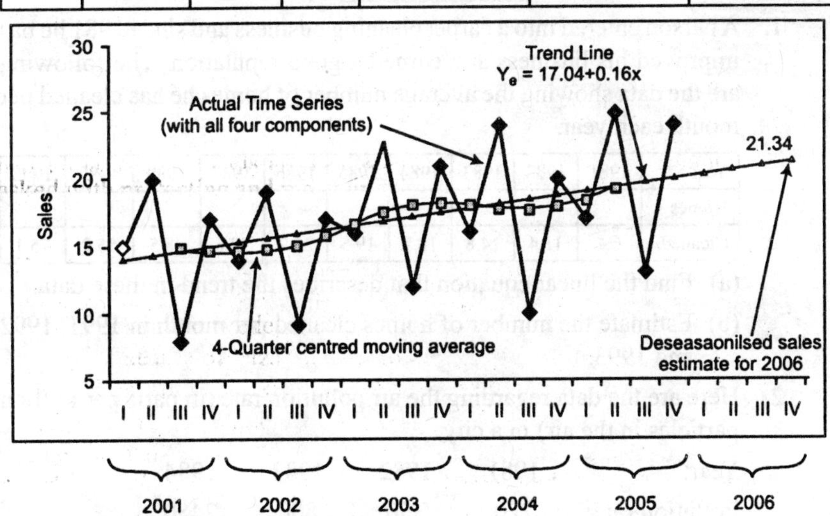

The coded figures for the four quarters of year 2006 (X-axis values) are 21, 23, 25 and 27 respectively. These values are put in the trend equation and the predictive deseasonalised values are- 20.39, 20.71, 21.03 and 21.34 for the four quarters of the year 2006 respectively (read from the graph). These figures are now to be seasonalised by multiplying with the respective seasonal indices expressed as fractions of 100, which are given below:-

Year	Quarter	De-seasonalised sales figures
2006	I	$94.0032 \times \frac{20.39}{100}$ = 19.167 thousand rupees
	II	$132.5008 \times \frac{20.71}{100}$ = 27.44 thousand rupees
	III	$59.71567 \times \frac{21.03}{100}$ = 12.558 thousand rupees
	IV	$113.7803 \times \frac{21.34}{100}$ = 24.280 thousand rupees

It is to be noted that these values of 2006 are estimates only, which do not take into account the cyclical and irregular components. Because, irregular component, due to its irregular nature, cannot be predicted mathematically and cyclical variation is descriptive of past behaviour, which cannot be used for future prediction.

EXERCISES

1. A person entered into a carpet cleaning business and since 1981 he has improved his business and earned a good reputation. The following are the data showing the average number of homes he has cleaned per month each year.

Year	1981	1982	1983	1984	1985	1986	1987	1989	1989	1990
Homes										
Cleaned:	6.4	11.4	14.8	18.5	19.5	25.7	32.6	48.5	55.3	75.1

   (a) Find the linear equation that describes the trend in these data.

   (b) Estimate the number of homes cleaned per month in 1991, 1992 and 1993.

2. Here are the data regarding the air pollution rate (in parts per million particles in the air) in a city.

Year:	1981	1982	1983	1984
Pollution rate:	220	350	800	2450

(a) Develop a linear estimating equation that best describes. Plot the line on a graph.

(b) Develop a second-degree estimating equation that best describes these data. Plot the curve on a graph.

3. An environmental agency has been watching a city's air quality over a 5-year period and has assembled the following seasonal data regarding amount of contaminants (in parts per million) in the air.

Year	Winter	Spring	Summer	Rainy
2001	450	380	330	385
2002	475	395	355	395
2003	495	405	375	415
2004	505	425	395	435
2005	525	455	425	485

(a) Determine the seasonal indices and deseasonalize these data (using a 4-quarter centered moving average)

(b) Calculate the least squares line that best describes these data.

(c) Identify the cyclic variation in these data by the relative cyclical residual method.

(d) Plot the original data, deseasonalized data and the trend.

4. The following data describes the marketing performance of a regional manager.

Sales by Quarter (in lakh rupees)

Year	I	II	III	IV
1991	20	25	39	26
1992	22	29	45	24
1993	24	32	42	24
1994	25	35	48	22

(a) Calculate the seasonal indices for these data. (Use a 4-quarter centred moving average)

(b) De-seasonalize these data using the indices from above.

(c) Find the least square line that best describes the trend in deseasonalized sales.

(d) Identify the cyclical component in this time series by computing the percent of trend.

INDEX NUMBERS

An index number measures how much a variable changes over time. For example- the police department may want to know the increase or decrease in road accidents, theft cases etc. in the city during the past 10 years. A man may want to know how much the price of daily consumables has increased so that he can adjust his budget accordingly. A manager of an industry may wish to compare the production cost of current month with that of the past 6 months. A medical expert team may want to compare the number of dengu fever cases reported this year with the number reported in previous year. In each of these phenomena, the degree of change is determined and defined. Characteristically, index numbers are used to measure such changes.

Usually, an index measures change in a variable over a period of time, such as a time series. However, it can also be used to measure the differences in a given variable in different locations. This is done by simultaneously collecting data in different locations. The comparative cost of living index, for example, shows that it is cheaper to live in Hyderabad than in Mumbai in terms of cost of goods and services.

The index number is calculated by finding the ratio of the current value to a base value. The year for which this base value is considered is called base period. Then multiplying this ratio with 100 gives the index as a percentage. This is called the *percentage relative*. The index number for the base period is always 100.

TYPES OF INDEX NUMBERS

There are four types of index numbers:

1. **Price Index Number:** This is the most frequently used index number. It compares the levels of prices from one period to another. The prices may be for one commodity or a group of commodities.

   Most familiar is the *Consumer Price Index* (CPI), tabulated by Labour Bureau, Ministry of Labour, Government of India. Consumer Price Index number, otherwise called as cost of living index number measures changes in cost of living of a particular class of people stationed at a particular time. Computation of CPI is done at the retail prices of commodities consumed by the class of people for whom it is meant. Though Government of India publishes the cost of living index numbers, State Governments have their own compilations too.

Retail Price Index Number measures changes in retail prices and is affected by economy of a country. Labour Bureau, Government of India compiles index numbers of retail prices both for urban and rural outlets.

Wholesale Price Index Number (WPI) measures the change in price level of a whole country. Prices are collected from wholesale dealers only and indications of price movements in all markets of the country other than retail markets are computed.

2. **Quantity Index Number:** This measures how much the number or quantity of a variable changes temporally or spatially. Quantity index numbers measure changes in the volume of production in different sectors of economy with respect to time period or space. Usually the commodities, which are subjected to considerable price fluctuations, are measured by quantity index numbers, e.g. agricultural produces. When inflation occurs, the quantity index number provides a more reliable measure of change in actual output and finished products than corresponding price or value index.
3. **Value Index Number:** This measures the changes in total monetary worth. As the name indicates, value means per unit price multiplied by quantity. In fact, the value index combines price and quantity changes. It does not distinguish between the effects of these two components and hence it presents a more informative index.
4. **Special Purpose Index Number:** This falls into a special category, where the index numbers are constructed for specific purposes like export indices, import indices, development indices, literacy indices etc.

WHOLESALE PRICE INDEX Vs. CONSUMER PRICE INDEX

- These two are different, because there is a lagging behind between the movement of wholesale and retail prices.
- Undeniably, the cost of living index for the daily labourer class people in Delhi is not the same as that in other parts of the country, whereas there is one wholesale price index for the whole country.

- Consumer price index numbers change erratically in comparison to wholesale price index numbers, because retail prices required for computation of the former increase at a faster rate than that of the later.
- The wholesale price index is based on different weighting systems and has a different coverage as regards the commodities compared to consumer price index number.

USED OF INEDX NUMBERS

The most common use of index number is as an end result.

- Index number measures the value of money, which is determined by its purchasing power, which in turn is determined by prices of various commodities (usually called as basket of goods) determined through a price index.
- Index numbers help in studying changes in volume of trade, production, sales, profits etc. over a period of time.
- The Consumer Price Index is often used as an indicator to cite the nation's economic condition.
- Management uses index numbers as part of an intermediate computation to understand other information in better ways.
- In chapter on time series, the seasonal indices were used to modify and improve the estimates of the future.
- Index numbers are being widely used not only in analysis of economic data and business situation, but also in many other social and natural sciences.
- The market analysts use index numbers to study market situations both with respect to time and place.
- Planning process of economic policies makes use of various kinds of index numbers such as wholesale price index numbers, cost of living index numbers, retail price index numbers, index numbers for profits, subsidies, credits, wage rates, index numbers for industrial and agricultural productivity.
- Consumer price index is used as a yardstick for an increase in wage rates and dearness allowances to employees.
- Price index numbers provide the simplest method to study the effect of the Government policies on the rising prices.

- Government often devises ways and means to control the general prices by observing the wholesale price index numbers.
- The direction in which the country's economy is moving can be known by studying the index numbers. For example- the index numbers of exports and imports for a span of say 20 years, indices of industrial and agricultural production etc.
- In health sector also, the health and family welfare indices help in evaluating fruitfulness of various Government launched health programs.
- For this reason, index numbers are rightly termed as *Economic Barometers* in a country's economy.

PROBLEMS RELATED TO INDEX NUMBERS

Several factors tend to distort index numbers. The most common causes of distortion are:

1. **Inappropriate Weighting Factor:** In developing a composite index, for example the consumer price index, care must be taken in considering the fact that changes in some variable are more important than changes in others. Take for example- An increase of petrol price by Rs.5.00 per liter at the retail end cannot be counterbalanced by decreasing the cost of car by Rs.5.00. So greater weight has to be assigned to the increased petrol price than to the decrease in the cost of cars.
2. **Improper Base Selection:** Sometimes the base period is chosen to suit the interest of a certain firm, which can distort the index numbers. For example- if the profit of a company is measured with respect to a base period when recession occurred, then this would produce an index showing that the profit has substantially increased. On the contrary, if the base period is chosen when profits are high, then this would produce an index indicating small increase in profit. So the base period should be chosen so as not to distort the index numbers. It should be a normal period free from abnormalities or extremities and not a distant past.
3. **Difficulty in Finding Suitable Data:** If it is required to find out an index describing seasonal variation in the sales of a company's products, then the annual sales data of the company will not be in much use.

4. **Incomparability of Indices:** This occurs when attempts are made to compare one index with another after there has been a basic change in what is being measured. If prices of personal computers for the period 2001 to 2010 are compared, then many fluctuations in prices are observed. But this comparison does not take into consideration about the technological advances in the electronic field over the time period in consideration.
5. **Purpose for construction:** The purpose for which the index number is to be found should be clearly defined. It will be futile to consider the wholesale prices of commodities, if cost of living/ consumer price index is to be found out, which takes into account only the retail prices. Similarly, if it is desired to construct the index number for low income group of agricultural workers, then commodities generally consumed by high income people e.g. refrigerators, air conditioners, four-wheelers are to be excluded from consideration.
6. **Selection of Items:** Index number based on inadequate number of items may make it unrepresentative whereas inclusion of large number of items may lead to heavy calculations and would be expensive and time consuming. So the sample of items to be included should be relevant, representative, reliable, adequate and comparable.

The various methods available for construction of index number may be classified as:

(A) Simple or Unweighted Index Numbers

1. Simple Aggregative Index
2. Simple Average of 'Relatives'

(B) Weighted Average Index Numbers

1. Weighted Aggregative Index
2. Weighted Average of 'Relatives'

Each of the above methods is discussed below:

1. Simple Aggregative Index

The table below shows a hypothetical example on prices of five commodities for 2001 and 2005. Price for the year 2001 (Base period) is denoted as p_0 and that in 2005 as p_1.

Commodity	Unit	2001		2005	
		Price Notation (p_0)	Price in Rs.	Price Notation (p_1)	Price in Rs.
Egg	Dozen	p_{0A}	25.00	p_{1A}	28.00
Milk	Kg	p_{0B}	8.00	p_{1B}	9.50
Rice	Quintal	p_{0C}	120.00	p_{1C}	200.00
Kerosene	Litre	p_{0D}	4.50	p_{1D}	7.00
Electric Wire	Metre	p_{0E}	11.50	p_{1E}	14.00
Total:			169.00		258.50

Now the index for the year 2005 taking 2001 as the base period will be

$I_{01} = \frac{\sum p_1}{\sum p_0} \times 100 = \frac{258.50}{169.00} \times 100 =$ 152.96 i.e. 52.96% overall price increase.

This indicates that there has been 52.96% increase in price from 2001 to 2005. But taking the commodity-wise index numbers we have,

$I_{01:A} = \frac{28.00}{25.00} \times 100 = 112.00$ i.e. 12% increase in price of egg

$I_{01:B} = \frac{9.50}{8.00} \times 100 = 118.75$ i.e. 18.75% increase in price of milk

$I_{01:C} = \frac{200.00}{120.00} \times 100 = 166.667$ i.e. 66.67% increase in price of rice

$I_{01:D} = \frac{7.00}{4.50} \times 100 = 155.56$ i.e. 55.56% increase in price of kerosene

$I_{01:E} = \frac{14.00}{11.50} \times 100 = 121.74$ i.e. 21.74% increase in price of electric wire.

It may be observed that there has been a price increase as low as 12% in case of egg and as high as 66.67% in case of rice, the index shows the overall increase of 52.96%. This wide variation has occurred because of dominance of the commodity rice.

Let us now add one more item, hand held calculator, to the above table for calculation of index number.

Commodity	Unit	2001		2005	
		Price Notation (p_0)	Price in Rs.	Price Notation (p_1)	Price in Rs.
Egg	Dozen	p_{0A}	25.00	p_{1A}	28.00
Milk	Kg	p_{0B}	8.00	p_{1B}	9.50
Rice	Quintal	p_{0C}	120.00	p_{1C}	200.00
Kerosene	Litre	p_{0D}	4.50	p_{1D}	7.00
Electric Wire	Metre	p_{0E}	11.50	p_{1E}	14.00
Hand held calculator	Number	p_{0F}	300.00	P_{1F}	199.00
Total:			469.00		457.50

Price Index will be: $\frac{457.50}{469.00} \times 100 = 97.55$

But intuitively the earlier price index of 152.95 is more accurate than the later value of 97.55. *So the major disadvantage of an unweighted index is, it does not attach greater importance, or weight, to the price change of high use item than it does to a low-volume item.* A family may purchase 50 dozens of egg a year, but it would be unusual for a family to own more than one or two calculators. So the distortion observed by using unweighted index can be rectified by using weighted index numbers, which is discussed at subsequent sections.

2. Simple Average of 'Relatives'

In this method, price relatives are calculated, which means that for each commodity, the price of the current year is expressed as percentage of the price in the base year. After this the mean value of all these percentages provides the index value. In most cases, arithmetic mean is taken, or else geometric mean is also taken as the method of averaging. Considering the arithmetic average method, the index number is: $I_{01} = \frac{\sum\left(\frac{p_1}{p_0} \times 100\right)}{n}$

Taking the previous example and using arithmetic average method:

Commodity	Unit	2001		2005		$\frac{p_1}{p_0}\times 100$
		Price Notation (p_0)	Price in Rs	Price Notation (p_1)	Price in Rs	
(A) Egg	Dozen	p_{0A}	25.00	p_{1A}	28.00	2825100 =112
(B) Milk	Kg	p_{0B}	8.00	p_{1B}	9.50	9.50 ÷ 8100 = 118.75
(C) Rice	Quintal	p_{0C}	120.00	p_{1C}	200.00	200 ÷ 120100 = 166.67
(C) Kerosene	Litre	p_{0D}	4.50	p_{1D}	7.00	7 ÷ 4.5100 = 155.56
(D) Electric Wire	Metre	p_{0E}	11.50	p_{1E}	14.00	14 ÷ 11.5100 =121.74
					Total:-	674.72

$$I_{01} = \frac{\sum\left(\frac{p_1}{p_0}\times 100\right)}{n} = \frac{674.72}{5} = 134.94,$$ showing there is 34.94% price increase in 2005.

Taking the previous example and using geometric average method:

Commodity	Unit	2001		2005		$I=\frac{p_1}{p_0}\times 100$	log(I)
		Price Notation (p_0)	Price in Rs	Price Notation (p_1)	Price in Rs		
(A) Egg	Dozen	p_{0A}	25.00	p_{1A}	28.00	28÷25×100 =112	log(112) =4.718
(B) Milk	Kg	p_{0B}	8.00	p_{1B}	9.50	9.50÷8×100 =118.75	log(118.75) =4.777
(C) Rice	Quintal	p_{0C}	120.00	p_{1C}	200.00	200÷120×100 =166.67	log(166.67) =5.116
(D) Kerosene	Litre	p_{0D}	4.50	p_{1D}	7.00	7÷4.5X100 =155.56	log(155.56) =5.047
(E) Electric Wire	Metre	p_{0E}	11.50	p_{1E}	14.00	14÷11.5X100 =121.74	log(121.74) =4.802
						Total:-	24.46

$$I_{01} = \text{Antilog}\left(\frac{\sum \log I}{n}\right) = \text{Antilog}\left(\frac{24.46}{5}\right) = 133.22;$$

It may be noted that geometric mean or harmonic mean is preferred to arithmetic mean in case the values of items vary widely between the two periods, as the later may yield biased results.

B1: Weighted Aggregative Index

As there are a number of formulae available using this method, we will discuss only a few of them for present purpose.

Various notations to be used in the formulae are:

p_{1i} = price of ith commodity in the current year

p_{0i} = price of ith commodity in the base year

q_{1i} = quantity of ith commodity in the current year

q_{0i} = quantity of ith commodity in the base year

q_a = quantity / weight of some other period other than the base year or current year or, average quantity based on two or more years.

S. No.	Name of Formula	Formula
(a)	Laspeyre's Index or L-formula	$L_{01} = \frac{\sum p_{1i}q_{0i}}{\sum p_{0i}q_{0i}} \times 100$
(b).	Paasche's Index or P-formula	$P_{01} = \frac{\sum p_{1i}q_{1i}}{\sum p_{0i}q_{1i}} \times 100$
(c)	Walsch's Index or W-formula	$W_{01} = \frac{\sum p_1\sqrt{q_0q_1}}{\sum p_0\sqrt{q_0q_1}} \times 100$
(d)	Marshall & Edgeworth's Index or ME-formula	$ME_{01} = \frac{\sum p_{1i}q_{0i} + \sum p_{1i}q_{1i}}{\sum p_{0i}q_{0i} + \sum p_{0i}q_{1i}} \times 100$
(e)	Fisher's Ideal Index or F-formula	$F_{01} = \sqrt{L_{01}P_{01}} = \sqrt{\frac{\sum p_1q_0}{\sum p_0q_0} . \frac{\sum p_1q_1}{\sum p_0q_1}} \times 100$
(f)	Dorbish & Bowley's Index or DB-formula	$DB_{01} = \frac{L_{01} + P_{01}}{2}$

S. No.	Name of Formula	Formula
(g)	Weighted Average Relative or I-formula $(=L_{01})$	$I_{01} = \dfrac{\sum\left(\dfrac{p_{1i}}{p_{0i}}\times 100\right)W_i}{\sum W_i}$
(h)	Kelley's Index or K-formula	$K_{01} = \dfrac{\sum p_1 q_a}{\sum p_0 q_a}\times 100$

Example 17.1: Considering the above example on prices of different commodities during 2001 and 2005, find the weighted aggregative index using different formulae.

Commodity	Unit	2001		2005	
		Price (p_0)	Quantity (q_0)	Price (p_1)	Quantity (q_1)
(A) Egg	Dozen	25.00	50	28.00	40
(B) Milk	Kg	8.00	60	9.50	50
(C) Rice	Quintal	120.00	100	200.00	90
(D) Kerosene	Litre	4.50	100	7.00	75
(E) Electric Wire	Metre	11.50	500	14.00	300

Solution:

Commodity	Unit	2001		2005		p_0q_0	p_1q_0	p_1q_1	p_0q_1
		Price (p_0)	Quantity (q_0)	Price (p_1)	Quantity (q_1)				
(A) Egg	Dozen	25.00	50	28.00	40	1250	1400	1120	1000
(B) Milk	Kg	8.00	60	9.50	50	480	570	475	400
(C) Rice	Quintal	120.00	100	200.00	90	12000	20000	18000	10800
(D) Kerosene	Litre	4.50	100	7.00	75	450	700	525	337.5
(E) Electric Wire	Metre	11.50	500	14.00	300	5750	7000	4200	3450
Total:				258.5	555	19930	29670	24320	15987.5

(a) $$L_{01} = \frac{\sum p_{1i}q_{0i}}{\sum p_{0i}q_{0i}}\times 100 = \frac{29670}{19930}\times 100 = 148.871$$

(b) $$P_{01} = \frac{\sum p_{1i}q_{1i}}{\sum p_{0i}q_{1i}}\times 100 = \frac{24320}{15987.5}\times 100 = 152.119$$

(c) $= W_{01} = \dfrac{\sum p_1\sqrt{q_0q_1}}{\sum p_0\sqrt{q_0q_1}} \times 100$

Commodity	Unit	2001		2005		q_0q_1	$\sqrt{q_0q_1}$	$P_1\sqrt{q_0q_1}$	$P_0\sqrt{q_0q_1}$
		Price (p_0)	Quantity (q_0)	Price (p_1)	Quantity (q_1)				
(a) Egg	Dozen	25.00	50	28.00	40	2000	44.72136	1252.198	1118.034
(b) Milk	Kg	8.00	60	9.50	50	3000	54.77226	520.3364	438.178
(c) Rice	Quintal	120.00	100	200.00	90	9000	94.86833	18973.67	11384.2
(d) Kerosene	Litre	4.50	100	7.00	75	7500	86.60254	606.2178	389.7114
(e) Electric Wire	Metre	11.50	500	14.00	300	150000	387.2983	5422.177	4453.931
Total:								26774.59	17784.05

Hence, $W_{01} = \dfrac{26774.59}{17784.05} \times 100 = 150.554$

(d) $ME_{01} = \dfrac{\sum p_{1i}q_{0i} + \sum p_{1i}q_{1i}}{\sum p_{0i}q_{0i} + \sum p_{0i}q_{1i}} \times 100 = \dfrac{29670 + 24320}{19930 + 15987.5} \times 100 = 150.317$

(e) $F_{01} = \sqrt{L_{01}P_{01}} = \sqrt{\dfrac{\sum p_1q_0}{\sum p_0q_0} \cdot \dfrac{\sum p_1q_1}{\sum p_0q_1}} = \sqrt{148.871 \times 152.119} = 150.486$

(f) $DB_{01} = \dfrac{L_{01} + P_{01}}{2} = \dfrac{148.871 + 152.119}{2} = 150.495$

(g) Weighted average of relatives:

Commodity	Unit	2001		2005		p_0q_0	W_i (Proportions)	$p_1/p_0 \times 100$	$p_1/p_0 \times 100\ W_i$
		Price (p_0)	Quantity (q_0)	Price (p_1)	Quantity (q_1)				
(a) Egg	Dozen	25.00	50	28.00	40	1250	125	112.00	14000
(b) Milk	Kg	8.00	60	9.50	50	480	48	118.75	5700
(c) Rice	Quintal	120.00	100	200.00	90	12000	1200	166.67	200000
(d) Kerosene	Litre	4.50	100	7.00	75	450	45	155.56	7000
(e) Electric Wire	Metre	11.50	500	14.00	300	5750	575	121.74	70000
Total:-					555		1993	674.71	29670

The amounts purchased or the values of commodities in the base year are in proportions of:

25 × 50 : 8 × 60 : 120 × 100 : 4.5 × 100 : 11.5 × 500

i.e. 1250 : 480 : 12000 : 450 : 5750

i.e. 125 : 48 : 1200 : 45 : 575

$$I_{01} = \frac{\sum\left(\frac{p_{1i}}{p_{0i}} \times 100\right)W_i}{\sum W_i} = \frac{29670}{1993} = 148.871 \text{ which is also equal to } L_{01}.$$

CONSIDERING A GOOD INDEX NUMBER

1. **Time Reversal Test:** This test signifies that the index number of year 'b' with base year 'a' should be reciprocal of the index number of year 'a' with base year 'b'. Hence,

$$I_{ab} \times I_{ba} = \frac{p_b}{p_a} \times \frac{p_a}{p_b} = 1$$

That means, the difference in the base year should not make any difference to the index numbers except one of scale. This property is always satisfied by any price relative. But the index number calculated by using any one of the above described methods does not guarantee that this test be satisfied.

The following Index Numbers satisfy the time-reversal test:

(i) Fishers' Ideal Index satisfies time reversal test. That means,

$$F_{01} = \sqrt{\frac{\sum p_1 q_0}{\sum p_0 q_0} . \frac{\sum p_1 q_1}{\sum p_0 q_1}} \times 100$$

$$F_{10} = \sqrt{\frac{\sum p_0 q_1}{\sum p_1 q_1} . \frac{\sum p_0 q_0}{\sum p_1 q_0}} \times 100$$

$$F_{01} \text{ X } F_{10} = \sqrt{\frac{\sum p_1 q_0}{\sum p_0 q_0} . \frac{\sum p_1 q_1}{\sum p_0 q_1}} \times \sqrt{\frac{\sum p_0 q_1}{\sum p_1 q_1} . \frac{\sum p_0 q_0}{\sum p_1 q_0}} = 1$$

(ii) Marshall-Edgeworth's Index satisfies time-reversal test. That means,

$$M_{01} = \frac{\sum p_1(q_0 + q_1)}{\sum p_0(q_0 + q_1)} \times 100$$

$$M_{10} = \frac{\sum p_0(q_0 + q_1)}{\sum p_1(q_0 + q_1)} \times 100$$

$$M_{01} \times M_{10} = \frac{\sum p_1(q_0+q_1)}{\sum p_0(q_0+q_1)} \times \frac{\sum p_0(q_0+q_1)}{\sum p_1(q_0+q_1)} = 1$$

(iii) Index number based on simple aggregative method satisfies time-reversal test. That means,

$$I_{01} \times I_{10} = \frac{\sum p_1}{\sum p_0} \times \frac{\sum p_0}{\sum p_1} = 1$$

(iv) Index number based on simple averages of relatives, where geometric mean of price relatives is used, satisfies time-reversal test. That means,

$$I_{01} \times I_{10} = \text{Antilog of}\left(\frac{\sum \log \frac{p_1}{p_0}}{n}\right) \times \text{Antilog of}\left(\frac{\sum \log \frac{p_0}{p_1}}{n}\right) = 1$$

Or, symbolically:

$$I_{01} \times I_{10} = \prod\left(\frac{p_1}{p_0}\right)^{\frac{1}{n}} \times \prod\left(\frac{p_0}{p_1}\right)^{\frac{1}{n}} = 1$$

(v) Weighted geometric mean of price relatives satisfies time-reversal test. That means,

$$I_{01} \times I_{10} = \left[\prod\left(\frac{p_1}{p_0}\right)^{W_i}\right]^{\frac{1}{\sum W_i}} \times \left[\prod\left(\frac{p_0}{p_1}\right)^{W_i}\right]^{\frac{1}{\sum W_i}} = 1$$

The following Index Numbers do not satisfy the time-reversal test:

(i) Laspeyre's Index number do not satisfy time-reversal test.

$$L_{01} \times L_{10} = \frac{\sum p_{1i}q_{0i}}{\sum p_{0i}q_{0i}} \times \frac{\sum p_{0i}q_{1i}}{\sum p_{1i}q_{1i}} \neq 1$$

(ii) Paasche's Index number do not satisfy time-reversal test.

$$P_{01} \times P_{10} = \frac{\sum p_{1i}q_{1i}}{\sum p_{0i}q_{1i}} \times \frac{\sum p_{0i}q_{0i}}{\sum p_{1i}q_{0i}} \neq 1$$

(iii) Index number based on simple averages of relatives, where arithmetic mean of price relatives is used do not satisfy time-reversal test.

$$I_{01} \times I_{10} = \left[\frac{1}{n}\sum\frac{p_1}{p_0}\right] \times \left[\frac{1}{n}\sum\frac{p_0}{p_1}\right] \neq 1$$

2. **Factor Reversal Test:** Factor-reversal test means, if the symbols for price and quantity in a formula for price index number are interchanged, then an index of quantity is obtained called as quantity index number. Multiplication of both price and quantity index numbers yie]ds an index that measures the change in the total value.

(i) Fisher's index number satisfies factor-reversal test. F_{01} is an ideal index number, because it satisfies both time and factor reversal test.

$$\overbrace{F_{01}}^{P} \times \overbrace{F_{01}}^{Q} = \sqrt{\overbrace{L_{01}}^{P} \times \overbrace{P_{01}}^{P}}\sqrt{\overbrace{L_{01}}^{Q}\,\overbrace{P_{01}}^{Q}} = \frac{\sum p_1q_1}{\sum p_0q_0} = \text{value ratio}$$

(ii) Laspeyre's and Paasche's index numbers do not satisfy the factor-reversal test.

$$\overbrace{L_{01}}^{P} \times \overbrace{L_{01}}^{Q} = \frac{\sum p_1q_0}{\sum p_0q_0} \times \frac{\sum q_1p_0}{\sum q_0p_0} \neq \frac{\sum p_1q_1}{\sum p_0q_0}(\text{value ratio});$$ P for price and Q for quantity index numbers.

$$\overbrace{P_{01}}^{P} \times \overbrace{P_{01}}^{Q} = \frac{\sum p_1q_1}{\sum p_0q_1} \times \frac{\sum q_1p_1}{\sum q_0p_1} \neq \frac{\sum p_1q_1}{\sum p_0q_0}$$

(iii) **Circular Test:** This test is an extension of time-reversal test and requires that $I_{ab} \times I_{bc} \times I_{ca} = 1$, where a, b and c stand for different years. The well known Laspeyre's, Paache's and Fisher's index numbers do not satisfy this test. But index numbers based on simple aggregative method and fixed weight aggregative method (Kelley's index number) satisfy the circular test. Circular test has its importance when it is required that comparisons are made over space instead of time. So the areas or periods, in which similar conditions do not prevail, are not to be compared.

(iv) **Unit Test:** As the name of the test signifies, according to this test, the index number formula should be such that it is independent of the units in which prices and quantities are measured. Except the simple aggregative formula, all other index number formulae discussed so far satisfy unit test.

Example 17.2: Taking data from the example 17.1 perform (1) Time-reversal test (2) Factor-reversal test.

Solution:

1. **Time Reversal Test:**

(i) Fisher's Ideal Index satisfies time reversal test

$$F_{01} \times F_{10} = \sqrt{\frac{\sum p_1q_0}{\sum p_0q_0} \times \frac{\sum p_1q_1}{\sum p_0q_1}} \times \sqrt{\frac{\sum p_0q_1}{\sum p_1q_1} \times \frac{\sum p_0q_0}{\sum p_1q_0}}$$

$$= \sqrt{\frac{29670}{19930} \times \frac{24320}{15987.5}} \times \sqrt{\frac{15987.5}{24320} \times \frac{19930}{29670}} = 1$$

(ii) Marshall-Edgeworth's Index satisfies time-reversal test

$$M_{01} \times M_{10} = \frac{\sum p_1(q_0 + q_1)}{\sum p_0(q_0 + q_1)} \times \frac{\sum p_0(q_0 + q_1)}{\sum p_1(q_0 + q_1)}$$

$$= \frac{29670 + 24320}{19930 + 15987.5} \times \frac{19930 + 15987.5}{29670 + 24320} = 1$$

(iii) Index number based on simple aggregative method satisfies time-reversal test

$$I_{01} \times I_{10} = \frac{\sum p_1}{\sum p_0} \times \frac{\sum p_0}{\sum p_1} = \frac{258.5}{169} \times \frac{169}{258.5} = 1$$

(iv) Index number based on simple averages of relatives, where geometric mean of price relatives is used, satisfies time-reversal test

Commodity	Unit	2001		2005		p_1/p_0	log (p_1/p_0)	p_0/p_1	log (p_0/p_1)
		Price (p_0)	Quantity (q_0)	Price (p_1)	Quantity (q_1)				
(a) Egg	Dozen	25.00	50	28.00	40	1.12	0.05	0.89	-0.05
(b) Milk	Kg	8.00	60	9.50	50	1.19	0.07	0.84	-0.07
(c) Rice	Quintal	120.00	100	200.00	90	1.67	0.22	0.60	-0.22
(d) Kerosene	Litre	4.50	100	7.00	75	1.56	0.19	0.64	-0.19
(e) Electric Wire	Metre	11.50	500	14.00	300	1.22	0.09	0.82	-0.09
Total:						6.76	0.62	3.79	-0.62

$$I_{01} \times I_{10} = \text{Antilog of}\left(\frac{\sum \log \frac{p_1}{p_0}}{n}\right) \times \text{Antilog of}\left(\frac{\sum \log \frac{p_0}{p_1}}{n}\right)$$

Antilog of (0.62 ÷ 5) = 1.330 × 0.750 = 0.998 ≅ 1 (discrepancy is due to rounding off of the decimal parts)

The following Index Numbers do not satisfy the time-reversal test:

(i) Laspeyre's Index number do not satisfy time-reversal test

$$L_{01} \times L_{10} = \frac{\sum p_{1i}q_{0i}}{\sum p_{0i}q_{0i}} \times \frac{\sum p_{0i}q_{1i}}{\sum p_{1i}q_{1i}}$$

$$= \frac{29670}{19930} \times \frac{15987.5}{24320} = 0.979 \neq 1;$$

(ii) Paasche's Index number do not satisfy time-reversal test

$$P_{01} \times P_{10} = \frac{\sum p_{1i}q_{1i}}{\sum p_{0i}q_{1i}} \times \frac{\sum p_{0i}q_{0i}}{\sum p_{1i}q_{0i}}$$

$$= \frac{24320}{15987.5} \times \frac{19930}{29670} = 1.022 \neq 1;$$

(iii) Index number based on simple averages of relatives, where arithmetic mean of price relatives is used do not satisfy time-reversal test

$$I_{01} \times I_{10} = \left[\frac{1}{n}\sum \frac{p_1}{p_0}\right] \times \left[\frac{1}{n}\sum \frac{p_0}{p_1}\right]$$

$$= \frac{1}{5} \times 6.76 \times \frac{1}{5} \times 3.79 = 1.0248 \neq 1$$

2. **Factor Reversal Test:** Fisher's index number satisfies factor-reversal test. F_{01} is an ideal index number, because it satisfies both time and factor reversal test.

Commodity	Unit	2001		2005		p_0q_0	p_1q_0	p_1q_1	p_0q_1
		Price (p_0)	Quantity (q_0)	Price (p_1)	Quantity (q_1)				
(a) Egg	Dozen	25.00	50	28.00	40	1250	1400	1120	1000
(b) Milk	Kg	8.00	60	9.50	50	480	570	475	400
(c) Rice	Quintal	120.00	100	200.00	90	12000	20000	18000	10800
(d) Kerosene	Litre	4.50	100	7.00	75	450	700	525	337.5
(e) Electric Wire	Metre	11.50	500	14.00	300	5750	7000	4200	3450
Total:				258.5	555	19930	29670	24320	15987.5

$$\overbrace{L_{01}}^{P}=\frac{\sum p_{1i}q_{0i}}{\sum p_{0i}q_{0i}}\times 100=\frac{29670}{19930}\times 100=148.871$$

$$\overbrace{P_{01}}^{P}=\frac{\sum p_{1i}q_{1i}}{\sum p_{0i}q_{1i}}\times 100=\frac{24320}{15987.5}\times 100=152.119$$

$$\overbrace{L_{01}}^{Q}=\frac{\sum q_{1i}p_{0i}}{\sum q_{0i}p_{0i}}\times 100=\frac{15987.5}{19930}\times 100=80.218$$

$$\overbrace{P_{01}}^{Q}=\frac{\sum q_{1i}p_{1i}}{\sum q_{0i}p_{1i}}\times 100=\frac{24320}{29670}\times 100=81.968$$

$$\overbrace{F_{01}}^{P}\times\overbrace{F_{01}}^{Q}=\sqrt{\overbrace{L_{01}}^{P}\times\overbrace{P_{01}}^{P}}\sqrt{\overbrace{L_{01}}^{Q}\,\overbrace{P_{01}}^{Q}}$$

$$=\sqrt{\frac{148.871\times 152.119}{100\times 100}}\cdot\sqrt{\frac{80.218\times 81.968}{100\times 100}}$$

$$=150.486\times\frac{1}{100}\times 81.09\times\frac{1}{100}=1.2202$$

$$\frac{\sum p_1q_1}{\sum p_0q_0}=\frac{24320}{19930}=1.2202=\text{value ratio}$$

But Laspeyre's and Paasche's index numbers do not satisfy the factor-reversal test

$$\overbrace{L_{01}}^{P}\times\overbrace{L_{01}}^{Q}=\frac{\sum p_1q_0}{\sum p_0q_0}\times\frac{\sum q_1p_0}{\sum q_0p_0}\neq\frac{\sum p_1q_1}{\sum p_0q_0}(\text{value ratio});\ \text{P for price and Q}$$

for quantity index numbers.

$$\overbrace{P_{01}}^{P}\times\overbrace{P_{01}}^{Q}=\frac{\sum p_1q_1}{\sum p_0q_1}\times\frac{\sum q_1p_1}{\sum q_0p_1}\neq\frac{\sum p_1q_1}{\sum p_0q_0}$$

The detailed calculation is left to the students to work out.

WHOLESALE PRICE INDEX NUMBER (WPI)

The Wholesale Price Index (WPI) is the price of a representative basket of wholesale goods. Some countries (like The Philippines) use WPI changes as a central measure of inflation. However, United States now report a producer price index instead. But on April 2, 2014 during the first bimonthly monetary policy for FY:2014-15, Mr. Raghuram Rajan (RBI Governor) declared that CPI would be the key index of inflation measurement going forward.

The Wholesale Price Index or WPI is "the price of a representative basket of wholesale goods". Some countries use the changes in this index to measure inflation in their economies, in particular India – the Indian WPI figure was earlier released weekly on every Thursday. But since 2009 it has been made monthly. It also influences stock and fixed price markets. The Indian WPI is now updated on a monthly basis. The Wholesale Price Index focuses on the price of goods traded between corporations, rather than goods bought by consumers, which is measured by the Consumer Price Index. The purpose of the WPI is to monitor price movements that reflect supply and demand in industry, manufacturing and construction. This helps in analyzing both macroeconomic and microeconomic conditions.

Calculation

The wholesale price index (WPI) is based on the wholesale price of a few relevant commodities of over 240 commodities available. The commodities chosen for the calculation are based on their importance in the region and the point of time the WPI is employed. For example in India about 435 items were used for calculating the WPI in base year 1993-94 while the advanced base year 2004-05 and which has now been changed to 2010-2011; uses 676 items. The indicator tracks the price movement of each commodity individually. Based on this individual movement, the WPI is determined through the averaging principle. The following methods are used to compute the WPI:

Laspeyre's Formula: It is the weighted arithmetic mean based on the fixed value-based weights for the base period.

Ten-Day Price Index: Under this method, "sample prices" with high intra-month fluctuations are selected and surveyed every ten days through phone. Utilizing the data retrieved by this procedure and with the assumption that other non-surveyed "sample prices" remain unchanged, a "ten-day price index" is compiled and released.

COST OF LIVING INDEX NUMBER

The cost of living index number, otherwise known as consumer price index number, is a measure of the changes in the level of prices of commodities and services that directly affect the cost of living of a particular group of people at a specified place. It does not measure the changes in cost of living due to causes other than changes in price level, but how much a particular class of people pays towards a specified basket of goods and services in a given period compared to the base period. That is why, the cost of living index number of a daily laborer Mumbai will be different from that of a film producer in Mumbai and also will be different from the cost of living index number of a labourer in Cuttack.

It is not easy, though not impossible, to construct the true cost of living index number. The first step is to clearly and unambiguously specify its purpose of construction, that is to say first define the class of people for whom it is to be found out. Then adequate random samples are surveyed as to their expenditure generally called as "family budget enquiry" for the commodities like (a) Food articles (b) Clothing (c) Fuel & Lighting (d) House rent and (e) Miscellaneous. Such commodities should be stable in quality without much seasonal variations in supply. Then retail prices of the commodities are collected from retail shops from where these families make their purchases. Thus the family budget enquiry reveals the average quantity of such commodities consumed by an average family. Weighted index numbers are preferred method of calculation as shown below:

1. Weighted aggregative method: Here Laspeyre's formula is used. Index number for current year = $\frac{\sum p_1 q_0}{\sum p_0 q_0} \times 100$; where the symbols have their usual notations.
2. Family budget method or Method of weighted relatives: Index number for current year = $\frac{\sum (p_1/q_0 \times 100) p_0 q_0}{\sum p_0 q_0} \times 100 = \frac{\sum IV}{\sum V}$.

   where, $I = \frac{p_1}{p_0} \times 100$ and $V = p_0 q_0$

It may be noted that, due to random sampling, which may not always be true representative of the actual class of people meant for the study, errors generally creep into the construction of consumer index number. Moreover faulty selection of commodities too affects it. So the utility consumer price index number poses a big question.

What is the difference between WPI and CPI Inflation?

Inflation is a general and sustained increase in overall price level of goods and services. The inflation rate is a key parameter basis which central government proposes its monetary and fiscal policy from time to time. The monetary policy primarily focuses on price stability and hence its concern for inflation is rather obvious.

Indicators of Measure of Inflation: Whole-sale price index (WPI) and Consumer price index (CPI) are the two primary measures of inflation.

WPI index reflects average price changes of goods that are bought and sold in the wholesale market. But whole sale price index (WPI) does not include the cost of services. Further, as WPI accounts for changes in general price level of goods at wholesale level, it fails to communicate actual burden borne by the end consumer. WPI is the primary measure that is used by the Indian central government for ascertaining inflation as WPI, whereas CPI accounts for changes in price at an early distribution stage. CPI is computed by executing a weighted average on a particular set of goods and services. The computation of CPI takes into account price changes and the actual inflation that affects the end consumer. CPI is thus a reflection of changes in the retail prices of specified goods and services over a time period, which are traded by particular consumer group. While earlier the Reserve Bank of India used WPI inflation to manage monetary policy expectations, it is now the CPI inflation which is largely taken into account. The RBI highlights its inflation expectations based on the CPI inflation data that comes in. For example, it sets targets on CPI Inflation and monitors it accordingly. Many analysts for long had suggested that the RBI should move to the CPI data vs the WPI data, which had now happened in the last couple of years.

Example 17.3: Construct the consumer price index number for 2013 on the basis of 2001 from the following data:

Goods	Quantity consumed in 2001	Unit	Price (Rs.) in 2001	Price (Rs.) in 2013
Food	6	Quintal	200	600
Clothing	45	Metre	20	40
Petrol	12	Quintal	40	80
Kerosene	42	Litre	2	6
House Rent	1	House	600	1200

Solution:

(a) Aggregative Expenditure Method:

Goods	Quantity consumed in 2001 (q_0)	Price (Rs.) in 2001 (p_0)	Price (Rs.) in 2013 (p_1)	p_1q_0	p_0q_0
Food	6	200	600	3600	1200
Clothing	45	20	40	1800	900
Petrol	12	40	80	960	480
Kerosene	42	2	6	252	84
House Rent	1	600	1200	1200	600
Total:				7812	3264

Cost of living index number for 2001

$$= \frac{\sum p_1 q_0}{\sum p_0 q_0} \times 100 = \frac{7812}{3264} \times 100 = 239.338$$

(b) Family Budget Method:

Goods	Quantity consumed in 2001 (q_0)	Price (Rs) in 2001 (p_0)	Price (Rs) in 2013 (p_1)	$I = p_1/p_0 \times 100$	$V = p_0 \times q_0$	I × V
Food	6	200	600	300.00	1200	360000
Clothing	45	20	40	200.00	900	180000
Petrol	12	40	80	200.00	480	96000
Kerosene	42	2	6	300.00	84	25200
House Rent	1	600	1200	200.00	600	120000
Total:					3264	781200

Cost of living index number for 2001= CLI $= \frac{\sum IV}{\sum V} = \frac{78100}{3264} = 239.388$

(Same as above)

QUANTITY INDEX NUMBER

The quantity or production indices measure the relative changes in quantities or volume of output compared to a base year. There are three types of quantity indices:

1	Unweighted Average of Relatives	Quantity Index $\overset{Q}{\hat{I}} = \frac{1}{n}\sum\left(\frac{q_1}{q_0}\right) \times 100$
2	Weighted Aggregative	Laspeyre's type $\overset{Q}{\overbrace{L_{01}}} = \frac{\sum q_1 p_0}{\sum q_0 p_0} \times 100$ Paasche's type $\overset{Q}{\overbrace{P_{01}}} = \frac{\sum q_1 p_1}{\sum q_0 p_1} \times 100$
3	Weighted Relatives Quantity Index	$\overset{Q}{\hat{I}} = \frac{\sum\left(\frac{q_1}{q_0}\right)W}{\sum W}$

CHAIN INDEX NUMBER

In course of time, the conditions of base period change and it becomes unreasonable to compare two distant periods keeping the base period fixed. This necessitates construction of chain index numbers, which use varying bases. First, link relatives which express the figures for each year as percentages of previous year are calculated. Then the link relatives are chained together by successive multiplication.

Chain index for current year = (Link relative for current year Chain Index for previous year) $\times \frac{1}{100}$

Example 17.4: Compute the chain indices for the average retail price (per kg) of wheat at Balasore district of Odisha (Hypothetical data).

Year	Per kg price of wheat (Rs)
2001	22
2002	25
2003	26
2004	27
2005	29
2006	30

Solution:

Year	Per kg price of wheat (Rs.)	Link Relatives		Chain Indices (2001=100)	
2001	22		100		100
2002	25	25÷22100=	113.64	113.64100÷100=	113.64
2003	26	26÷25100=	104.00	104113.64÷100=	118.18
2004	27	27÷26100=	103.85	103.85118.18÷100=	122.73
2005	29	29÷27100=	107.41	107.41122.73÷100=	131.83
2006	30	30÷29100=	103.45	103.45131.83÷100=	136.38

SPLICING AND SHIFTING THE BASE OF AN INDEX NUMBER

The problem of combining two or more overlapping series of index numbers is called splicing. Suppose index numbers for the year 2000, 2001 and 2002 are given with base year 2000. Again index numbers for the years 2002, 2003 and 2004 are given with base year as 2002. Then how to combine the two such series? Splicing is needed when the weights of an index number get changed through years and at a later occasion it is desired to combine the two overlapping series into a continuous one.

This is achieved by proportional distribution. Consider the following example:

Year	Series of Index Numbers	
	P	Q
2000	100	
2001	130	
2002	150	100
2003	x_1	160
2004	x_2	170

If x_1 and x_2 are the index numbers for the years 2003 and 2004 respectively of P-series, then,

$$\frac{150}{x_1} = \frac{100}{160}$$

$$x_1 = \frac{150 \times 160}{100} = 240$$

And

$$\frac{150}{x_2} = \frac{100}{170}$$

$$x_2 = \frac{150 \times 170}{100} = 255$$

The shifting of the base of an index number from one period to another is carried out by the method of proportions. The following example explains the detailed calculation procedure:

Year	Index Number	Revised Index Numbers (Base shifted to 2003)	Revised Index Numbers (Base shifted to 2004)
2001	100	$x_1 = \frac{100}{108} \times 100 = 92.593$	$x_1 = \frac{100}{112} \times 100 = 89.286$
2002	106	$x_2 = \frac{106}{108} \times 100 = 98.148$	$x_2 = \frac{106}{112} \times 100 = 94.643$

Year	Index Number	Revised Index Numbers (Base shifted to 2003)	Revised Index Numbers (Base shifted to 2004)
2003	108	100	$x_3 = \frac{108}{112} \times 100 = 96.429$
2004	112	$x_4 = \frac{112}{108} \times 100 = 103.704$	100

DEFLATING PRICES

Deflating price is the adjusted price of a commodity in terms of base year. Adjustment is done through price index only. Suppose there are two commodities P and Q whose prices in 2000 are 35 and 62 respectively. Their prices in 2005 are 45 and 65 respectively. The price index during 2005 is 130 with base year as 2000. So the deflated prices are calculated as below:

Year	Price (Rs.)		Price Index	Deflated Prices (Rs.)	
	Commodity P	Commodity Q		Commodity P	Commodity Q
2000	35	45	100	—	—
2005	42	65	130	$\frac{42}{130} \times 100 = 32.308$	$\frac{65}{130} \times 100 = 50$

Deflated price of commodity P = 32.308 – 35 = Rs. –2.692. That means the real value of commodity P has decreased during the period.

Deflated price of commodity Q = 50 – 45 = Rs. 5.00. That means the real value of commodity Q has increased during the period.

Moreover, demand of the commodities also affects the increase or decrease in real value of the commodities.

EXERCISES

1. What is index number? Briefly describe the problems that are encountered in construction of an index number of prices.
2. Explain the concept and uses of an index number. Discuss the conditions that a good index number should satisfy.
3. Explain base shifting and splicing of an index number.
4. What do you mean by deflating prices? Explain.

5. Compute the index number based on aggregate of actual prices (simple aggregative) index for the following data. Compare it with the index number calculated by the method of simple average (arithmetic mean) of relatives.

Commodity	Unit	Price (Rs.)	
		2000	2005
P	Per quintal	145	195
Q	Per quintal	40	55
R	Per kg	3	5
S	Per meter run	80	95

6. Compute the index number from the following data by the method of simple average of 'relatives' using arithmetic mean and geometric mean.

Commodity	Price (Rs.)	
	2000	2005
A	30	25
B	40	55
C	6	8
D	15	25
E	10	12

7. Calculate the index number from the following data using (a) Laspeyre's formula (b) Paasche's formula (c) Fisher's ideal formula (d) Dorbish-Bowley's method (e) Marshall-Edgeworth method.

Commodity	2011		2013	
	Quantity	Price	Quantity	Price
A	12	10	10	18
B	20	8	18	14
C	8	3	10	6
D	4	16	6	25

8. Compute Fisher's ideal index for 2012 from the following data and show numerically that it satisfies time-reversal and factor-reversal tests. Also find if Laspeyre's and Paasche's index satisfy these tests.

Year	Commodity A		Commodity B		Commodity C	
	p	q	p	q	p	q
2000	5	12	8	6	6	4
2012	12	6	16	4	8	3

9. Compute the index number for the year 2012 using the method of weighted average of price relatives.

Commodity	2010		2012	
	Price	Quantity	Price	Quantity
A	145	25	180	20
B	40	8	50	8
C	12	6	15	5
D	4	18	6	15

10. From the following data on cosmetics prices, show that the arithmetic mean of relatives (unweighted) does not satisfy the time-reversal test.

Items	Price (Rs.)	
	2005	2010
Hair Oil	10	15
Nail Polish	20	30
Shampoo	30.	45

11. Calculate the chain indices for the price of commodity 'Q' in different periods:

Year	2008	2009	2010	2011	2012
Price	32	34	37	41	45

12. Compute the chain index number with 2002 price as base from the following hypothetical data:

Commodity	Price (Rs.) per unit		
	2002	2003	2004
A	16	18	22
B	18	25	35
C	14	16	20
D	4	6	7

13. Construct the chain index numbers from the following link relatives:

Year	2001	2002	2003	2004
Link relative	100	120	98	115

14. Calculate chain base index numbers:

Year	2001	2002	2003	2004	2005
Fixed base index	100	145	135	150	155

15. Calculate fixed base index numbers from the following chain base index numbers:

Year	2005	2006	2007	2008	2009
Chain base index	100	110	90	95	115

16. Splice the following two series:

Year	Series of Index Numbers	
	A	B
2005	100	
2006	110	
2007	130	
2008	150	100
2009		130
2010		155
2011		170

17. Shift the base from 2000 to 2010 and construct the new series:

Year	1980	1985	1990	1995	2000	2005
Index Number	100	110	125	135	160	175

18. Given the following data, deflate the wage series to construct the real wage index:

Year	1980	1985	1990	1995	2000
Wage (W)	300	510	650	1000	1400
Cost of living index (I)	100	110	170	220	250

19. Construct the cost of living (consumer price) index number from the following data:

Items of consumption	Quantity consumed in 2000	Price (Rs.) per unit	
		2000	2010
Food	80 kg	1.42	2.80
Clothing	30 kg	2.40	6.80
Education	2 children	26.00	50.00
Fuel	75 kg	22.00	55.00
Miscellaneous	12 units	12.00	14.00

STATISTICAL MODELS IN ENVIRONMENTAL STUDIES

INTRODUCTION

Statistics and mathematics provide effective tools in developing various models, which can be conveniently used in environmental studies. Some of the few models are discussed in this chapter to get an insight of the techniques adopted in developing the models.

(A) Air Pollutant Concentration Models

Atmospheric dispersion modeling is the mathematical simulation of how air pollutants disperse in the ambient atmosphere. The dispersion models are used to estimate or to predict the downwind concentration of air pollutants or toxins emitted from sources such as industrial plants, vehicular traffic or accidental chemical releases. The perfect air concentration model will estimate the concentration emitted from any specified set of pollutant, at any location, at any period of time and with high confidence levels in our prediction.

Practically, it is performed using computer programs that solved the mathematical equations and algorithms which simulate the pollutant dispersion. Such models are important especially to governmental agencies tasked with protecting and managing the ambient air quality. The models also supplied information to assist in the design of effective control strategies to reduce emissions of harmful air pollutants.

Air dispersion models are also used for emergency planning of accidental chemical releases. Models are used to determine the consequences of accidental releases of hazardous or toxic materials. Accidental releases may result in fires, spills or explosions that involve hazardous materials, such as chemicals or radionuclides. The results of dispersion modeling, using worst case accidental release source terms and meteorological conditions, can provide an estimation of location impacted areas and ambient concentrations.

Thus, it and can be used to determine protective actions appropriate to the accident. In this chapter we will learn from the simplest model (least reliable) to the most complex (most reliable) model so that the theories behind the mathematical simulation program can be revealed.

All the models studied here will be using a general mass balance equation as given below:

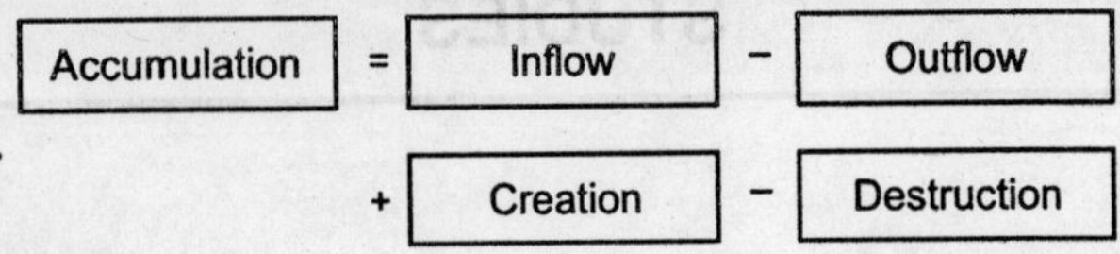

FIXED BOX MODEL

The box model is the simplest of the model types. It assumes the volume of atmospheric air is in the shape of a box. It also assumes that the air pollutants inside the box are homogeneously distributed. Although useful, this model is very limited in its ability to accurately predict dispersion of air pollutants o because the assumption of homogeneous pollutant distribution is much too simple. To start adopting box model, the following major simplifying assumptions must be made according to the illustrated figure below:

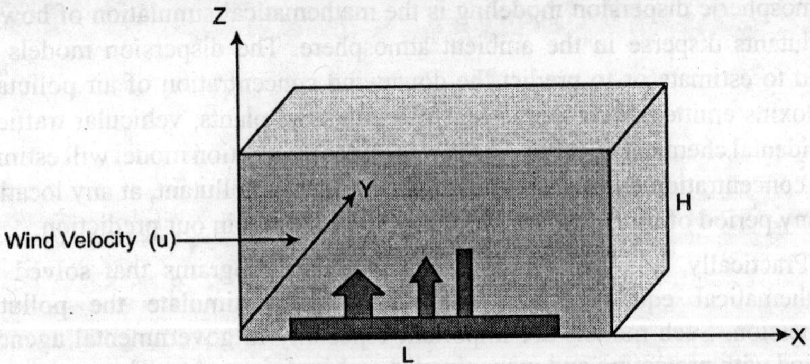

Figure Rectangular city, showing the symbol used in the fixed box model

- The city is a rectangle with dimension W and L and one side is parallel with wind direction.
- Complete atmospheric turbulence is produced and total mixing of pollutants up to H and no mixing above this height.
- The turbulence is strong enough in the upwind direction that the pollutant concentration is uniform in the whole volume of air over the city and not higher at downwind side than upwind side. [This emission rate is constant and unchanging with time (in steady state condition)]

- The wind blows in x direction with velocity u. This velocity is constant and is independent of time, location or elevation above the ground (steady state condition). We use average u between at the ground and at H.
- The concentration of pollutant entering the city (at $x = 0$) is constant and is equal to b (background concentration). The units are g/m^3 or $\mu g/m^3$.
- The air pollution rate of the city is Q (g/s). This is usually given as emission rate per unit area, q, $g/s\text{-}m^2$. The conversion can be made by $Q = qA$
  where A is the area (W L).
- No pollutant enters or leaves through the top or side of the box.
- The pollutant is long-lived in the atmosphere (destruction rate = 0).
- Emission rate is constant and unchanging with time (steady state condition)

With all these assumption, the general mass balance equation is simplified as below

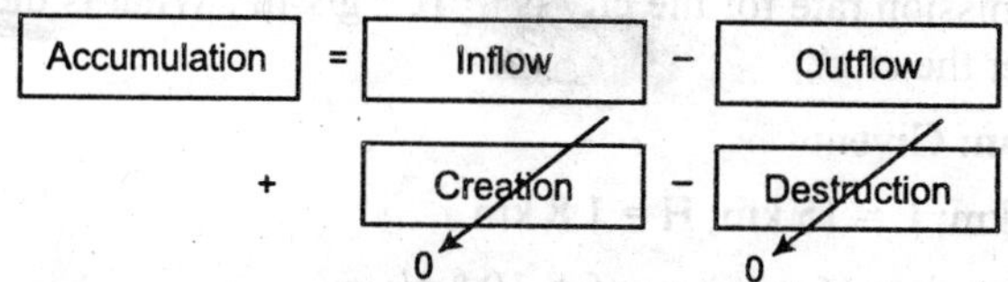

So, that the equation is simplified as below

Inflow	=	Outflow

There are two flow rates. The flow rate of the pollutant into the upwind side is:

Flow rate-in = (uWH)b = (volume per time) (mass per volume) = mass per time

The second flow rate-in is that the pollutant emitted by the city into the lower boundary:

Flow rate-in $Q = qWL$

According to the assumption, the concentration of entire city is a constant and equal to c.

Flow rate-out = (uWH)c

Then, combining all these equations we get:

$$0 = (uWHb + qWL) - uWHc$$

$$0 = [W(uHb + qL) - uHc]$$

$$uHc = uHb + qL$$

$$\therefore \quad c = b + \frac{qL}{uH}$$

The above equation on Box Model is widely used in chemical engineering.

The model indicates that the upwind concentration for a long lived pollutant is additive to the concentration produced by the city. Clearly from equation 3, the concentration, c increases with the increase of q and L and decreases with the increases in u and H. Most of the models predict concentrations for only one specific meteorological condition. To calculate the annual average concentration of some pollutant, we need to calculate the concentration for each meteorological condition and then multiply by the frequency of occurrence for that meteorological condition. Mathematically speaking:

Annual Average Concentration = $\sum${(concentration for that meteorology) × (frequency of occurrence of that meteorology) }

Example 18.1: A city with dimensions W × L × H (8 km × 15 km × 1.8 km) had a wind velocity of 5 m/s. The upwind concentration of SO_2 is 15 μg/m³. The emission rate for the city is 6 10^{-6} g/s-m². What is the concentration of SO_2 over the city?

Solution: Given:

W = 8 km; L = 15 km; H = 1.8 km

u = 5 m/s; b = 15 g/m³; q = 6 × 10^{-6} g/s-m²

$$c = b + \frac{qL}{uH} = \frac{15\mu g}{m^3} + \left(6 \times 10^{-6} \frac{g}{s - m^2}\right) \times \left(\frac{15000 \text{ m}}{(5 \text{ m/s}) \times (1800 \text{ m})}\right)$$

$$c = \frac{15\mu g}{m^3} + \frac{10 \times 10^{-6} g}{m^3} = \frac{15\mu g}{m^3} + \frac{10\mu g}{m^3} = 25\frac{\mu g}{m^3}$$

Example 18.2: For the city in Example 18.1, the meteorological conditions described (u = 5 m/s; H = 1.8 km) occur 40% of the time. For the remaining period, the wind blows from L-direction shown in the Box model figure at a velocity of 10 m/s and the same mixing height. What is the annual concentration of SO_2 in this city ?

Solution: In this case, wind blows from L-direction, that means, wind crosses L-H plane. So, L = 8 km.

$$c = b + \frac{qL}{uH} = \frac{15\mu g}{m^3} + \left(6 \times 10^{-6} \frac{g}{s - m^2}\right) \times \left(\frac{8000 \text{ m}}{(5 \text{ m/s}) \times (1800 \text{ m})}\right)$$

$$c = \frac{15\mu g}{m^3} + \frac{5.33\times10^{-6} g}{m^3} = \frac{15\mu g}{m^3} + \frac{5.33\mu g}{m^3} = 20.33\frac{\mu g}{m^3}$$

This occurs for 60% of the time, as the wind from W-direction blows for 40% of the time, as per the question.

So annual average concentration

$$\left(25\frac{\mu g}{m^3}\times 0.4\right)+\left(20.33\frac{\mu g}{m^3}\times 0.6\right)=22.198\frac{\mu g}{m^3}$$

GAUSSIAN PLUME MODEL

Gaussian plume model is usually called as diffusion model and sometimes dispersion model. Most of diffusion model used Gaussian plume idea which is also a material balance model. In this model, a point source such as a factory smoke stack is selected and the downwind concentration resulting from this point source is calculated.

Consider a point source somewhere in the air where a pollutant is released at a constant rate Q(kg/s). The wind is blowing continuously in a direction x (measured in meters from the source) with a speed U (m/s). The plume spreads as it moves in the x-direction such that the local concentrations C(x,y,z) at any point in space form distributions which have shapes that are "Gaussian" or "Normal" in planes normal to the x-direction.

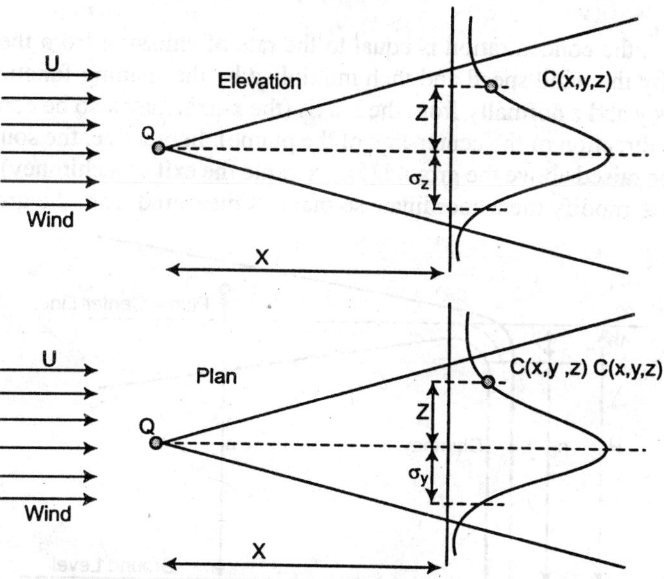

In the lateral or Y-direction, the profile shape is given by, $\frac{1}{\sigma_y\sqrt{2\pi}} \times e^{\left(\frac{-y^2}{2\sigma_y^2}\right)}$

Whilst in the vertical or Z-direction, it is given by, $\frac{1}{\sigma_z\sqrt{2\pi}} \times e^{\left(\frac{-z^2}{2\sigma_z^2}\right)}$

The parameters σ_y and σ_z are the standard deviations of these Gaussian distributions, which indicate the spread of the plume in the Y and Z-directions respectively. They increase with the distance x from the source. The area under the distribution, determined by integration of the functions given above between $-\infty$ and $+\infty$, is equal to unity.

Combining these two-dimensional shape distributions by multiplying the functions together gives us the function for the shape of the distribution in three-dimensions (a kind of 'hill' of pollutant) = $\frac{1}{2\pi\sigma_y\sigma_z} \times e^{\left(\frac{-y^2}{2\sigma_y^2}\right)} \times e^{\left(\frac{-z^2}{2\sigma_z^2}\right)}$

The concentration at any point is given by

$$C_{(x,y,z)} = \frac{Q}{U} \times \frac{1}{2\pi\sigma_y\sigma_z} \times e^{\left(\frac{-y^2}{2\sigma_y^2}\right)} \times e^{\left(\frac{-z^2}{2\sigma_z^2}\right)}$$

Hence, the concentration is equal to the rate of emission from the source divided by the wind speed and then multiplied by the shaping function. This measures y and z normally from the x-axis (the x-axis may also be considered to be the direction of the center-line of the plume). In practice, the source will always be raised above the ground (for example the exit of a chimney). Hence we need to modify the z coordinate so that it is measured from the ground.

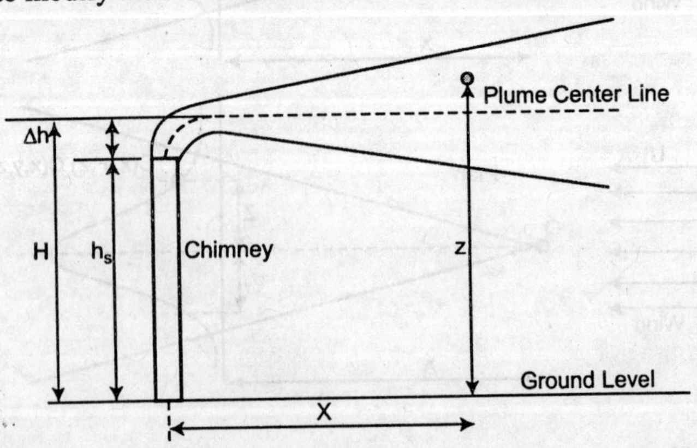

H = effective height of plume center line (m)

h_s = height of source above ground (m)

Δh = initial plume rise (m)

z = coordinate measured vertically from the ground to a point in the plume (m)

Hence, relative to the plume axis the new vertical coordinate is (z – H) giving

$$C_{(x,y,z)} = \frac{Q}{U} \times \frac{1}{2\pi\sigma_y\sigma_z} \times e^{\left(\frac{-y^2}{2\sigma_y^2}\right)} \times e^{\left(\frac{-(z-H)^2}{2\sigma_z^2}\right)}$$

One further modification is to be made to this equation. Unlike a plume spreading in free air, most plumes will be emitted close to the ground, like the case of the chimney shown above. Hence, as the plume spreads downwards (as well as upwards) as it moves downwind from the source, it will eventually hit the ground. But the plume cannot continue to spread into the ground. Instead it is reflected back into the air above the ground. The effect of the ground boundary is included in the concentration equation mathematically by using a fictitious mirror-image source (S_i = S) of the same strength (Q_i = Q) placed at the same distance from the ground (h_s) but on the other side of the boundary.

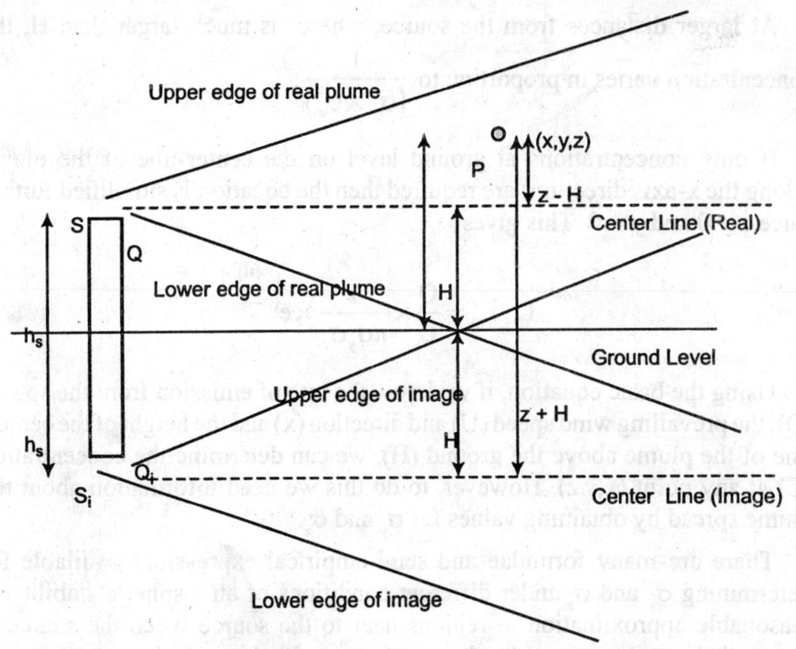

Hence, at any point P there is a contribution to the concentration C(x,y,z) from both the real source (S) and the imaginary source (S_i). The vertical distance to P from the center-line of the real plume is (z-H). The vertical distance to P from the center-line of the imaginary source is (z+H). The lateral distance into the page is the same for both sources.

Hence, the total concentration at P is:

$$C_{(x,y,z)} = \frac{Q}{U} \times \frac{1}{2\pi\sigma_y\sigma_z} \times e^{\left(\frac{-y^2}{2\sigma_y^2}\right)} \times \left[e^{\left(\frac{-(z-H)^2}{2\sigma_z^2}\right)} + e^{\left(\frac{-(z+H)^2}{2\sigma_z^2}\right)} \right]$$

If only concentrations at ground level is required (for example in assessing the exposure of crops or humans to the pollutant) then we can simplify the equation by setting z = 0. This gives

$$C_{(x,y,0)} = \frac{Q}{U} \times \frac{1}{\pi\sigma_y\sigma_z} \times e^{\left(\frac{-y^2}{2\sigma_y^2}\right)} \times e^{\left(\frac{-(H)^2}{2\sigma_z^2}\right)}$$

It should be noted that the maximum concentration occurs when $\sigma_z = \frac{H}{\sqrt{2}}$

At larger distances from the source, where is much larger than H, the concentration varies in proportion to $\frac{1}{(\sigma_y \times \sigma_z)}$.

If only concentrations at ground level on the center-line of the plume (along the x-axis direction) are required then the equation is simplified further since z = 0 and y = 0. This gives

$$C_{(x,0,0)} = \frac{Q}{U} \times \frac{1}{\pi\sigma_y\sigma_z} \times e^{\left(\frac{-(H)^2}{2\sigma_z^2}\right)}$$

Using the basic equation, if we know the rate of emission from the source (Q), the prevailing wind speed (U) and direction (x) and the height of the center-line of the plume above the ground (H), we can determine the concentration (C) at any point (x,y,z). However, to do this we need information about the plume spread by obtaining values for σ_y and σ_z.

There are many formulae and semi-empirical expressions available for determining σ_y and σ_z under different conditions of atmospheric stability. A reasonable approximation in regions near to the source when the source is elevated above the ground (such as at the top of a chimney) is:

$\sigma_y = I_y \times x$ and $\sigma_z = I_z \times x$

Where,

I_y = turbulent wind speed fluctuations (turbulence intensities) in the y-direction

I_z = turbulent wind speed fluctuations (turbulence intensities) in the z-direction

Under neutral atmospheric conditions, it has been found that, over a range of heights corresponding to the vertical plume spread, centered at approximately h_s, I_y and I_z may be estimated as:

$$= I_y = \frac{0.88}{\ln\left(\frac{h_s}{z_0}\right) - 1} \text{ and } I_z = \frac{0.50}{\ln\left(\frac{h_s}{z_0}\right) - 1}$$

ln = natural log

h_s =release height

z_0 = aerodynamic roughness, which varies according to various topographic ground conditions.

For general cases of different atmospheric conditions the following typical values apply.

Thermal Stratification	Lateral Intensity (I_y)	Vertical Intensity (I_z)
Extremely Unstable	0.40 – 0.55	0.15 – 0.55
Moderately Unstable	0.25 – 0.40	0.10 – 0.15
Near Neutral	0.10 – 0.25	0.05 – 0.08
Moderately Stable	0.08 – 0.25	0.03 – 0.07
Extremely Stable	0.03 – 0.25	0.00 – 0.03

Different values of σ_y and σ_z mean that the spreading in the vertical and horizontal directions is not equal. Usually the values of . It may be seen that the turbulence intensities, specially the vertical wind speed fluctuations, increase as atmospheric conditions become more unstable. Another factor that needs to be considered in practice is the plume rise (Δh). This is the path or trajectory of the plume center-line after it leaves the source. Its course depends upon atmospheric conditions and the amount of buoyancy and vertical momentum in the initial plume at the source.

Buoyance force causes the plume rise to vary with $x^{2/3}$

Momentum force causes the plume rise to vary with $x^{1/3}$

Hence, the shape of the trajectory will depend on which forces dominate the plume. If the chimney plume is buoyant the plume rises to a maximum level of $\Delta h = 0.2h_s$ to $\Delta h = 0.6h_s$ above the source, depending on atmospheric conditions. If the initial momentum of the plume dominates then an approximate expression for the final plume rise is:

$$\Delta h = \frac{3DW}{U}$$

Where

D = Diameter of the source e.g. chimney exit diameter (m)

W = Initial vertical velocity of the plume (m/s)

In practice, the initial vertical velocity is often similar to the prevailing wind speed. So, for a typical chimney diameter of, say, 4m the final plume rise is of the order of 8m. In many cases the value of the plume rise is so small, relative to the magnitude of the release height and the size of the plume, that the plume center-line may be considered to be horizontal for simplicity.

Example 18.3: A factory emits 40 g/s of SO_2 at H height. The wind speed is 5 m/s. At a distance of 1 km downwind, the values of σ_y and σ_z are 25 m and 35 m respectively. Compute the concentration of SO_2 (Neglect ground effect)

(a) At the center-line of the plume.

(b) At a point 50 m to the side and 10 m below the center-line.

Solution:

Given, Q = 40 g/s; U = 5 m/s; x = 1000 m; σ_y = 25 m; σ_z = 35 m

The general equation for concentration, without considering ground effect, at any point is

$$C_{(x,y,z)} = \frac{Q}{U} \times \frac{1}{2\pi\sigma_y\sigma_z} \times e^{\left(\frac{-y^2}{2\sigma_y^2}\right)} \times \left[e^{\left(\frac{-(z-H)^2}{2\sigma_z^2}\right)} \right]$$

(a) At the center-line of plume, y = 0 and z = H i.e. z-H =0

$$C_{(1000,0,H)} = \frac{40}{5} \times \frac{1}{2 \times \pi \times 25 \times 35} \times e^0 \times e^0 = 1454.54 \frac{\mu g}{m^3}$$

(b) At a point away from the center-line, the 2nd and 3rd term are to be calculated, where y = 50 m and z – H = –10 m

$$= C_{(1000,50,-20)} = \frac{40}{5} \times \frac{1}{25 \times 35 \times \quad \times} \times e^{\left(\frac{-(50)^2}{2\times25^2}\right)} \times \left[e^{\left(\frac{-(-10)^2}{2\times35^2}\right)} \right]$$

$$= 1454.54\ 0.1353\ 0.2396 = 47.175\ \frac{\mu g}{m^3}$$

Example 18.4 A factory emits 30 g/s of CO at a height of 25 m. The wind speed is 4 m/s at a distance of 0.6 km downwind, where $\sigma_y = 40$ m and $\sigma_z =$ 25 m. What are the CO concentration at a point of 20 m to the side and 3 m above the ground?

Solution:

Given that, H = 25 m; Q = 30 g/s; U = 4 m/s; $\sigma_y = 50$ m; $\sigma_z = 25$ m; x = 600 m; y = 20 m; z = 3 m

The general equation for concentration, considering ground effect, at any point is:

$$= C_{(x,y,z)} = \frac{Q}{U} \times \frac{1}{2\pi\sigma_y\sigma_z} \times e^{\left(\frac{-y^2}{2\sigma_y^2}\right)} \times \left[e^{\left(\frac{-(z-H)^2}{2\sigma_z^2}\right)} + e^{\left(\frac{-(z+H)^2}{2\sigma_z^2}\right)} \right]$$

$$= C_{(600,20,3)} = \frac{30}{4} \times \frac{1}{2 \times \pi \times 50 \times 25} \times e^{\left(\frac{-(20)^2}{2\times(50)^2}\right)} \times \left[e^{\left(\frac{-(3-25)^2}{2\times(25)^2}\right)} + e^{\left(\frac{-(3+25)^2}{2\times(25)^2}\right)} \right]$$

$$= 0.00193 \times 0.8825 \times (1.4728 + 1.8723) = 0.00569 \text{ g/m}^3$$

(B) Population Model (Leslie Matrix)

Before any discussion is made regarding population projection, a careful look at the census figure can be had in order to appreciate the male and female population, which seem to maintain an equal proportion as shown below: -

Vensus of India 2011: Primary Census Abstract Figure at a Glance: Odisha

	Absolute			Percentage		
	Total	Rural	Urban	Total	Rural	Urban
Persons	41,974,218	34,970,562	7,003,656	100	83.3	16.7
Males	21,212,136	17,586,203	3,625,933	100	82.9	17.1
Females	20,762,082	17,384,359	3,377,723	100	83.7	16.3

It can be seen that the percentages of male and female are almost same. Hence, if female population only is considered for estimation of future population projection, it will serve the purpose of finding total future population by simply doubling it. Female population is used, as they have that gifted quality of reproduction of off-springs, which is otherwise known as fecundity level. In many cases, half the population will be females. In addition, survival rates of males and females are generally almost the same. For most animals, the breeding rate depends on the number of females of breeding age. Although the breeding rates may be different for females of different ages, females will normally form a fixed proportion of the population. This concept can be applied for various species like fishes, owl or any animal.

For these reasons, we can ignore the males and concentrate on the female half of the population. Even if the proportion of females is not half, we can still do this, as long as the proportion remains fairly constant for any particular age group.

In ecology, matrices may be used to model the changes in a population of animals over a period of time. In order to apply the modeling method, we need some information about the reproduction of the population and the actual numbers of animals present. The method makes some assumptions about reproduction in general, but these are generally fulfilled, so the model has wide applicability.

When we examine a population, the age-specific data is of particular interest. The Fisheries Department want to know what catch limits and sizes they should set to ensure that fishing does not deplete a population. National Parks and Wildlife officers want to know what culling should be undertaken to ensure that particular species, if left mismanaged, may lead to extinction by disease or starvation etc.

Consider a population of a particular species in a national park. After a study of the population, the following data was collected. No animals lived longer than 5 years.

Age-Specific Population Distribution

Age (years)	0-1	1-2	2-3	3-4	4-5	Total
Population	403	157	102	52	11	725
Female population	194	82	55	22	6	359

Clearly, the assumption that half the population is female is reasonable. The breeding rate was estimated for each age group as the number of live offspring in a year divided by the number of such species in the age group at the start of the year. To concentrate only on females, we actually want the breeding rate for females. This will be the number of live females produced, divided by the number of females at the start of the year. Since both numbers are half the population totals, the breeding rate for females is the same as the breeding rate for the population. The following data was obtained by previous study:

Age-Specific Breeding Rates: Female Only

Age (years)	0-1	1-2	2-3	3-4	4-5
Breeding rate	0	1.3	1.8	0.9	0.2

Survival rates were also studied during the tagging process. The survival rate is the proportion of the age group that survives into the next year. This is also the same for females as for the whole population. The data for that part of the study was as follows:

Age-Specific Survival Rates of The Populat Ion (Calculated As: 1 – Death Rate)

Age (years)	0-1	1-2	2-3	3-4	4-5
Survival rate	0.6	0.8	0.8	0.4	0

We want to know how to arrange the information and calculate the number of females in each age group in the next year. For each of the older ages, the number next year will be the number this year multiplied by the survival rate. For the 0 – 1 group, the number will be the number born in the year, which will be the total of products of the breeding rates and the number in each age group. We will have:

NEXT YEAR:

$0\text{-}1 = 0.0 \times 194 + 1.3 \times 82 + 1.8 \times 55 + 0.9 \times 22 + 0.2 \times 6 = 226.6 \cong 227$

$1\text{-}2 = 0.6 \times 194 = 116.4 \cong 116$

$2\text{-}3 = 0.8 \times 82 = 66$

$3\text{-}4 = 0.8 \times 55 = 44$

$4\text{-}5 = 0.4 \times 22 = 8.8 \cong 9$

Total = 462

Total population will be double of this female population = 2 × 462 = 924

If we arrange the initial age-specific population of females as a column matrix, then the age specific breeding rates and age-specific survival rates may be arranged in a square matrix to produce the age-specific female population after one year as:

$$= \begin{bmatrix} 0.0 & 1.3 & 1.8 & 0.8 & 0.2 \\ 0.6 & 0.0 & 0.0 & 0.0 & 0.0 \\ 0.0 & 0.8 & 0.0 & 0.0 & 0.0 \\ 0.0 & 0.0 & 0.8 & 0.0 & 0.0 \\ 0.0 & 0.0 & 0.0 & 0.4 & 0.0 \end{bmatrix} \times \begin{bmatrix} 194 \\ 85 \\ 55 \\ 22 \\ 6 \end{bmatrix} = \begin{bmatrix} 226.6 \\ 116.4 \\ 65.6 \\ 44.0 \\ 8.8 \end{bmatrix}$$

$$L \times X_0 = X_1$$

This is normally written as the equation $X_1 = LX_0$, where X_0 is the initial age-specific female population, X_1 is the age-specific female population after 1 year and L is the Leslie Matrix, which incorporates the age-specific breeding rates and survival rates.

$$= \begin{bmatrix} B1 & B2 & B3 & ... & Bn-1 & Bn \\ S1 & 0 & 0 & ... & 0 & 0 \\ 0 & S2 & 0 & ... & 0 & 0 \\ ... & ... & ... & ... & ... & ... \\ 0 & 0 & 0 & ... & Sn-1 & 0 \end{bmatrix}$$

The age grouping and associated rates are chosen for convenience. It may be 3 months, 1 year, 5 years, 20 years etc. However, best are achieved when the age grouping is chosen to be the generation gap- that is, the time between birth and breeding.B_i is the breeding rate of the ith age group and S_i is the survival rate of the ith age group. The survival rate, S_n does not appear because it must be zero.

Example 18.5: Using the above data, find the population after (a) 2 years, (b) 3 years, (c) 4 years, (d) 5 years.

Solution: To find the population after 2 years, we simply multiply the female population distribution after 1 year by the Leslie matrix again to get the new female population distribution.

(a) After 2 years:

$$X_2 = \begin{bmatrix} 0.0 & 1.3 & 1.8 & 0.9 & 0.2 \\ 0.6 & 0.0 & 0.0 & 0.0 & 0.0 \\ 0.0 & 0.8 & 0.0 & 0.0 & 0.0 \\ 0.0 & 0.0 & 0.8 & 0.0 & 0.0 \\ 0.0 & 0.0 & 0.0 & 0.4 & 0.0 \end{bmatrix} \times \begin{bmatrix} 226.6 \\ 116.4 \\ 65.6 \\ 44.0 \\ 8.8 \end{bmatrix} = \begin{bmatrix} 310.8 \\ 136.0 \\ 93.1 \\ 52.5 \\ 17.6 \end{bmatrix}$$

Total = 610 (Only female)

Total Population will be 2 × 610 = 1220

Percentage increase is (1220-924) ÷ 924 × 100 = 32.46%

(b) After 3 years:

$$X_3 = \begin{bmatrix} 0.0 & 1.3 & 1.8 & 0.9 & 0.2 \\ 0.6 & 0.0 & 0.0 & 0.0 & 0.0 \\ 0.0 & 0.8 & 0.0 & 0.0 & 0.0 \\ 0.0 & 0.0 & 0.8 & 0.0 & 0.0 \\ 0.0 & 0.0 & 0.0 & 0.4 & 0.0 \end{bmatrix} \times \begin{bmatrix} 310.8 \\ 136.0 \\ 93.1 \\ 52.5 \\ 17.6 \end{bmatrix} = \begin{bmatrix} 395.2 \\ 186.5 \\ 108.8 \\ 74.5 \\ 21.0 \end{bmatrix}$$

Total = 786 (Only female)

Total Population will be 2 × 786 = 1572

Percentage increase is (1572-1220) ÷ 1220 × 100 = 28.85%

(c) After 4 years:

$$X_4 = \begin{bmatrix} 0.0 & 1.3 & 1.8 & 0.9 & 0.2 \\ 0.6 & 0.0 & 0.0 & 0.0 & 0.0 \\ 0.0 & 0.8 & 0.0 & 0.0 & 0.0 \\ 0.0 & 0.0 & 0.8 & 0.0 & 0.0 \\ 0.0 & 0.0 & 0.0 & 0.4 & 0.0 \end{bmatrix} \times \begin{bmatrix} 395.2 \\ 186.5 \\ 108.8 \\ 74.5 \\ 21.0 \end{bmatrix} = \begin{bmatrix} 509.5 \\ 237.1 \\ 149.2 \\ 87.0 \\ 29.8 \end{bmatrix}$$

Total = 1012.6 (Only female)

Total Population will be 2 × 1012.6 = 2025

Percentage increase is (2025-1572) ÷ 1572 × 100=28.81%

(d) After 5 years:

$$X_5 = \begin{bmatrix} 0.0 & 1.3 & 1.8 & 0.9 & 0.2 \\ 0.6 & 0.0 & 0.0 & 0.0 & 0.0 \\ 0.0 & 0.8 & 0.0 & 0.0 & 0.0 \\ 0.0 & 0.0 & 0.8 & 0.0 & 0.0 \\ 0.0 & 0.0 & 0.0 & 0.4 & 0.0 \end{bmatrix} \times \begin{bmatrix} 509.5 \\ 237.1 \\ 149. \\ 87.0 \\ 29.8 \end{bmatrix} = \begin{bmatrix} 661.1 \\ 305.7 \\ 189.7 \\ 119.4 \\ 34.8 \end{bmatrix}$$

Total = 1310.7 (Only female)

Total Population will be 2 × 1310.7 = 2621

Percentage increase is (2621-2025) ÷ 2025 × 100 = 29.473%

The otal population will be double the total female population in each year.

In geleral, we can write that, if conditions do not change, the female populatioı distribution after n years, X_n, is given by the expression $X_n = L^n X_0$, when the ppulation growth is modeled by a Leslie matrix. The population increases from 725 animals to 923, 1220, 1572, 2025, 2621 and so on. The percentage increase in population in each year varies (32.46%, 28.85%, 28.81%, 29.437%), but eventually settles to about 29%. In fact, a general property of the Leslie matrix model as shown is that it eventually produces a stable rate of increase or decrease.

This is more easily seen using computer programs to perform the calculation. The age distribution of the population also stabilize. We eventually find that we can write $X_n = kX_{n-1}$ for any Leslie matrix model. In the case above, k = 1.2922 and the age proportions stabilize, so that 50% of the population is aged 0-1, 23% is aged 1-2, 15%is aged 2-3, 9%is aged 3-4, and 3%is aged 4-5.

The increase in species population in this example is such that the authority may consider that there is likely to be a problem of population pressure from the species. In this case they may want to consider the effects of culling.

These can be introduced into the model in various ways. If a particular number of animals is culled at a particular time of the year (by *size, bag limits* and *licensing,* for example) then the column matrix for the number in each age group that is culled is just subtracted from X before it is multiplied by L. Policing of this method is expensive for wild animals, so a more common method is to have a *season* during which the animals are culled. The length of the season is adjusted by experience to maintain a stable population. In this case, hunting is more easily and cheaply policed and changes to the model are also simplified. We assume that the same proportion of all animals is culled (or *harvested*), and simply multiply the original Leslie matrix by an appropriate factor.

Suppose that, for every age group, the proportion removed from the population is h. Then the proportion, remaining is (1 - h). Each survival rate must be multiplied by (1 - h) and so must the birth rates, because there will be fewer animals to breed. Thus, when harvesting/culling is introduced, we obtain a new Leslie matrix L' = (1 – h)L. Every entry is multiplied by (1 – h), using scalar multiplication. If 15% of the animals were culled, then we would have h = 0.15 and L' = 0.85L.

Example 18.6: In the above example, suggest a culling factor to gradually reduce the population after 10 years.

Solution: In the previous example, it is seen that although the pecentage increase fluctuates wildly at first, it gradually settles down to a figre 29%. Continuation of the process for many generations shows that the stable increase is about 25.5%.

Thus we have that $X_n = 1.255X_{n-1}$ for large values of n.

To choose a culling factor h to keep the population stable, we use the fact that $X_n = L'X_{n-1}$ with $L' = (1 - h)L$, we have

$X_n = L'X_{n-1} = (1 - h)LX_{n-1} = (1 - h)\,1.255X_{n-1}$

Thus, for a stable population, we would like $1.255(1 - h) = 1$.

$1.255(1 - h) = 1$

$= 1 - h = \dfrac{1}{1.255}$

$h = 1 - \dfrac{1}{1.255} = 0.2032$ Say 20.32%

Since we want to reduce the population, the culling factor should be more than 20.32%. We should use a culling factor of 25% or more to gradually reduce the population.

(C) Predator ~ Prey Model (Lokta-Voltera Equation)

The Lotka–Volterra equations, also known as the predator–prey equations, are a pair of first-order, non-linear, differential equations frequently used to describe the dynamics of biological systems in which two species interact, one as a predator and the other as prey. The populations change through time according to the pair of equations:

$$\frac{dx}{dt} = \alpha x - \beta xy$$

$$\frac{dy}{dt} = \delta xy - \gamma y$$

Where,

x is the number of prey (for example, rabbits)

y is the number of some predator (for example, foxes)

$\frac{dx}{dt}$ and $\frac{dy}{dt}$ represent the growth rates of the two populations over time

t represents time and

α, β, γ, δ are positive real parameters describing the interaction of the two species.

αx = prey growth rate

βx = predation of the predator

δ = Conversion efficiency

γ = Starvation rate of predator

Physical Meanings of the Equations

The Lotka-Volterra model makes a number of assumptions about the environment and evolution of the predator and prey populations:

- The prey population finds ample food at all times.
- The food supply of the predator population depends entirely on the size of the prey population.
- The rate of change of population is proportional to its size.
- During the process, the environment does not change in favour of one species and genetic adaptation is inconsequential.
- Predators have limitless appetite.
- As differential equations are used, the solution is deterministic and continuous. This, in turn, implies that the generations of both the predator and prey are continually overlapping.

Prey

The prey equation is:

$$\frac{dx}{dt} = \alpha x - \beta xy$$

But in absence of predators, the equation becomes, $\frac{dx}{dt} = \acute{a}x$; as the second term vanishes (y = 0). Integrating both sides, we get, $\int \frac{1}{x} dx = \int \alpha\, dt$

$\Rightarrow \log x = \alpha t$

$\Rightarrow x = e^{\alpha t}$

That means, the prey are assumed to have an unlimited food supply and to *reproduce exponentially* unless subject to predation; this exponential growth is represented in the equation above by the term αx. The rate of predation upon the prey is assumed to be proportional to the rate at which the predators and the prey meet; this is represented above by βxy. If either x or y is zero then there can be no predation.

With these two terms, the equation above can be interpreted as: the change in the prey's numbers is given by its own growth minus the rate at which it is preyed upon.

Predators

The predator equation is:

$$\frac{dy}{dt} = \delta xy - \gamma y$$

In this equation, represents the growth of the predator population. (Note the similarity to the predation rate; however, a different constant is used as the rate at which the predator population grows is not necessarily equal to the rate at which it consumes the prey). represents the loss rate of the predators due to either natural death or emigration; it leads to an *exponential decay* in the absence of prey, as can be derived in the following manner:

But in the absence of prey, the equation becomes, $\frac{dy}{dt} = -\gamma y$; as the first term vanishes (x = 0). Integrating both sides, we get, $\int \frac{1}{y} dy = \int \gamma \, dt$

$\Rightarrow \log y = -\gamma t$

$\Rightarrow y = e^{-\gamma t}$

Hence the equation expresses the change in the predator population as growth fueled by the food supply, minus natural death.

Solutions to the Equations

The equations have periodic solutions and do not have a simple expression in terms of the usual trigonometric functions, although they are quite tractable.

If none of the non-negative parameters α,β,γ,δ vanishes, three can be absorbed into the normalization of variables to leave but merely one behind: Since the first equation is homogeneous in x, and the second one in y, the parameters β/α and δ/γ are absorbable in the normalizations of y and x, respectively, and γ into the normalization of t, so that only α/γ remains arbitrary. It is the only parameter affecting the nature of the solutions.

A linearization of the equations yields a solution similar to simple harmonic motion with the population of predators trailing that of prey by 90° in the cycle.

Frequency Plot

x changes at y = 1

y changes at x = 1

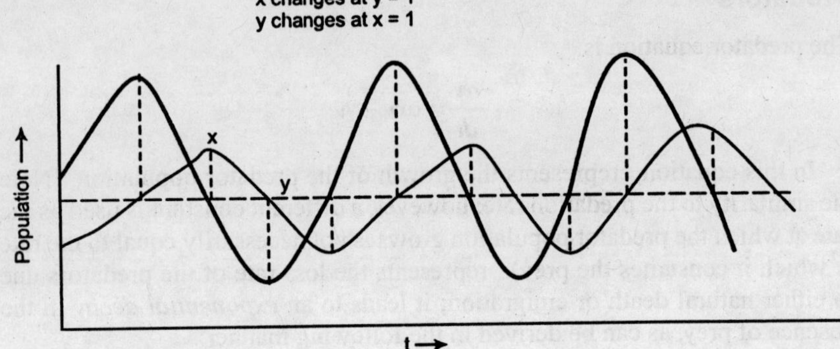

A Simple Example

One may also plot solutions parametrically as orbits in "phase-space", without representing time, but with one axis representing the number of prey and the other axis representing the number of predators for all times.

This is to say, eliminating time from the two differential equations above results in only one such,

$$\frac{dy}{dx} = -\frac{y}{x} \times \frac{\delta x - \gamma}{\beta y - \alpha}$$

or, $\frac{1}{y}dy(\beta y - \alpha) = -\frac{1}{x}dx(\delta x - \gamma)$

or, $\beta dy - \alpha\frac{1}{y}dy = \gamma\frac{1}{x}dx - \delta dx$

$$\int\left(\beta dy - \alpha\frac{1}{y}dy\right) = \int\left(\gamma\frac{1}{x}dx - \delta dx\right)$$

or, $\int\beta dy - \int\alpha\frac{1}{y}dy = \int\gamma\frac{1}{x}dx - \int\delta dx$

or, $\beta y - \alpha \log y + C = \gamma \log x - \delta x + D$

or, $C - D = -\delta x + \gamma \log x - \beta y + \alpha \log y$

or, $V = -\delta x + \gamma \log x - \beta y + \alpha \log y$

$e^V = e^{-\delta x + \gamma \log x - \beta y + \alpha \log y} = K$ (Say)

or, $K = \frac{x^\gamma \times y^\alpha}{e^{\delta x + \beta y}}$

whose solutions are closed curves.

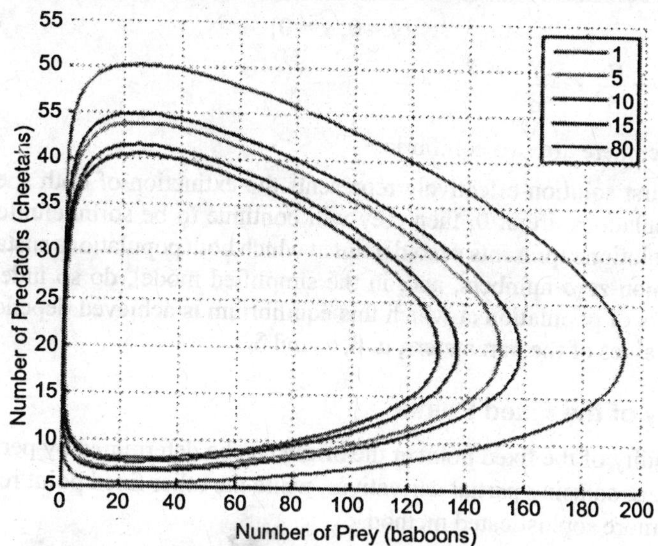

An aside These graphs illustrate a serious potential problem with this *as a biological model*: For this specific choice of parameters, in each cycle, the baboon population is reduced to extremely low numbers, yet recovers (while the cheetah population remains sizeable at the lowest baboon density). In real-life situations, however, chance fluctuations of the discrete numbers of individuals, as well as the family structure and life-cycle of baboons, might cause the baboons to actually go extinct, and, by consequence, the cheetahs as well. This modeling problem has been called the "atto-fox problem", an atto-fox being a notional 10^{-18} of a fox, in the context of rabies modeling.

Dynamics of the System

In the model system, the predators thrive when there are plentiful prey but, ultimately, outstrip their food supply and decline. As the predator population is low the prey population will increase again. These dynamics continue in a cycle of growth and decline.

Population Equilibrium

Population equilibrium occurs in the model when neither of the population levels is changing, i.e. when both of the derivatives are equal to 0.

$$x(\alpha - \beta y) = 0$$

$$-y(\gamma - \delta x) = 0$$

When solved for x and y the above system of equations yields

$$\{y = 0, x = 0\}$$

and $y = \frac{\alpha}{\beta}, x = \frac{\gamma}{\delta}$

Hence, there are two equilibria.

The first solution effectively represents the extinction of both species. If both populations are at 0, then they will continue to be so indefinitely. The second solution represents a fixed point at which both populations sustain their current, non-zero numbers, and, in the simplified model, do so indefinitely. The levels of population at which this equilibrium is achieved depend on the chosen values of the parameters, α, β, γ, and δ.

Stability of the Fixed Points

The stability of the fixed point at the origin can be determined by performing a linearization using partial derivatives, while the other fixed point requires a slightly more sophisticated method.

The Jacobian matrix of the predator-prey model is

$$J(x, y) = \quad J(x,y) = \begin{bmatrix} \alpha - \beta y & -\beta x \\ \delta y & \delta x - \end{bmatrix}$$

First Fixed Point (Extinction)

When evaluated at the steady state of (0, 0) the Jacobian matrix J becomes

$$J(0,0) = \begin{bmatrix} \alpha & 0 \\ 0 & -\gamma \end{bmatrix}$$

The eigen values of this matrix are $\lambda_1 = \alpha, \lambda_2 = -\gamma$

In the model α and γ are always greater than zero, and as such the sign of the eigenvalues above will always differ. Hence the fixed point at the origin is a saddle point.

The stability of this fixed point is of significance. If it were stable, non-zero populations might be attracted towards it, and as such the dynamics of the system might lead towards the extinction of both species for many cases of initial population levels. However, as the fixed point at the origin is a saddle point, and hence unstable, it follows that the extinction of both species is difficult in the model. (In fact, this could only occur if the prey were artificially completely eradicated, causing the predators to die of starvation. If the predators were eradicated, the prey population would grow without bound in this simple model): The populations of prey and predator can get infinitesimally close to zero and still recover.

Second Fixed Point (Oscillations)

Evaluating J at the second fixed point leads to

$$J\left(\frac{\gamma}{\delta},\frac{\alpha}{\beta}\right)=\begin{bmatrix} 0 & -\frac{\beta\gamma}{\delta} \\ \frac{\alpha\delta}{\beta} & 0 \end{bmatrix}$$

The eigenvalues of this matrix are

$$\lambda_1 = i\sqrt{(\alpha\gamma)}, \lambda_2 = -i\sqrt{(\alpha\gamma)}$$

As the eigenvalues are both purely imaginary, this fixed point is not hyperbolic, so no conclusions can be drawn from the linear analysis. However, as illustrated above, the system admits a constant of motion *V*, or, equivalently, exp(*V*),

$$K = y^{\alpha} e^{-\beta y} x^{\gamma} e^{-\delta x}$$

and the level curves, for each constant K, are closed orbits surrounding the fixed point: the levels of the predator and prey populations cycle, and oscillate around this fixed point.

Increasing K moves a closed orbit closer to the fixed point. The largest value of the constant K is obtained by solving the optimization problem.

$$y^{\alpha}e^{-\beta\psi}x^{\gamma}e^{-\delta x}=\frac{x^{\gamma}\times y^{\alpha}}{e^{\delta x+\beta y}}\rightarrow \max(x,y>0)$$

The maximal value of *K* is thus attained at the stationary (fixed) point $\left(\frac{\gamma}{\delta},\frac{\alpha}{\beta}\right)$ and amounts to

$$K^{*}=\left(\frac{\alpha}{\beta e}\right)^{\alpha}\times\left(\frac{\gamma}{\delta e}\right)^{\gamma},$$

where *e* is Euler's Number.

GLOSSARY OF TERMS

TERMS USED

Alpha (α)	The probability of a Type-I error
Alternative Hypothesis	The conclusion that is accepted when the data fail to support the null hypothesis.
Analysis of Variance (ANOVA)	A statistical technique used to test the equality of three or more sample means and thus make inferences as to whether the samples come from population having the same mean.
Base Year	A normal year which is taken as a reference period to compare the price of the current year.
Bayes' Theorem	The formula for conditional probability under statistical dependence.
Bernoulli Process	A process in which each trial has only two possible outcomes, the probability of the outcome of any trial remains fixed over time, and the trials are statistically independent.
Beta (β)	The probability of a Type-II error
Between-Column Variance	An estimate of the population variance derived from the variance among the sample means.
Bimodal Distribution	A distribution of data points in which two values occur more frequently than the rest of the values in the data set.
Binomial Distribution	A discrete distribution describing the results of an experiment known as a Bernoulli process.
Census	The measurement or examination of every element in the population.
Central Limit Theorem	A result assuring that the sampling distribution of the mean approaches normality as the sample size increases, regardless of the shape of the population distribution from which the sample is selected.
Chain Index Number	This index number uses varying base year through the computation of link relatives which expresses the figures for each year as percentages of the previous year.

Chi-Square Distribution	A family of probability distributions, differentiated by their degrees of freedom, used to test a number of different hypotheses about variances, proportions, and distributional goodness of fit.
Cluster Sampling	A method of random sampling in which the population is divided into groups, or clusters of elements, and then a random sample of these clusters is selected.
Clusters	Within a population, groups that are essentially similar to each other, although the groups themselves have wide internal variation.
Coding	A method of converting traditional measures of time to a form that simplifies computation (often called translating)
Coefficient of Correlation	The square root of the coefficient of determination. Its sign indicates the direction of the relationship between two variables, direct or inverse.
Coefficient of Determination	A measure of the proportion of variation in Y, the dependent variable, that is explained by the regression line, that is, by Y's relationship with the independent variable.
Coefficient of Variation	A relative measure of dispersion, comparable across distributions, that expresses the standard deviation as a percentage of the mean.
Conditional Probability	The probability of one event occurring, given that another event has occurred.
Confidence Interval	A range of values that has some designated probability of including the true population parameter value.
Confidence Level	The probability that statisticians associate with an interval estimate of a population parameter, indicating how confident they are that the interval estimate will include the population parameter.
Confidence Limits	The upper and lower boundaries of a confidence interval.
Consistent Estimator	The estimator that yields values more closely approaching the population parameter as the sample size increases.
Consumer Price Index	An index that measures the price changes in a representative set of consumer goods over time.
Contingency Table	A table having R rows and C columns. Each row corresponds to a level of one variable, each column to a level of another variable. Entries in the body of the table are the frequencies with which each variable combination occurred.
Continuous Data	Data that may progress from one class to the next without a break and may be expressed by either whole numbers or fractions.

Continuous Probability Distribution	A probability distribution in which the variable is allowed to take on any value within a given range.
Correlation Analysis	A technique to determine the degree to which variables are linearly related.
Cost of Living Index	It measures the changes in the level of prices of commodities and services directly affecting the cost of living of a specified group of persons at a specified place.
Critical Value	The value of standard statistic (z or t *etc*) beyond which the null hypothesis is rejected; the boundary between the acceptance and rejection regions.
Cumulative Frequency Distribution	A tabular display of data showing how many observations lie above, or below certain values.
Curvilinear Relationship	An association between two variables that is described by a curved line.
Cyclical Fluctuation	A type of variation in a time series, in which the value of the variable fluctuates above and below a secular trend line.
Data	A collection of any number of related observations of one or more variables.
Data Array	The arrangement of raw data by observations in either ascending or descending order.
Data Point	A single observation from a data set.
Data Set	Allocation of data. collection of any number of related observations of one or more variables.
Degrees of Freedom	The number of values in a sample that can be specified freely once something about the sample is known.
Dependent variable	The variable that is predicted in regression analysis.
Deseasonalization	A statistical process used to remove the effects of seasonality from a time series.
Direct Relationship	A relationship between two variables in which, as the independent variable's value increases, so does the value of the dependent variable.
Discrete Class	Data that do not progress from one class to the next without a break; that is, where classes represent distinct categories or counts and may be represented by whole numbers.
Discrete Probability Distribution	A probability distribution in which the variable is allowed to take on only a limited number of values, which can be listed.
Discrete Random Variable	A random variable that is allowed to take on only a limited number of values, which can be listed.
Dispersion	The spread or variability in a set of data.

Dummy variable	A variable taking a value 0 or 1, enabling us to include in a regression model qualitative factors such as sex, marital status and education level etc.
Efficient Estimator	An estimator with a smaller standard error than some other estimator of the population parameter; that is, the smaller the standard error of an estimator, the more efficient that estimator is.
Estimate	A specific observed value of an estimator.
Estimating Equation	A mathematical formula that relates the unknown variable to the known variables in regression analysis.
Estimator	A sample statistic used to estimate a population parameter.
Event	One or more of the possible outcomes of doing something, or one of the possible outcomes from conducting an experiment.
Exhaustive Events	A list of events that represents all the possible outcomes of an experiment.
Expected Frequencies	The frequencies those are expected to see in a contingency table or frequency distribution if the null hypothesis is true.
Expected Value	A weighted average of the outcomes of an experiment.
Expected Value of a Random Variable	The sum of the products of each value of the random variable with the value's probability of occurrence.
Experiment	The activity that results in, or produces, an event.
F Distribution	A family of distributions differentiated by two parameters (df-numerator, df-denominator), used primarily to test hypothesis regarding variances.
F Ratio	A ratio used in the analysis of variance, among other tests, to compare the magnitude of two estimates of the population variance to determine whether the two estimates are approximately equal; in ANOVA, the ratio of between-column variance to within-column variance is used.
Factor Reversal Test	When price index and quantity index are multiplied together, it should measure the change in total value.
Factorial Experiment	An experiment in which each factor involved is used once with each other factor. In a complete factorial experiment, every level of each factor is used with each level of every other factor.
Finite Population	A population having a stated or limited size.
Finite Population Multiplier	A factor used to correct the standard error of the mean for studying a population of finite size that is small in relation to the size of the sample.

Free-Hand Curve	A highly subjective method to fit a line or a curve to a set of points plotted in a graph using one's own judgement.
Frequency Curve	A frequency polygon smoothed by adding classes and data points to a data set.
Frequency Distribution	An organized display of data that shows the numbers of observations from the data set that falls into each of a set of mutually exclusive and collectively exhaustive classes.
Frequency Polygon	A line graph connecting the midpoints of each class in a data set, plotted at a height corresponding to the frequency of the class.
Geometric Mean	A measure of central tendency used to measure the average rate of change or growth for some quantity, computed by taking the nth root of the product of *n* values representing change.
Goodness-of-Fit Test	A statistical test used for determining whether there is a significant difference between an observed frequency distribution and a theoretical probability distribution hypothesised to describe the observed distribution.
Grand Mean	The mean for the entire group of subjects from all the samples in the experiment.
Histogram	A graph of a data set, composed of a series of rectangles, each proportional in width to the range of values in a class and proportional in height to the number of items falling in the class, or the fraction of items in the class.
Hypothesis	An assumption or speculation that is made about a population parameter.
Independent Variable	The variable or variables that take values independently in regression analysis.
Index Number	It is a ratio of two numbers which have intrinsic characteristic of measuring change in the values of a variable with respect to some other variables like time or space etc.
Infinite Population	A population in which it is theoretically impossible to observe all the elements.
Interval Estimate	A range of values to estimate an unknown population parameter.
Inverse Relationship	A relationship between two variables in which, as the independent variable increases, the dependent variable decreases.
Irregular Variation	A condition in a time series in which the value of a variable is completely unpredictable.
Joint Probability	The probability of two events occurring together or in succession.

Judgement Sampling	A method of selecting a sample from a population in which personal knowledge or expertise is used to identify the items from the population that are to be included in the sample.
Kolmogorov-Smirnov Test	A non-parametric test, which does not require that data be grouped in any way, for determining whether there is a significant difference between an observed frequency distribution and a theoretical frequency distribution.
Kruskal-Wallis Test	A non-parametric method for testing whether three or more independent samples have been drawn from populations with the same distribution. It is a non-parametric version of ANOVA test.
Kurtosis	The degree of peakedness of a distribution of points.
Latin Square	An efficient experimental design that makes it unnecessary to use a complete factorial experiment.
Least-Squared Method	A technique for fitting a straight line through a set of points in such a way that the sum of the squared vertical distances from the *n* points to the line is minimised.
Linear Relationship	A particular type of association between two variables that can be described mathematically by a straight line.
Link Relative Method	This is a method to compute seasonal indices. The link relatives are calculated by dividing the figures for each period by the figure for the previous period.
Lower-Tailed Test	A one-tailed hypothesis test in which a sample value significantly below the hypothesisized population value that leads to reject the null hypothesis.
Mann-Whitney U Test	A non-parametric method used to determine whether two independent samples have been drawn from populations with the same distribution.
Mean	A central tendency measure representing the arithmetic average of a set of observations.
Measure of Central Tendency	A measure indicating the value to be expected of a typical or middle data point.
Measure of Dispersion	A measure describing how the observations in a data set are scattered or spread out.
Median	The middle point of a data set, a measure of location that divides the data set into halves.
Median Class	The class in a frequency distribution that contains the median value for a data set.
Mode	The value most often repeated in the data set. It is represented by the highest point in the distribution curve of a data set.

Modelling Technique	Methods for deciding which variables to include in a regression model and the different ways in which they can be included.
Moving Average Method	This is a method for smoothening the data to depict the trend. Moving averages are successive simple or weighted averages of a specified number of consecutive values in a time series data.
Multiple Regression	A statistical process by which several variables are used to predict another variable.
Mutually Exclusive Events	Events that cannot happen together.
Nonparametric Tests	Statistical techniques that do not make restrictive assumptions about the shape of a population distribution when performing a hypothesis test.
Normal Distribution	A distribution of a continuous random variable with a single-peaked, bell-shaped curve. The mean lies at the center of the distribution, and the curve is symmetrical around a vertical line erected at the mean. The two tails extend indefinitely, never touching the horizontal axis.
Null Hypothesis	The hypothesis, or assumption, about a population parameter under test, usually an assumption of the status quo.
Ogive	A graph of a cumulative frequency distribution.
One-Tailed Test	A hypothesis in which there is only one rejection region; that is, where the only concern is whether the observed value deviates from the hypothesised value in one direction.
Open-Ended Class	A class that allows either the upper or lower end of a quantitative classification scheme to be limitless.
Parameters	Numerical values that describe the characteristics of a whole population, commonly represented by Greek letters.
Percentile	Fractiles that divide the data into 100 equal parts.
Point Estimate	A single number used to estimate an unknown population parameter.
Poisson Distribution	A discrete distribution in which the probability of the occurrence of an event within a very small time period is a very small number, the probability that two or more such events will occur within the same time interval is effectively 0, and the probability of the occurrence of the event within one time period is independent of where that time period is.
Population	A collection of all elements under study and about which conclusions are drawn.

Posterior Probability	A probability that has been revised after additional information was obtained.
Power of Hypothesis Test	The probability of rejecting the null hypothesis when it is false, that is, a measure of how well the hypothesis test is working.
Precision	The degree of accuracy with which the sample mean can estimate the population mean, as revealed by the standard error of the mean.
Price Index	Index number showing the changes in price level of a commodity or a group of commodities.
Priory Probability	Probability estimate made prior to receiving new information.
Probability	The chance that something will happen.
Probability Distribution	A list of the outcomes of an experiment with the probabilities expected to be associated with these outcomes.
Quantity Index	It measures changes in the volume of production in different sectors of economy with respect to time or space.
Quartiles	Fractiles that divide the data into four equal parts.
Random or Probability Sampling	A method of selecting a sample from a population in which all the items in the population have an equal chance of being chosen in the sample.
Random variable	A variable that takes on different values as a result of the outcomes of a random experiment.
Range	The distance between the highest and lowest values in a data set.
Rank Correlation	A method for doing correlation analysis when the data are not available to use in numerical form, but when information is sufficient to rank the data.
Rank-Correlation Coefficient	A measure of the degree of association between two variables that is based on the ranks of observations, not their numerical values.
Ratio-to-Moving-Average Method	A statistical method used to measure seasonal variation. It uses an index describing the degree of that variation.
Raw Data	Information before it is arranged or analyzed by statistical methods.
Regression	The general process of predicting one variable from another by statistical means, using previous data.
Regression Line	A line fitted to a set of data points to estimate the relationship between two variables.

Relative Cyclical Residual	**A measure of cyclical variation, it uses the percentage from the trend for each value in the series.**
Relative Frequency Distribution	**The display of a data set that shows the fraction or percentage of the total data set that falls into each of a set of mutually exclusive and collectively exhaustive classes.**
Representative Sample	**A sample that contains the relevant characteristics of the population in the same proportions as they are included in that population.**
Residual Method	**A method of describing the cyclical component of a time series. It assumes that most of the variation in the series not explained by the secular trend is cyclical variation.**
Sample	**A collection of some, but not all, of the elements of the population under study, used to describe the population. A portion of the elements in a population chosen for direct examination or measurement.**
Sample Space	**The set of all possible outcomes of an experiment.**
Sampling Distribution of the Mean	**A probability distribution of all the possible means of samples of a given size, *n*, from a population.**
Sampling Error	**Error or variation among sample statistics due to chance, that is, differences between each sample and the population, and among several samples, which are due to the elements chosen for the sample.**
Sampling Fraction	**The fraction or proportion of the population contained in a sample.**
Sampling With Replacement	**A sampling procedure in which sampled items are not returned to the population after being picked, so that no member of the population can appear in the sample more than once.**
Scatter Diagram	**A graph of points on a rectangular grid; the X and Y coordinates of each point correspond to the two measurements made on some particular sample element, and the pattern of points illustrates the relationship between the two variables.**
Seasonal Variation	**Pattern of change in a time series within a year; pattern that tend to be repeated from year to year.**
Second-Degree Equation	**A mathematical form used to describe a parabolic curve that may be used in time-series trend analysis.**
Secular Trend	**A type of variation in a time series, the value of the variable tending to increase or decrease over a long period of time.**
Sign Test	**A test for the difference between paired observations where + and - signs are substituted for quantitative values.**

Significant Level	A value indicating the percentage of sample values that is outside certain limits, assuming the null hypothesis is correct, that is, the probability of rejecting the null hypothesis when it is true.
Simple Random Sampling	Methods of selecting samples that allow each possible sample an equal probability of being picked and each item in the entire population an equal chance of being included in the sample.
Skewness	The extent to which a distribution of data points is concentrated at one end or the other; the lack of symmetry.
Slope	A constant for any given straight line whose value represents how much each unit change of the independent variable changes the dependent variable.
Splicing of Index Numbers	When two or more overlapping series of index numbers are combined into one continuous series, the process is called splicing.
Standard Deviation	The positive square root of the variance; a measure of dispersion in the same units as the original data, rather than in the squared units of the variance.
Standard Error	The standard deviation of the sampling distribution of a statistic.
Standard Error of Estimate	A measure of the reliability of the estimating equation, indicating the variability of the observed points around the regression line, that is, the extent to which observed values differ from their predicted values on the regression line.
Standard Error of Regression Coefficient	A measure of uncertainty about the exact value of a regression coefficient.
Standard Error of the Mean	The standard deviation of the sampling distribution of the mean; a measure of the extent to which the means from different samples are expected to varry from the population mean, owing to the chance error in the sampling process.
Standard Error of the Regression Coefficient	A measure of the variability of sample regression coefficient around the true population regression coefficient.
Standard Normal Probability Distribution	A normal probability distribution, with mean $\mu = 0$ and standard deviation $\sigma = 1$.
Strata	Groups within a population formed in such a way that each group is relatively homogeneous, but wider variability exists among the separate groups.
Statistical Inference	The process of making inferences about population from information contained in samples.

Statistic	Numerical measures describing the characteristics of a sample. Represented by Roman letters.
Stratified Sampling	A method of random sampling in which the population is divided into homogeneous groups, or strata, and elements within each stratum are selected at random according to one of two rules. (1) A specified number of elements is drawn from each stratum in the population, or (2) equal numbers of elements are drawn from each stratum, and the results are weighted according to the stratum's proportion of the total population.
Student's *t* Distribution	A family of probability distributions distinguished by their individual degrees of freedom, similar in form to the normal distribution, and used when the population standard deviation is unknown and the sample size is relatively small (*n* 30)
Sufficient Estimator	An estimator that uses all the information available in the data concerning a parameter.
Symmetrical Distribution	A characteristic of a distribution in which each half is the mirror image of the other half.
Systematic Sampling	A method of sampling in which elements to be sampled are selected from the population at a uniform interval that is measured in time, order, or space.
Test of Independence	A statistical test of proportions of frequencies to determine whether membership in categories of one variable is different as a function of membership in the categories of a second variable.
Time Reversal Test	The index number of year 'b' with base year 'a' is reciprocal of the index number of year 'a' with base year 'b'.
Time Series	Information accumulated at regular intervals and the statistical methods used to determine patterns in such data.
Transformations	Mathematical manipulations for converting one variable into a different form so we can fit curves as well as lines by regression.
Two-Tailed Test	A hypothesis test in which the null hypothesis is rejected if the sample value is significantly higher or lower than the hypothesised value of the population parameter; a test involving two rejection regions.
Type-I Error	Rejecting a null hypothesis when it is true.(α)
Type-II Error	Accepting a null hypothesis when it is false.()
Unbiased Estimator	An estimator of a population parameter that, on the average, assumes values above the population parameter as often, and to the same extent, as it tends to assume values below the population parameter.

Unit Test	According to this test, the index number formula should be such that it is independent of the units in which prices and quantities are measured.
Value Index	It measures the change in the value of commodities between two different time periods.
Variance	A measure of the average squared distance between the mean and each item in the population.
Venn Diagram	A pictorial representation of probability concepts in which the sample space is represented as a rectangle and the events in the sample space as portions of that rectangle.
Weighted Aggregate Index	Weights either related to base year or current year or both are used to form the aggregate price for both the current year and base year in the construction of an index number.
Weighted Average of Relatives	Weights are used for averaging the price/quantity relatives.
Weighted Mean	An average calculated to take into account the importance of each value to the overall total, that is, an average in which each observation value is weighted by some index of its importance.
Wholesale Price Index	It measures the change in general price level in a country as a whole. Economic Adviser's index number is a wholesale price index number.
Within-Column Variance	An estimate of the population variance based on the variances within the *k* samples, using a weighted average of the *k* sample variances.
Y-Intercept	A constant for any given straight line whose value represents the value of the *Y* variable when the *X* variable has a value 0.